Food Processing and Its Impact on Phenolic and Other Bioactive Constituents in Food – Second Edition

Food Processing and Its Impact on Phenolic and Other Bioactive Constituents in Food – Second Edition

Editor

Sabina Lachowicz-Wiśniewska

Basel • Beijing • Wuhan • Barcelona • Belgrade • Novi Sad • Cluj • Manchester

Editor
Sabina Lachowicz-Wiśniewska
Department of Health Sciences
Calisia University
Kalisz
Poland

Editorial Office
MDPI
St. Alban-Anlage 66
4052 Basel, Switzerland

This is a reprint of articles from the Special Issue published online in the open access journal *Molecules* (ISSN 1420-3049) (available at: www.mdpi.com/journal/molecules/special_issues/food_phenol_second_ed).

For citation purposes, cite each article independently as indicated on the article page online and as indicated below:

Lastname, A.A.; Lastname, B.B. Article Title. *Journal Name* **Year**, *Volume Number*, Page Range.

ISBN 978-3-0365-9035-6 (Hbk)
ISBN 978-3-0365-9034-9 (PDF)
doi.org/10.3390/books978-3-0365-9034-9

© 2023 by the authors. Articles in this book are Open Access and distributed under the Creative Commons Attribution (CC BY) license. The book as a whole is distributed by MDPI under the terms and conditions of the Creative Commons Attribution-NonCommercial-NoDerivs (CC BY-NC-ND) license.

Contents

About the Editor vii

Preface ix

Anna Bieniek, Sabina Lachowicz-Wiśniewska and Justyna Bojarska
The Bioactive Profile, Nutritional Value, Health Benefits and Agronomic Requirements of Cherry Silverberry (*Elaeagnus multiflora* Thunb.): A Review
Reprinted from: *Molecules* **2022**, *27*, 2719, doi:10.3390/molecules27092719 1

Łukasz Woźniak, Anna Szakiel, Agnieszka Głowacka, Elżbieta Rozpara, Krystian Marszałek and Sylwia Skąpska
Triterpenoids of Three Apple Cultivars—Biosynthesis, Antioxidative and Anti-Inflammatory Properties, and Fate during Processing
Reprinted from: *Molecules* **2023**, *28*, 2584, doi:10.3390/molecules28062584 17

Ji Hye Kim, You Jin Lim, Shucheng Duan, Tae Jung Park and Seok Hyun Eom
Accumulation of Antioxidative Phenolics and Carotenoids Using Thermal Processing in Different Stages of *Momordica charantia* Fruit
Reprinted from: *Molecules* **2023**, *28*, 1500, doi:10.3390/molecules28031500 32

Gabriela Salazar-Orbea, Rocío García-Villalba, Luis M. Sánchez-Siles, Francisco A. Tomás-Barberán and Carlos J. García
Untargeted Metabolomics Reveals New Markers of Food Processing for Strawberry and Apple Purees
Reprinted from: *Molecules* **2022**, *27*, 7275, doi:10.3390/molecules27217275 46

Maria Clara de Moraes Motta Machado, Bárbara Morandi Lepaus, Patrícia Campos Bernardes and Jackline Freitas Brilhante de São José
Ultrasound, Acetic Acid, and Peracetic Acid as Alternatives Sanitizers to Chlorine Compounds for Fresh-Cut Kale Decontamination
Reprinted from: *Molecules* **2022**, *27*, 7019, doi:10.3390/molecules27207019 59

Venelina Popova, Zhana Petkova, Nadezhda Mazova, Tanya Ivanova, Nadezhda Petkova and Magdalena Stoyanova et al.
Chemical Composition Assessment of Structural Parts (Seeds, Peel, Pulp) of *Physalis alkekengi* L. Fruits
Reprinted from: *Molecules* **2022**, *27*, 5787, doi:10.3390/molecules27185787 74

Magdalena Błaszak, Barbara Jakubowska, Sabina Lachowicz-Wiśniewska, Wojciech Migdał, Urszula Gryczka and Ireneusz Ochmian
Effectiveness of E-Beam Radiation against *Saccharomyces cerevisiae*, *Brettanomyces bruxellensis*, and Wild Yeast and Their Influence on Wine Quality
Reprinted from: *Molecules* **2023**, *28*, 4867, doi:10.3390/molecules28124867 90

Mirjana B. Pešić, Milica M. Pešić, Jelena Bezbradica, Anđela B. Stanojević, Petra Ivković and Danijel D. Milinčić et al.
Okara-Enriched Gluten-Free Bread: Nutritional, Antioxidant and Sensory Properties
Reprinted from: *Molecules* **2023**, *28*, 4098, doi:10.3390/molecules28104098 106

Józef Gorzelany, Michał Patyna, Stanisław Pluta, Ireneusz Kapusta, Maciej Balawejder and Justyna Belcar
The Effect of the Addition of Ozonated and Non-Ozonated Fruits of the Saskatoon Berry (*Amelanchier alnifolia* Nutt.) on the Quality and Pro-Healthy Profile of Craft Wheat Beers
Reprinted from: *Molecules* **2022**, *27*, 4544, doi:10.3390/molecules27144544 . . . . **126**

Justyna Belcar and Józef Gorzelany
Feasibility of Defatted Juice from Sea-Buckthorn Berries (*Hippophae rhamnoides* L.) as a Wheat Beer Enhancer
Reprinted from: *Molecules* **2022**, *27*, 3916, doi:10.3390/molecules27123916 . . . . **142**

About the Editor

Sabina Lachowicz-Wiśniewska

Sabina Lachowicz-Wiśniewska, Associate Professor, has a Doctor of Science degree received at the Institute of Rural Health in the area of natural products and dietetics. She completed her PhD degree at Wrocław University of Environmental and Life Sciences under the direction of Professor Jan Oszmianski in the area of natural product technology and chemistry. Nowadays, she carries out research at Calisia University in the Kalisz Microbiota Research Team in the area of prebiotic functional foods and the bioavailability of their bioactive compounds for the prevention of oxidative stress and inflammation. Her past research is in the general area of functional food and pharmaceutical and medicinal plant chemistry, with over 74 peer-reviewed scientific papers covering research fields such as (i) the production of innovative functional food designed to have health-promoting properties; (ii) the bioavailability and digestibility of bioactive compounds in the simulated digestive system by in vitro method; (iii) the determination of antioxidant, anti-diabetic, anti-obesity, and anti-inflammatory potential; and (iv) the identification and assessment of the health-promoting properties of bioactive compounds from plant materials based on chromatographic techniques. Research collaborations have been established within Dekaban Foundation with Prof. Anubhav Pratap Singh (Faculty of Land and Food Systems (LFS), University of British Columbia), Prof. Antonio J. Melendez-Martınez in the Food Color and Quality Laboratory, Universidad de Sevilla, and also with many European research centers.

Preface

Special Issue of *Molecules* entitled "Food Processing and Its Impact on Phenolic and Other Bioactive Constituents in Food – Second Edition". Bioactive compounds, including phenolic ingredients, have long been used as important constituents of a healthy diet. As a result, consumer awareness about the important role of high-quality products rich in bioactive compounds—especially phenolic compounds—in human nutrition, health, and prevention against diseases has increased. Additionally, methods for food processing, regardless of the technology used, have a huge impact on the quality of the final products. Therefore, the big challenges for scientists lie in the monitoring of changes during food processing and the optimization of technology to achieve minimal degradation of nutrients (including phenolic compounds). Thus, for this Special Issue, we published the latest scientific news, insights, and advances in the field of food processing and its impact on bioactive constituents in food, especially phenolic compounds. The information presented will certainly attract considerable interest among a large group of our readers from different disciplines and research fields.

Sabina Lachowicz-Wiśniewska
Editor

Review

The Bioactive Profile, Nutritional Value, Health Benefits and Agronomic Requirements of Cherry Silverberry (*Elaeagnus multiflora* Thunb.): A Review

Anna Bieniek [1], Sabina Lachowicz-Wiśniewska [2,3,*] and Justyna Bojarska [4]

1. Department of Agroecosystems and Horticulture, Faculty of Agriculture and Forestry, University of Warmia and Mazury in Olsztyn, Prawocheńskiego 21 Street, 10-720 Olsztyn, Poland; anna.bieniek@uwm.edu.pl
2. Department of Food and Nutrition, Calisia University, Nowy Świat 4 Street, 62-800 Kalisz, Poland
3. Department of Horticulture, West Pomeranian University of Technology Szczecin, Słowackiego 17 Street, 71-434 Szczecin, Poland
4. Chair of Food Plant Chemistry and Processing, Faculty of Food Sciences, University of Warmia and Mazury in Olsztyn, Cieszyński Sq. 1 Street, 10-726 Olsztyn, Poland; justyna.bojarska@uwm.edu.pl
* Correspondence: s.lachowicz-wisniewska@akademiakaliska.edu.pl or slachowiczwisniewska@zut.edu.pl

Citation: Bieniek, A.; Lachowicz-Wiśniewska, S.; Bojarska, J. The Bioactive Profile, Nutritional Value, Health Benefits and Agronomic Requirements of Cherry Silverberry (*Elaeagnus multiflora* Thunb.): A Review. *Molecules* 2022, 27, 2719. https://doi.org/10.3390/molecules27092719

Academic Editor: Francesco Cacciola

Received: 28 February 2022
Accepted: 20 April 2022
Published: 23 April 2022

Publisher's Note: MDPI stays neutral with regard to jurisdictional claims in published maps and institutional affiliations.

Copyright: © 2022 by the authors. Licensee MDPI, Basel, Switzerland. This article is an open access article distributed under the terms and conditions of the Creative Commons Attribution (CC BY) license (https://creativecommons.org/licenses/by/4.0/).

Abstract: The cherry silverberry (*Elaeagnus multiflora* Thunb.) is a lesser-known plant species with high nutritional and therapeutic potential. Cherry silverberry contains numerous biologically active compounds. The cherry silverberry is a shrub growing up to 3 m. Its drupe-like fruit is ellipsoidal, up to 1 cm long, and set on stems. It is red in color, juicy, and sour, and its taste resembles that of red currants. According to the literature, cherry silverberry fruit contains carbohydrates, organic acids, and amino acids, as well as vitamin C, in addition to biominerals, polyphenols, flavonoids, carotenoids, chlorophylls, and tocopherols, which contribute to its high nutritional value. New biotypes of cherry silverberry cultivated in Poland can be used for the production of functional foods and direct consumption. In China, the cherry silverberry, known as goumi, has been used as a medicinal plant and a natural remedy for cough, diarrhea, itch, foul sores, and, even, cancer. This review article summarizes the scant research findings on the nutritional and therapeutic benefits of cherry silverberry.

Keywords: cherry elaeagnus; chemical composition; biologically active compounds; antioxidant activity; cultivation

1. Introduction

Bioactive compounds are widespread in the vegetal world. They exert protective effects on plants, as well as human and animal health. Bioactive substances can act as natural antioxidants, whose presence in the body may help prevent a wide variety of lifestyle diseases [1]. Plant species that are rich sources of bioactive substances have been extensively researched around the world [2]. Particular attention has been paid to lesser-known plant species such as kiwiberry, cornelian cherry, honeysuckle, hawthorn, chokeberry, rowanberry, elderberry, medlar, bilberry, seabuckthorn, and silverberry, which grow in different climatic zones and have been introduced to cultivation outside their natural geographic ranges. Novel fruits and berries are increasingly being introduced into local and global food systems [3,4]. Some of them can be eaten raw, while others require processing [5,6]. Neglected and underutilized edible plant species can also boost the livelihoods of small-scale farmers and local producers [7]. This group of plants includes *Elaeagnus multiflora* Thunb. (*Elaeagnaceae*), also known as cherry silverberry, cherry elaeagnus, and goumi. The cherry silverberry belongs to the genus *Elaeagnus* L. and the family *Elaeagnaceae* Juss., which also includes the more popular common seabuckthorn (*Hippophaë rhamnoides* L.) [8–11]. According to the literature [12–19], *E. multiflora* fruit, which is suitable for direct consumption and processing, can be classified as a "superfood" due to its high content of carotenoids, exogenous amino acids, macronutrients, micronutrients, unsaturated fatty acids, and vitamin C. Fresh and processed silverberries are a valuable source of lycopene, the most potent antioxidant among common carotenoids, which is renowned for its anticarcinogenic effects [18–21]. The cherry silverberry is native to China, Korea, and Japan [9]. In traditional Chinese medicine, the species is known as a phytosterol-rich plant [20–24]. The fruit, leaves, and young branches of *E. multiflora* can be used as phenolic antioxidant additives and dietary supplements [2,8,22,25–39] as well as natural remedies for cough, diarrhea, gastrointestinal disorders, itch, cancer, and bone diseases [8,12,19]. Cherry silverberry seeds are used in dietary therapy and as a functional food for cancer prevention [22,28]. According to Kim et al. [29], *E. multiflora* fruit extract can be applied as a whitening functional cosmetic material, due to the suppression of melanin biosynthesis. Cherry silverberries can be processed at home to prepare juice, compote, jam and jelly, herbal tea, wine, soup, sauces, desserts, candies, pudding, ice-cream topping, fruit leather, and other food products [2,9,11]. Today, this species is grown not only in China but also in the eastern United States and in Europe, including Poland [8,21]. As demonstrated by Bieniek et al. [9], the cherry silverberry thrives in the temperate climate of Poland, as it is easy to cultivate and resistant to diseases.

Elaeagnus multiflora is a thorny shrub, growing up to 3 m (Figure 1). The leaves are typical of the genus *Elaeagnus*—the upper part of the leaf blade is green, whereas its bottom is silvery. Figure 2 presents the flowers and fruit with seeds of *E. multiflora*. The flowers are solitary or in pairs in the leaf axils, fragrant, with a four-lobed pale-yellowish-white corolla 1.5 cm long; flowering occurs in mid-spring. Since silverberry flowers give off a strong aroma, resembling that of cinnamon and vanilla, this plant can be used for flavoring cakes and other desserts [9]. Its drupe-like fruit is ellipsoidal, up to 1 cm long, and set on stems. It is red in color, juicy, and sour, while its taste resembles that of red currants. In Poland, silverberries ripen at the end of June or at the beginning of July [2]. This species is currently being introduced to Russia and the USA, while it has not yet been commercially produced in Poland. Since the 1990s, research has been carried out at the Department of Horticulture, University of Warmia and Mazury in Olsztyn (formerly: University of Agriculture and Technology), to select the most suitable biotypes for cultivation in Poland [2,8,9,26,35–37]. According to Lachowicz et al. [2], the cherry silverberry biotypes grown in north-eastern Poland constitute a highly interesting material and could be an excellent source of functional foods. This species also deserves special attention as a fruit plant for organic cultivation.

Figure 1. *Elaeagnus multiflora* Thunb. with fruit.

Figure 2. Flowers and fruit with seeds of *Elaeagnus multiflora* Thunb.

The aim of this article was to review the latest research findings regarding the cherry silverberry.

2. Selection of Varieties and Cultivation Characteristics

The cherry silverberry has been cultivated as a fruit plant since 1974. The first variety of the cherry silverberry, Sakhalinsky pervyi, was bred in the Far Eastern Research Institute of Agriculture in Russia. In 1999, it was entered into the State Register of Breeding Achievements Approved for Use. Other varieties, including Moneron and Taisa (2002), Krilon (2006), Shikotan (2009), Yuzhnyi (2009), Kunashi (2011), Cunai (2015), and Paramushir (2016) were also registered in Russia (State Register of Breeding Achievements Approved for Use, 2016) [9].

A collection of *E. multiflora* was created at the M.M. Gryshko National Botanical Garden (NBG) of the National Academy of Sciences of Ukraine in Kyiv in 1980–1982. The primary material (seeds from free pollination) was imported from Sakhalin (Sakhalin Scientific Research Institute of Agriculture). At present, the *E. multiflora* collection includes 45 genotypes. Grygorieva et al. [22] analyzed the morphometric parameters of fruit in selected genotypes of cherry silverberry grown in the "Forest-Steppe of Ukraine" geographic plot in the M.M. Gryshko NBG. The results of this preliminary study have contributed to increasing interest in *E. multiflora* cultivation among farmers, which can be followed by the domestication and introduction of this species to the agricultural production system in Ukraine and other countries.

In Poland, research into *E. multiflora* was initiated in 1995 at the Department of Horticulture, University of Agriculture and Technology (presently: University of Warmia and Mazury in Olsztyn), when three-year-old plants were obtained from the Institute for Fruit Growing in Samokhvalovitchy in Belarus. [9]. At present, experiments involving several dozen seedlings are being carried out to select the optimal biotypes that could be grown in Poland and other countries [2,8,35–37]. Lachowicz et al. [8] noted considerable differences in the chemical composition and antioxidant activity of the *E. multiflora* varieties and biotypes selected at the University of Warmia and Mazury in Olsztyn. Lachowicz et al. [8] found that the fruit of biotypes Si1 and Si2 contained high concentrations of vitamin C, linoleic acid, and α-linolenic acid. The fruit of biotypes Si5 and Si4 was characterized by the highest content of glucose, fructose, and ash, whereas the fruit of biotypes Si0 and Si3 contained the highest levels of the remaining fatty acids as palmitic, oleic, stearic, and organic acids, exhibiting the highest antioxidant activity. Moreover, biotype Si0 had a high content of total polyphenolics, organic acids, and palmitoleic acid, and demonstrated higher antioxidant activity than the remaining biotypes. The above authors concluded that new biotypes of cherry silverberry grown in north-eastern Poland are highly promising and can be consumed raw or used in the production of functional foods.

Elaeagnus multiflora varieties 'Sweet Scarlet' and 'SSP' (seedlings obtained from Austria) can be purchased from Polish nurseries. 'Sweet Scarlet' is the earliest-maturing variety. The fruit begins to ripen in the first half of June; the berries remain on the stems for four weeks and, then, fall down. This variety has darker and sweeter fruit than other varieties. 'Sweet Scarlet' is an allogamous variety, which requires pollen from another variety for fruit setting. 'SSP' is an autogamous variety, with a slower growth rate than 'Sweet Scarlet'. The fruit ripens at the beginning of July, and it has a sweet taste. Another *E. multiflora* variety is 'Jahidka', which produces much shorter shrubs (up to 1.5 m) and red oval fruit weighing 1–1.5 g that ripens in early July [31].

Elaeagnus multiflora is often confused with *E. umbellata* because both species have similar leaves and flowers. However, *E. umbellata* produces round fruit with short petioles, typically ripening in September [35].

Cultivation of Elaeagnus multiflora

The cherry silverberry has low nutritional requirements, and it thrives on dry, sandy, and poor soils. However, the species requires large amounts of sunlight. Cherry silverberry shrubs can grow in the same site for 25 years [9,40]. A symbiosis with nitrogen-fixing actinomycetes makes the cherry silverberry a pioneer soil-fertilizing species [9,40].

In commercial plantations, cherry silverberry shrubs should be planted at 4×2 m spacing, 5–8 cm deeper than in the seed bed (Figure 3). The species has similar fertilizer requirements to currants and gooseberries. *Elaeagnus multiflora* is highly resistant to drought. Due to its high-quality fruit, it is a promising fruit plant that can be recommended for organic cultivation. Most seedlings begin to bear fruit in the fourth year after planting [21,31]. According to Kołbasina [41], 5-year-old plants can yield 3–4 kg fruit per shrub, 10-year-old plants up to 15 kg, and 20-year-old plants up to 30 kg. Cultivation conditions, as well as climatic factors during the growing season, regardless of genetic factors, have a significant effect on the yield and qualitative characteristics of fruit [9,21]. *Elaeagnus multiflora* can be grown on a small scale and cultivated commercially with the use of combine harvesters [26].

Figure 3. *Elaeagnus multiflora* growing in the Experimental Garden of the University of Warmia and Mazury in Olsztyn (north-eastern Poland).

3. Biologically Active Compounds in *Elaeagnus multiflora* Thunb.

Cherry silverberry fruit is abundant in bioactive components that are responsible for its health-promoting properties [8,9]. These substances can be divided into primary and secondary metabolites. Primary metabolites are a source of nutrients, energy, and structural components in plants with limited bioactive properties, whereas secondary metabolites are metabolic products in plants that deliver a wide range of health-promoting effects.

Primary metabolites include, among others, carbohydrates, organic acids, and amino acids. Secondary metabolites include, among others, vitamin C, biominerals, polyphenols, flavonoids, carotenoids, chlorophylls, and tocopherols (Table 1) [42,43].

Table 1. The basic chemical composition of cherry silverberry fruit.

Components	Contents	Ref.	Components	Contents	Ref.
Dry weight [%]	12.64–15.55	[9,44]	Amino acids [mg/100 g FW]	89.68	[44]
Total saccharides [%]	5.34–6.30	[9]	serine	13.93	[44]
Monosaccharides [%]	1.54–1.96	[9]	phosphoethanolamine	13.93	[44]
Total free sugars [mg/100 g FW *]	781.44	[44]	alanine	13.16	[44]
fructose	370.34	[44]	β-alanine	13.16	[44]
glucose	401.96	[44]	aspartic acid	4.62	[44]
sucrose	5.80	[44]	phosphoserine	4.62	[44]
trehalose	3.34	[44]	cystine	4.45	[44]
Crude protein [%]	1.29	[44]	methionine	3.89	[44]
Soluble protein [g/100 g FW]	0.48	[44]	phenylalanine	2.85	[44]
pH	3.29	[44]	threonine	2.63	[44]
Crude ash [%]	0.46–0.62	[2,44]	taurine	2.63	[44]
Biominerals [mg/100 g FW]	1353.70–1855.94	[17,44]	tyrosine	2.17	[44]
potassium	1627.44	[44]	leucine	1.41	[44]
magnesium	140.28	[44]	isoleucine	1.16	[44]
sodium	56.70	[44]	valine	1.12	[44]
calcium	14.70	[44]	β-aminoisobutyric acid	1.12	[44]
iron	7.98	[44]	α-aminoisobutyric acid	0.62	[44]
manganese	5.53	[44]	ornithine	0.57	[44]
zinc	2.89	[44]	glutamic acid	0.51	[44]
copper	0.10	[44]	sarcosine	0.51	[44]
lithium	0.20	[44]	Polyphenolic compounds [mg/100 g DW]	417.02–1268.90	[2,8,37]
nickel	0.12	[44]	phenolic acids	1.22–3.80	[2,8,37]
Lipids [g/100 g]	1.40	[1,9]	flavonols	37.29–56.25	[2,8,37]
unsaturated fatty acids account [%], of which	48.70–54.50	[1,9]	hydrolyzable tannins	3.07–10.60	[2,8,37]
α-linolenic acid [%]	17.50–20.80	[1,9]	stilbenes	0.91–1.71	[2,8,37]
linolinic acid [%]	21.80–25.90	[1,9]	polymeric procyanidins	861.36–1197.34	[2,8,37]
oleic acid [%]	19.30–22.70	[9]	Carotenoids [mg/100 g DW]	40.09–170.00	[2,8,37]
Organic acids [g/100 g DW **], of which	18.48–34.11	[2,36]	phytoene	0.93–0.97	[2,8,37]
malic acid account [%]	55–60	[2]	lycopene	39.16–169.00	[2,8,37]
quinic account [%]	11–15	[2]	β-carotene	0.21–0.31	[2,8,37]
tartaric acid account [%]	9–18	[2]	Tocopherols [mg/100 g DW]	2.00–9.93	[37]
Vitamin C [mg/100 g]	4.22–562.72	[9,44]	Chlorophylls [mg/100 g DW]	393.00	[2,37]

* FW, fresh weight; ** DW dry weight.

Sugars, organic acids, and their ratio can affect the sensory and chemical attributes of the food matrix, including sweetness, microbiological stability, total acidity, pH, and overall sensory acceptability [38]. Therefore, the palatability of cherry silverberry fruit, mainly its sweet and sour taste, is determined by the content of sugars and organic acids. The average content of organic acids in the fruit of *E. multiflora* Thunb. biotypes grown in Poland range from 0.78% to 1.20% [9], or 18.48 to 34.11 g/100 g of dry weight (DW) [2,36], which implies that cherry silverberries are abundant in these compounds. A liquid chromatography analysis revealed the presence of seven organic acids in cherry silverberry fruit: malic, quinic, tartaric, oxalic, citric, isocitric, and succinic acid. The predominant organic acids were malic (55–60% of total organic acids), quinic (11–15%), and tartaric (9–18%) acids [2]. Kim et al. [44] identified four organic acids in cherry silverberry fruit and determined their total content at 294.44 mg/100 g of fresh weight (FW). According to Mikulic-Petkovsek et al. [45], citric and malic acids account for 30–95% of all organic acids in berries. Fruits that are low in citric acid include cherry silverberry as well as chokeberry, rowanberry, and eastern shadbush. Five organic acids with a total content of 167.8 g/100 g FW were identified in cherry silverberry leaves. Malic acid was the predominant compound (66% of total organic acids), followed by acetic (13.7%), citric (8.1%), lactic (6.3%), and succinic acid (5.3%) [17].

Another study demonstrated that cherry silverberry fruit contained 1.54–1.96% of monosaccharides and 5.34–6.30% of total sugars on a fresh weight (FW) basis [9]. Total sugar content was determined at 9.77 to 11.50 ° Brix by Hong et al. [46]. An analysis involving the high-pressure liquid chromatography with refractive index detectors (HPLC-RI) method revealed the presence of two sugars, fructose and glucose. Fructose accounted for around 57–59% and glucose for 41–43% of the total sugars in cherry silverberry fruit [2]. Kim et al. [44] identified five free sugars with a total content of 781 mg/100 g FW in cherry silverberry fruit. Fructose and glucose were the predominant sugars, whereas sucrose, maltose, and trehalose were detected in trace amounts [44]. Cherry silverberry leaves were found to contain five sugars: arabinose, fructose, glucose, maltose, and trehalose. Similar to the fruit, the predominant sugar in the leaves was fructose (46.9% of total sugars), followed by arabinose (27.2%) [17]. According to Mikulic-Petkovsek et al. [45], berries contain mainly fructose and glucose, and fructose accounts for up to 75% of the total sugars. However, some exceptions have been noted, such as kiwifruit, where sucrose represents 71.9% of the total sugars [45].

The sugar–acid ratio denotes the relative content of sugars and acids, which are responsible for the taste and aroma of fruit [45]. Sweet-tasting berries are not always rich in sugar, and they may be low in organic acids, mainly malic acid [45,47]. The sugar–acid ratio affects the perception of sweetness [48], and it ranges from 5.25 to 7.40 in cherry silverberry fruit [16]. In a study by Mikulic-Petkovsek et al. [45], white gooseberries and red, black, and white currants were the most acidic fruits with a sugar–acid ratio of around two. The sweetest-tasting fruits were black mulberries, brambles, and goji berries, with a sugar–acid ratio above 12.9 [45].

Vitamin C (ascorbic acid) is yet another bioactive substance that plays a very important role in fruit. Vitamin C has antioxidant, anticarcinogenic, anti-inflammatory, and antisclerotic properties; it lowers blood glucose levels and reduces the risk of cardiovascular diseases [49,50]. Cherry silverberries are abundant in vitamin C, although the content can vary depending on variety, genotype, growing conditions, weather, and ripeness [9]. In the work of Sakamura et al. [24], vitamin C concentration decreased in successive stages of fruit ripening. In contrast, in *Rubus sieboldi*, *Ribis nigrum*, pears, peaches, and papayas, the content of L-ascorbic acid increased with ripening [24]. In a study by Kim et al. [44], cherry silverberries grown in Korea contained 131.35 mg/100 g FW of ascorbic acid and 431.37 mg/100 g FW of dehydroascorbic acid, and the total content of vitamin C was determined at 562.72 mg/100 g FW. These results indicate that cherry silverberry fruit is an excellent source of vitamin C. In a study conducted by Bieniek et al. [9], the concentration of vitamin C in the fruit of cherry silverberry grown in Poland ranged from 4.22 to 7.70 mg/100 g FW. Vitamin C levels reached 15.8–33.1 mg/100 g in cherry silverberry fruit grown in Ukraine [51] and 27.8 mg/100 g in the fruit grown in Pakistan [52]. In other fruit, vitamin C concentrations were 30 mg/100 g in elderberries, 35–90 mg/100 g in blackcurrants, and 16–32 mg/100 mg in raspberries [53].

Cherry silverberries are also abundant in biominerals, mainly potassium (1627.44 mg/100 g FW), magnesium (140.28 mg/100 g FW), sodium (56.70 mg/100 g FW), calcium (14.70 mg/100 g FW), iron (7.98 mg/100 g FW), manganese (5.53 mg/100 g FW), zinc (2.89 mg/100 g FW), copper, lithium, and nickel (0.10–0.20 mg/100 g FW) [44]. According to Polish Standards [54], 100 g of cherry silverberry fruit provide approximately 65% of the recommended daily intake of potassium, 33–43% of magnesium, 53–79% of iron, 26–32% of zinc, and 240% of manganese, for healthy middle-aged adults [54]. Bal et al. [55] found that seabuckthorn is also a rich source of potassium, whose content was determined at 1012–1484 mg/100 g FW in fruit flesh and at 933–1342 mg/100 g FW in seeds. Cherry silverberry leaves can be used as functional food additives [8], and they have been found to contain 14 minerals with a total content of 1353.70 mg/100 g FW [17]. Similar to the fruit, 100 g of cherry silverberry leaves provided 33% of the recommended daily intake of potassium, 36% of calcium, 35–63% of iron, 22% of copper, 18–24% of magnesium, around

250% of manganese, and around 180% of selenium [54]. Other elements, including Li, Na, Al., Fe, Co, Ni, Cu, Zn, and Ge, were detected in trace amounts [17].

Free and bound amino acids and their derivatives are yet another important group of biologically active compounds. According to Kim et al. [44], cherry silverberries are abundant in amino acids, whose total content was determined at 89.68 mg/100 g FW. The content of serine, alanine, phosphoethanolamine, and β-alanine exceeded 10 mg/100 g FW, whereas aspartic acid, cystine, methionine, phosphoserine, threonine, glutamic acid, glycine, valine, isoleucine, leucine, tyrosine, phenylalanine, taurine, sarcosine, α-aminoisobutyric acid, β-aminoisobutiryc acid, and ornithine were detected at concentrations below 5 mg/100 g FW. In turn, cherry silverberry leaves contained 7 essential amino acids, 10 non-essential amino acids, and 11 amino acid derivatives, with a total content of 943 mg/100 g FW. The following amino acids were identified at concentrations higher than 50 mg/100 FW: threonine, valine, isoleucine, leucine, phenylalanine, glutamic acid, alanine, and tyrosine. Lysine, aspartic acid, serine, cystine, histidine, proline, glycine, tyrosine, arginine, phosphoserine, sarcosine, α-aminoadipic acid, β-aminoisobutyric acid, γ-aminoisobutyric acid, and anserine were detected at concentrations below 20 mg/100 g FW. Trace amounts of carnosine, β-alanine, cystathionine, and α-aminoisobutyric acid were also identified in cherry silverberry leaves [17]. The content of amino acids was similar in medlar leaves, but it was 10 times higher in ripe medlar fruit [56]. Amino acid concentrations in Saskatoon berries were estimated at 490 mg/100 g [57]. According to Zhang et al. [58], the content of free amino acids in fruits is determined mainly by ripeness, growing conditions, position on a plant, genotype, and the applied analytical methods.

Cherry silverberries are abundant in bioactive components, with antioxidant properties that deliver numerous health benefits, including polyphenols and isoprenoids [8]. These compounds promote a healthy oxidant/antioxidant balance and lower the risk of chronic non-infectious diseases, such as cardiovascular diseases, cancer, neurodegenerative disorders, diabetes, and obesity [59]. The total content of polyphenolic compounds in *E. multiflora* fruit, expressed in gallic acid equivalents (GAE), was determined at 280 mg/100 g FW by Kim et al. [44], at 12.21 mg% by Hong et al. [46], and at 568 mg GAE/100 g DW by Lachowicz et al. [2]. Polyphenol concentrations are similar in seabuckthorn fruit, where they range from 128.66 to 407.48 mg GAE/100 g [60,61]. High-performance liquid chromatography methods have been applied to assess the content and qualitative composition of polyphenols in cherry silverberry fruit [8,13,37]. Total polyphenol content was determined at 904.65–1268.90 mg/100 g DW in the fruit of the 'Jahidka' and 'Sweet Scarlet' varieties grown in Poland, after extraction with 30% ethanol [8]; 353 mg/100 g FW in Korean-grown fruit, after extraction with 50% ethanol [13]; and 417.02–819.04 mg/100 g DW in the fruit of Polish-grown biotypes, after extraction with 30% ethanol [37]. According to Cho et al. [62], differences in polyphenol concentrations may be attributed to variety, species, growing conditions, extraction methods, analytical methods, technological process, or the analyzed materials. Lee et al. [13] identified 13 polyphenolic compounds that were classified as flavan-3-ols (epigallocatechin, catechin, epicatechin, epigallocatechin gallate, epicatechin gallate, catechin gallate) and phenolic acids (gallic acid, protocatechuic acid, tannic acid, *p*-hydroxybenzoic acid, vanillic acid, *p*-coumaric acid, ferulic acid). Epicatechin gallate was the dominant flavan-3-ol (66%), whereas gallic acid and *p*-coumaric acid accounted for 26% and 23% of the total phenolic acids, respectively [13]. Lachowicz-Wiśniewska et al. [2,37] identified 16 polyphenols in the fruit of the cherry silverberry varieties 'Jahidka' and 'Sweet Scarlet', including one phenolic acid, one hydrolysable tannin, one stilbene, and 13 flavanols, as well as polymeric procyanidins. Polymeric procyanidins were the predominant compounds that accounted for 66.0–95.0% of total polyphenols, as evidenced by a mildly astringent taste [63]. In *E. umbellata* fruit, flavonols were the predominant polyphenols (78.8%) [64]. In *E. multiflora* fruit, flavonols—quercetin derivatives, kaempferol, and isorhamnetin—accounted for 5% of the total polyphenols, whereas the content of phenolic acids (sinapic acid derivatives) was determined at 0.2%, hydrolyzable tannins (galloyl derivatives) at 0.3%, and stilbenes (glucosylphloretin derivatives) at 0.2%. Kaempferol-pentoside-rutinoside was the predominant flavonol [37]. In turn, cherry silverberry leaves were found to contain 38 polyphenolic compounds, including three phenolic acids, 35 flavonols, and polymeric procyanidins. Polymeric procyanidins were also dominant and accounted for around 81% of the total polyphenols [2].

Isoprenoids, including carotenoids, chlorophylls, and tocopherols, are indirectly responsible for the color, taste, and aroma of fruits. Cherry silverberry fruit contains carotenoids, whereas chlorophylls have been identified in leaves. Carotenoids are highly biologically active compounds that boost immunity and prevent inflammations caused by excessive formation of reactive oxygen species (ROS) [65–68]. Chlorophylls stimulate intestinal peristalsis, lower blood pressure, and decrease the risk of anemia [65,67]; plants can hardly bear to live without chlorophyll. Lachowicz et al. [2,37] were the first research team to examine the content as well as the qualitative and quantitative composition of carotenoids and chlorophylls in cherry silverberry fruit [2,37]. Isoprenoid concentrations ranged from 95.69 to 170 mg/100 g DW in the fruit of Polish-grown biotypes, and from 40.09 to 97.15 mg/100 g DW in the fruit of the 'Sweet Scarlet' and 'Jahidka' varieties. The content of carotenoids ranged from 66.20 to 71.26 mg/100 g DW, and the content of chlorophyll ranged from 1634 to 1694 mg/100 g DW in the 'Sweet Scarlet' and 'Jahidka' varieties, whereas in the analyzed biotypes, carotenoid concentration was determined at 81 mg/100 g DW and chlorophyll concentration at 393 mg/100 g DW [2,37]. The carotenoid content of seabuckthorn fruit, which belongs to the same family as the cherry silverberry, ranged from 10 to 120 mg/100 g FW [69]. Sixteen carotenoid compounds were identified in cherry silverberry fruit, including eight lycopene derivatives, α- and β-carotene (provitamin A), their two derivatives, lutein, two violaxanthins, and neoxanthin [2,37]. Lycopene delivers numerous health benefits [70], and it was the dominant carotenoid (80%) in cherry silverberry fruit. The remaining carotenoids also have health-promoting properties [66,67,71]. Phytoene is a valuable, but rarely identified, carotenoid. This colorless compound is characterized by high dietary bioavailability, and recent research has shown that phytoene exhibits high levels of biological activity and exerts protective effects on the skin [66]. Seabuckthorn berries are more abundant in β-carotene (0.9–18 mg/100 g FW) than fruits and vegetables that are regarded as the richest sources of this compound [72]. Cherry silverberry fruit contains even more β-carotene (37–42 mg/100 g DW) [2]. Significant differences in carotenoid levels in the analyzed fruits could be related to numerous factors, such as climate, genotype, and agrotechnology [66,67,71]. The fruit of the studied cherry silverberry biotypes also contained α-tocopherol at 3.31–7.07 mg/100 mg DW. In turn, the content of α-tocopherol in cherry silverberry seeds ranged from 2.0 to 3.3 mg/100 g DW [37]. α-tocopherol and its derivatives, known as vitamin E, are powerful antioxidants that delay cell aging [41]. In a study by Piłat et al. [72], tocopherol levels in seabuckthorn berries ranged from 3.35 to 6.27 mg/100 g FW, and α-tocopherol was the predominant compound that accounted for 62–67% of the total tocopherols [73].

4. Health-Promoting Properties of *Elaeagnus multiflora* Thunb.

For thousands of years, plants have been used to treat various human and animal diseases [74,75]. Plants of the family *Elaeagnaceae* have gained popularity in recent years due to their exceptional chemical composition as well as health benefits. Seabuckthorn (*Hippophaë rhamnoides* L.) is the most researched representative of this family. Its berries contain more than 190 bioactive compounds, and it is considered a wonder of nature. The cherry silverberry is referred to as a "wonder berry" in the Far East [31,35]. Not only the fruit, but also other plant parts such as flowers, leaves, roots, and stems have been utilized in traditional medicine. Scientific studies have confirmed the antioxidant [13,25,76–80], anti-inflammatory [13,30,33,34], antiproliferative [12,32,81], anticancer [12,81,82], antimicrobial [19], antidiabetic [31,37,80] anti-fatigue [83], and alleviating [82] properties of *E. multiflora*.

Jung et al. [83] examined the effect of *Elaeagnus multiflora* fruits (EFM) on fatigue and exercise performance in BALB/c mice. These results suggest that EMF can be utilized as an efficacious natural resource for its anti-fatigue effects. Subsequent studies Jung et al. [83] conducted on aging male rats suggest that *Elaeagnus multiflora* and *Cynanchum wilfordii* can be effectively used to alleviate testosterone deficiency syndrome (TDS).

4.1. Antioxidant Activity

High antioxidant activity has become a topic of numerous studies [76]. The consumption of food that is rich in antioxidants reduces the risk of developing chronic diseases and oxidative stress [77,78]. The development of chronic, autoimmune, neurodegenerative, and metabolic diseases, as well as cancer, is positively correlated with oxidative stress [79]. Phenolics, as metabolites, possess antioxidant activity and can protect the body from damage caused by free-radical-induced oxidative stress (ROS) [24]. Oxidative stress, that is, the imbalance of antioxidants and prooxidants in favor of prooxidants, is caused by high levels of reactive ROS. In free radical processes, ROS react with cellular components, which leads to their modification and damage. A study investigating the total phenolics from different parts of *E. multiflora* from Gilgit-Baltistan (Pakistan) [25] revealed that this plant species is a good candidate for a natural antioxidant. It contains nutritional and functional material in its fruit, leaves, and young branches, and is able to repair damage caused to cells by ROS [25]. The results of the cited study indicate that the concentrations of phenolics in medicinal plant species vary across different plant parts and are affected by the nature of solvents. Lee et al. [13] demonstrated that the 50% ethanol extract of *E. multiflora* fruit displayed the highest antioxidant activities in ABTS$^{\bullet+}$ (2,2′-azino-bis(3-ethylbenzothiazoline-6-sulfonic acid) and DPPH$^{\bullet}$ (1,1-diphenyl-2-picrylhydrazyl) radical scavenging and power-reducing assays. The cited authors suggested that this extract may be used as a natural source for food supplements and pharmaceuticals, due to its strong biological activities and high phytochemical content. According to Ismail et al. [25], *E. multiflora* is rich in bioactive phenolic compounds, which should be isolated for further investigations.

According to Lizardo et al. [80], extracts of cherry silverberry fruits fermented by pure cultures of *Lactobacillus plantarum* KCTC 33131 and *L. casei* KCTC 13086 exhibited favorable physicochemical properties and enhanced phytochemical content, antioxidant properties (DPPH radical scavenging activity, reducing power, superoxide dismutase-like property and hydrogen peroxide scavenging activity), and α-glucosidase and tyrosinase enzyme inhibitory activity, as compared with unfermented fruits. Despite a decrease in the specific phenolic acid contents among the fermented samples, the cherry silverberry fruit, fermented by mixed cultures of *Lactobacillus plantarum* and *L. casei*, contained superior total polyphenols and total individual flavonoid contents in comparison with fruits fermented by single cultures and unfermented ones

4.2. Antimicrobial Properties of Elaeagnus

Microbes (such as bacteria, fungi, and viruses) are the major causative agents of infectious diseases, which pose threats to public health [74,75]. The search for plants with antimicrobial activity has gained importance in recent years, due to a growing concern about the increasing rates of infections caused by antibiotic-resistant microorganisms [84–89]. Several plant-derived products, such as essential oils and extracts, have been used as traditional antiseptics and have been reported to possess moderate to significant levels of antimicrobial properties. Extracts from plants of the genus *Elaeagnus* were found to be more active against Gram-positive than Gram-negative bacteria [74,90,91]. The antimicrobial activities of selected *Elaeagnus* species, namely *E. angustifolia* [92–94], *E. macrophylla* [95], *E. mollis* [58,96], *E. kologa* [97], *E. umbellate* [19,98,99], *E. maritime*, *E. submacrophylla* [100], and *E. indica* [74,101], have also been documented. In the work of Ismail et al. [25] and Nikolaeva et al. [102], epigallocatechin from *Elaeagnus galabra* has been recognized as an antibacterial agent. According to Zargari [103], the leaves and fruit of *E. angustifolia* and

E. *multiflora* exhibit antipyretic activity. Bacterial and fungal strains that are inhibited by E. *multiflora* extracts should be analyzed and characterized in more detail. Mubasher et al. [99] studied the antibacterial activity of E. *umbellate*, which is often confused with E. *multiflora* due to similarities in leaf and fruit morphology. The objective of their study was to evaluate the biological activity of E. *umbellata* extracts against standard microbial strains as well as multi-drug-resistant bacteria isolated from hospitals. Flowers, leaves, and berries were extracted in different solvents and were tested for their antibacterial activity by the disc diffusion method on selected organisms, such as the methicillin-resistant *Staphylococcus aureus* (*S. aureus*), multi-drug resistant *Pseudomonas aeruginosa* (*P. aeruginosa*), and enterohemorrhagic *Escherichia coli* (*E. coli*). Most of the extracts displayed broad-spectrum activity, since Gram-positive bacteria, including *S. aureus* and *B. subtilis*, as well as Gram-negative bacteria, including *E. coli* and *P. aeruginosa*, were inhibited. Srinivasan et al. [74] demonstrated that the leaf extracts of E. *indica* possess potent antimicrobial activities. They exerted varied inhibitory effects on the tested microbes. Most polar extracts exhibited strong antimicrobial activities [74,101]. The extracts of E. *umbellata* [104] and E. *indica* exerted greater inhibitory effects on bacteria than fungi. According to Piłat and Zadernowski [72], seabuckthorn leaves contain compounds that inhibit the growth of microorganisms such as *Bacillus cereus*, *Pseudomonas aeruginosa*, *Staphylococcus aureus*, and *Enterococcus faecalis* [105]. Moreover, seabuckthorn seed oil exhibits antibacterial activity against *Escherichia coli* [106].

The above findings indicate that E. *multiflora* can be used in the treatment of infectious diseases. The antimicrobial efficacy of various *Elaeganus* species has already been documented, but further research is needed to identify all of their bioactive compounds [19].

4.3. Antidiabetic Activity

Type 2 diabetes impairs insulin synthesis by the pancreas, thereby leading to hyperglycemia. The absorption of simple sugars should be controlled by the inhibitors of enzymes responsible for sugar hydrolysis in the gastrointestinal tract. In turn, obesity and lipid absorption are controlled by pancreatic lipase inhibitors [37,107,108]. Therefore, the antidiabetic activity of E. *multiflora* fruit parts was measured as the inhibitory activity against α-amylase, α-glucosidase, and pancreatic lipase [37]. The authors of the cited study tested six new biotypes of goumi, which were selected in the Experimental Garden of the University of Warmia and Mazury in Olsztyn (north-eastern Poland). The inhibitory activity against α-amylase and α-glucosidase in the fruit skin and pulp of E. *multiflora* reached 24.6 and 32.3 IC_{50} (mg/mL) on average, respectively, whereas the inhibitory activity against pancreatic lipase was 74.9 IC_{50} (mg/mL) on average, implying that the antidiabetic activity of the fruit skin and pulp was three-fold stronger than the antidiabetic activity of seeds and leaves. The highest inhibition of the tested enzymes was noted for the fruit skin and pulp of biotype Si5 (17.0 and 23.7 mg/mL against α-amylase and α-glucosidase, respectively), whereas obesity was most effectively controlled by the fruit skin and pulp of biotype Si4 (69.0 mg/mL against pancreatic lipase). The antidiabetic activity of E. *multiflora* fruit skin and pulp was similar to that noted for the extract of E. *umbellata* [107]. Lee et al. [13] found that the 50% ethanol extract of E. *multiflora* fruit has potent α-glucosidase inhibitory activity and could be an effective antidiabetic agent. α-glucosidase inhibitors can be used in the treatment of many diseases such as diabetes, cancer, and HIV [13,109–111], which has contributed to the increasing popularity of cherry silverberry. In a study by Lachowicz et al. [37], the fruit skin and pulp of E. *multiflora* exhibited the strongest antidiabetic properties because their components migrate to juice during pressing. Cherry silverberry juice can be used to produce a functional powdered additive. Furthermore, sugars can be removed from the juice to enhance its antidiabetic effect.

4.4. Anticancer Activity

In developed countries, cancer has emerged as the leading cause of premature death. Therefore, effective cancer prevention strategies are being sought. The results of epidemiologic studies have prompted food manufacturers to incorporate plant raw materials containing anticarcinogenic substances in their products [112]. This group of compounds includes lycopene, whose anticarcinogenic properties are associated with its high antioxidant activity. Cherry silverberry fruit is a valuable source of lycopene, which appears to be the most potent antioxidant among common carotenoids, known for its anticarcinogenic effects [14–19]. Studies involving cell lines, animals, and human subjects have shown that dietary lycopene can decrease the risk and growth of prostate cancer, ovarian cancer, cervical cancer, breast cancer, esophageal cancer, liver cancer, gallbladder cancer, brain tumors, and cardiovascular disease [109–117], as well as tumors of the upper respiratory tract [118].

Cancer is a disease in which some of the body's cells grow uncontrollably and spread to other parts of the body. The anticancer activity of *E. multiflora* has been confirmed by experiments, with in vitro as well as in vivo models. The mechanisms underlying tumor-suppressing properties, including the ability to remove ROS, interfere with cell division, and modulate the signal transduction pathway, are being investigated [2].

Lee et al. [12] examined the potential of cherry silverberry as a cancer-preventive agent through regulating inflammatory signals, including cyclooxygenase-2 (COX-2) and Akt. Extracts from the seeds and flesh of *E. multiflora* berries were obtained, and COX-2 and Akt activities were analyzed in cherry-silverberry-extract-treated HT-29 colon cancer cells. The study revealed that the analyzed seed extracts reduced cell viability at concentrations above 1600 mg/mL, and, effectively, reduced COX-2 and p-Akt expression. Both seed and flesh extracts inhibited cell growth and induced apoptosis in HT-29 cells. Lee et al. [30] confirmed that cherry silverberry extracts effectively scavenged 1,1-diphenyl-2-picrylhydrazyl (DPPH) radical in vitro, reduced nitric oxide production in LPS-treated macrophages, and inhibited cell proliferation in MCF7, Hela, and SNU-639 cancer cells. According to Lee et al. [12], further research is needed to elucidate the exact molecular mechanism, by which *E. multiflora* fruit induced apoptosis in colon cancer cells.

Several epidemiological studies [15] have suggested the presence of a positive correlation between inflammation and cancer, in particular a strong association between inflammatory bowel disease and a higher incidence rate of colon cancer. Oh and Lee [32] demonstrated that cherry silverberry seeds, in contrast to its flesh, are believed to exert a possible anticancer effect. *Elaeagnus multiflora* seeds are considered to be a candidate for an anticancer functional food in preventive nutrition programs.

Lizardo et al. [81] explored the possibility of adding value to an underutilized fruit, cherry silverberry, through the process of fermentation, which makes it a potential source of functional food and an ingredient for the prevention of colorectal cancer.

5. Conclusions

Similar to seabuckthorn, *E. multiflora* has many potential applications in human nutrition, food technology as an ingredient of functional food, cosmetics (including skin cosmetics), and pharmaceuticals as a component of nutraceuticals, medicine, manufacture, and animal nutrition. Cherry silverberry is a promising fruit plant, which perfectly matches the current trends in horticulture by promoting the cultivation of plants with edible fruit that is attractive to both consumers and food producers, on account of its high nutritional value, medicinal properties, and biological activity. Plant species that can be grown without chemicals and constitute rich sources of bioactive substances have attracted considerable interest from researchers worldwide. The identified bioactive compounds can be used to design new functional foods with specific properties. They are found not only in the fruit but also in other plant parts such as the bark, leaves, flowers, and seeds. The seeds are considered to be a candidate for an anticancer functional food in preventive nutrition programs. Nowadays, a healthy lifestyle is gaining increasing popularity, therefore, the health-promoting potential of plants should be further explored.

Author Contributions: Conceptualization, A.B. and S.L.-W.; investigation, A.B. and J.B.; writing–original draft preparation, A.B. (abstract, Sections 1, 2, 4 and 5) and S.L.-W. (Section 3); writing–review and editing, A.B. and S.L.-W.; project administration, A.B. and J.B.; funding acquisition, A.B. All authors have read and agreed to the published version of the manuscript.

Funding: The results presented in this paper were obtained as part of a comprehensive study financed by the University of Warmia and Mazury in Olsztyn, Faculty of Agriculture and Forestry, Department of Agroecosystems and Horticulture, 30.610.016-110. This project was financially supported by the Minister of Education and Science under the program entitled "Regional Initiative of Excellence" for the years 2019–2022, Project No. 010/RID/2018/19, amount of funding PLN 12,000,000.

Institutional Review Board Statement: Not applicable.

Informed Consent Statement: Not applicable.

Data Availability Statement: MDPI Research Data Policies.

Acknowledgments: Supported by the Foundation for Polish Science (FNP) and the scholarship for young scientists of the Ministry of Education and Science (MEIN) for Sabina Lachowicz-Wiśniewska.

Conflicts of Interest: The authors declare no conflict of interest.

Sample Availability: Samples of the compounds are available from the authors.

References

1. Piłat, B.; Zadernowski, R. Bioactive Substances—Positive and Negative Effects of their Addition to Foodstuffs. *Przemysł Spożywczy* **2017**, *71*, 24–27.
2. Lachowicz, S.; Bieniek, A.; Gil, Z.; Bielska, N.; Markuszewski, B. Phytochemical parameters and antioxidant activity of new cherry silverberry biotypes (*Elaeagnus multiflora* Thunb.). *Eur. Food Res. Technol.* **2019**, *245*, 1997–2005. [CrossRef]
3. Bieniek, A.; Dragańska, E.; Prancketis, V. Assessment of climatic conditions for *Actinidia arguta* cultivation in north-eastern Poland. *Zemdirb. Agric.* **2016**, *103*, 311–318. [CrossRef]
4. Latocha, P. The Nutritional and Health Benefits of Kiwiberry (*Actinidia arguta*)—A Review. *Plant Foods Hum. Nutr.* **2017**, *72*, 325–334. [CrossRef] [PubMed]
5. Czaplicki, S.; Ogrodowska, D.; Zadernowski, R.; Konopka, I. Effect of sea-buckthorn (*Hippophaë rhamnoides* L.) pulp oil consumption on fatty acids and vitamin A and E accumulation in adipose tissue and liver of rats. *Plant Foods Hum. Nutr.* **2017**, *72*, 198–204. [CrossRef] [PubMed]
6. Viapiana, A.; Wesolowski, M. The phenolic contents and antioxidant activities of infusions of *Sambucus nigra* L. *Plant Foods Hum. Nutr.* **2017**, *72*, 82–87. [CrossRef]
7. FAO. *The State of the World's Biodiversity for Food and Agriculture*; Bélanger, J.J., Pilling, D., Eds.; FAO Commission on Genetic Resources for Food and Agriculture Assessments: Rome, Italy, 2019.
8. Lachowicz, S.; Kapusta, I.; Świeca, M.; Stinco, C.M.; Meléndez-Martínez, A.J.; Bieniek, A. In vitro Antioxidant and Antidiabetic potency of fruits and leaves of *Elaeagnus multiflora* Thunb. and their isoprenoids and polyphenolics profile. *Antioxidants* **2020**, *9*, 436. [CrossRef]
9. Bieniek, A.; Piłat, B.; Szałkiewicz, M.; Markuszewski, B.; Gojło, E. Evaluation of yield, morphology and quality of (*Elaeagnus multiflora* Thunb.) biotypes under conditions of north-eastern Poland. *Pol. J. Nat. Sci.* **2017**, *32*, 61–70.
10. Wani, T.A.; Wani, S.M.; Ahmad, M.; Ahmad, M.; Ganil, A.; Masoodi, F.A. Bioacrive profile, health benefits and safety evaluation of sea buckthorn (*Hippophaë rhamnoides* L.): A review. *Cogent Food Agric.* **2016**, *2*, 1128519.
11. Bieniek, A.; Kawecki, Z.; Piotrowicz-Cieślak, A.I. The content of some organic ingredients in the fruit of less known fruit plants. *Biul. Nauk.* **2002**, *14*, 11–17. (In Polish)
12. Lee, M.S.; Lee, Y.K.; Park, O.J. Cherry silverberry (*Elaeagnus multiflora*) extracts exere anti-inflammatory effects by inhibiting COX-2 and Akt signals in HT-29 colon cancer cells. *Food Sci Biotechnol.* **2010**, *19*, 1673–1677. [CrossRef]
13. Lee, J.H.; Seo, W.T.; Cho, K.M. Determination of phytochemical contents and biological activities from the fruits of *Elaeagnus multiflora*. *Int. J. Food Sci. Nutr.* **2011**, *16*, 29–36. [CrossRef]
14. Nowak, K.W.; Mielnik, P.; Sięda, M.; Staniszewska, I.; Bieniek, A. The effect of ultrasound treatment on the extraction of lycopene and β-carotene from cherry silverberry fruits. *AIMS Agric. Food* **2021**, *6*, 247–254. [CrossRef]
15. Przybylska, S. Lycopene-a bioactive carotenoid offering multiple health benefits: A review. *Int. J. Food Sci. Technol.* **2020**, *55*, 11–32. [CrossRef]
16. Di Mascio, P.; Kaiser, S.; Sies, H. Lycopene as the most efficient biological carotenoid singlet oxygen quencher. *Arch. Biochem. Biophys.* **1989**, *274*, 532–538. [CrossRef]
17. Yoon, K.Y.; Hong, J.Y.; Shin, S.R. Analysis on the Components of the *Elaeagnus multiflora* Thunb. Leaves. *Korean J. Food Preserv.* **2007**, *14*, 639–644.
18. Stahl, W.; Sies, H. Antioxidant activity of carotenoids. *Mol. Aspects. Med.* **2003**, *24*, 345–351. [CrossRef]

19. Patel, S. Plant genus Elaeagnus: Underutilized lycopene and linoleic acid reserve with permaculture potential. *Fruits* **2015**, *70*, 191–199. [CrossRef]
20. Ahmadiani, A.; Hosseiny, J.; Semnanian, S.; Javan, M.; Saeedi, F.; Kamalinejad, M.; Saremi, S. Antinociceptive and antiinflammatory effects of *Elaeagnus angustifolia* fruit extract. *J. Ethnopharmacol.* **2000**, *72*, 287–292. [CrossRef]
21. Szałkiewicz, M.; Kawecki, Z. Oliwnik wielokwiatowy (*Elaeagnus multiflora* Thunb.)—Nowa roślina sadownicza. *Biul. Nauk.* **2003**, *22*, 285–290. (In Polish)
22. Grygorieva, O.; Klymenko, S.; Ilinska, A.; Brindza, J. Variation of fruits morphometric parameters of *Elaeagnus multiflora* Thunb., germplasm collection. *Potravin. Slovak J. Food Sci.* **2018**, *12*, 527–532. [CrossRef]
23. You, Y.H.; Kim, K.B.; An Ch, S.; Kim, J.H.; Song, S.D. Geographical Distribution and Soil Characteristics of Elaeagnus Plants in Korea. *Korean J. Ecol.* **1994**, *17*, 159–170.
24. Sakamura, F.; Suga, T. Changes in chemical components of ripening oleaster fruits. *Phytochemistry* **1987**, *26*, 2481–2484. [CrossRef]
25. Ismail, M.; Hussain, M.; Mahar, S.; Iqbal, S. Investigation on Total Phenolic Contents of *Elaeagnus Multiflora*. *Asian J. Chemstry* **2015**, *27*, 4587–4590. [CrossRef]
26. Bieniek, A.; Lachowicz, S. Oliwnik wielokwiatowy—Alternatywa dla produkcji ekologicznej. In Proceedings of the Conference materials X FairFruit and Vegetable Industry of TSW, Warsaw Expo, Nadarzyn, Poland, 15–16 January 2020. (In Polish)
27. Shin, S.R.; Hong, J.Y.; Yoon, K.Y. Antioxidant properties and total phenolic contents of cherry Elaeagnus (*Elaeagnus multiflora* Thunb.) leaf extracts. *Food Sci. Biotechnol.* **2008**, *17*, 608–612.
28. Kim, S.A.; Oh, S.I.; Lee, M.S. Antioxidative and cytotoxic effects of solvent fractions from *Elaeagnus multiflora*. *Korean J. Food Nutr.* **2007**, *20*, 134–142.
29. Kim, S.T.; Kim, S.W.; Ha, J.; Gal, S.W. *Elaeagnus multiflora* fruit extract inhibits melanin biosynthesis via regulation of tyrosinase gene on translational level. *Res. J. Biotechnol.* **2014**, *9*, 1–6.
30. Lee, Y.S.; Chang, Z.Q.; Oh, B.C.; Park, S.C.; Shin, S.R.; Kim, N.W. Antioxidant activity, anti-inflammatory activity, and whitening effects of extracts of *Elaeagnus multiflora* Thunb. *J. Med. Food* **2007**, *10*, 126–133. [CrossRef]
31. Bieniek, A. Oliwnik szansa na zwiększenie bioróżnorodności w sadownictwie. *Truskawka Malina Jagody* **2021**, *1*, 51–53. (In Polish)
32. Kim, S.; OH, S.; Lee, M. Antioxidative and Cytoxic Effects of Ethanol Extracts from *Elaeagnus multiflora*. *Korean J. Food Nutr.* **2008**, *21*, 403–409.
33. Houng, J.Y.; Nam, H.S.; Lee, Y.S.; Yoon, K.Y.; Kim, N.W.; Shin, S.R. Study on the antioxidant activity of extracts from the fruit of *Elaeagnus multiflora* Thunb. *Korean J. Food Preserv.* **2006**, *13*, 413–419.
34. Chang, Z.Q.; Park, S.C.; Oh, B.C.; Lee, Y.S.; Shin, S.R.; Kim, N. Antiplatet aggregation and antiinflammatory activity for extracts of *Elaeagnus multiflora*. *Korean J. Med. Crop Sci.* **2006**, *51*, 516–517.
35. Bieniek, A. „Cud—Jagoda, czyli oliwnik wielokwiatowy. *Szkółkarstwo* **2016**, *6*, 48–53. (In Polish)
36. Lachowicz, S.; Bieniek, A.; Wiśniewski, R.; Gil, Z.; Bielska, N.; Markuszewski, B. Profil parametrów fitochemicznych i właściwości przeciwoksydacyjne owoców oliwnika wielokwiatowego (Elaeagnusmultiflora Thunb.). *Materiały z konf. Naukowej "Miejsce ogrodnictwa we współczesnym życiu człowieka I ochronie środowiska Warszawa* **2019**, 35. (In Polish)
37. Lachowicz-Wiśniewska, S.; Kapusta, I.; Stinco, C.M.; Meléndez-Martínez, A.J.; Bieniek, A.; Ochmian, I.; Gil, Z. Distribution of Polyphenolic and Isoprenoid Compounds and Biological Activity Differences between in the Fruit Skin+ Pulp, Seeds, and Leaves of New Biotypes of *Elaeagnusmultiflora* Thunb. *Antioxidants* **2021**, *10*, 849. [CrossRef]
38. Chinnici, F.; Spinabelli, U.; Riponi, C.; Amati, A. Optimization of the determination of organic acids and sugars in fruit juices by ion-exclusion liquid chromatography. *J. Food Compos. Anal.* **2005**, *18*, 121–130. [CrossRef]
39. Zielińska, A.; Nowak, I. Tokoferole i tokotrienole jako witamina E. *Chemik* **2014**, *68*, 585–591. (In Polish)
40. Hryniewski, T. Drzewa i krzewy. In *Vademecum Miłośnika Przyrody*; Wyd. Mulico Oficyna Wydawnicza: Warsaw, Poland, 2008. (In Polish)
41. Kołbasina, E. *Jagodnyje Liany i Redkije Kustarniki*; Izdatielskij Dom MSP: Moscow, Russia, 2003; p. 112. (In Russion)
42. Kozioł, A. Anti-aging active substances and application methods based on nanotechnology. *Kosmetologia Estetyczna* **2020**, *2*, 213–218.
43. Pawlowski, R. Substancje czynne w ziołach. *Hod. Trzody Chlewnej* **2013**, 11–12. (In Polish)
44. Kim, N.W.; Yoo, E.Y.; Kim, S.L. Analysis on the Components of the Emit of *Elaeagnus multiflora* Thumb. *Korean J. Food Preserv.* **2003**, *10*, 534–539.
45. Mikulic-Petkovsek, M.; Schmitzer, V.; Slatnar, A.; Stampar, F.; Veberic, R. Composition of sugars, organic acids, and total phenolics in 25 wild or cultivated berry species. *J. Food Sci.* **2012**, *77*, C1064–C1070. [CrossRef] [PubMed]
46. Hong, J.Y.; Cha, H.S.; Shin, S.R.; Jeong, Y.J.; Youn, K.S.; Kim, M.H.; Kim, N.W. Optimization of manufacturing condition and physicochemical properties for mixing beverage added extract of *Elaeagnus multiflora* Thunb. fruits. *Korean J. Food Preserv.* **2007**, *14*, 269–275.
47. Mikulic-Petkovsek, M.M.; Stampar, F.; Veberic, R. Parameters of inner quality of the apple scab resistant and susceptible apple cultivars (*Malus domestica* Borkh.). *Sci. Hortic.* **2007**, *114*, 37–44. [CrossRef]
48. Keutgen, A.; Pawelzik, E. Modifications of taste-relevant compounds in strawberry fruit under NaCl salinity. *Food Chem.* **2007**, *105*, 1487–1494. [CrossRef]
49. Janda, K.; Kasprzak, M.; Wolska, J. Witamina C–budowa, właściwości, funkcje i występowanie. *Pom. J. Life Sci.* **2015**, *61*, 419–425. (In Polish) [CrossRef]

50. Yew, W.W.; Chang, K.C.; Leung, C.C.; Chan, D.P.; Zhang, Y. Vitamin C and Mycobacterium tuberculosis persisters. *Antimicrob. Agents Chemother.* **2018**, *62*, e01641-18. [CrossRef]
51. Wasiuk, E. Łoch mnogocwietkowyj kak płodowaja kultura. Materiały z VIII Międzynarodowej konferencji sadowniczej pt. Sowremennyje naucznyje issliedowanija w sadowodstwie. *Jałta* **2000**, *2*, 34–36.
52. Khattak, K.F. Free radical scavenging activity, phytochemical composition and nutrient analysis of *Elaeagnus umbellata* berry. *J Medic Plants Res.* **2012**, *6*, 5196–5203.
53. Senica, M.; Stampar, F.; Mikulic-Petkovsek, M. Blue honeysuckle (*Lonicera cearulea* L. subs. edulis) berry; A rich source of some nutrients and their differences among four different cultivars. *Sci. Hortic.* **2018**, *238*, 215–221. [CrossRef]
54. Jarosz, M.; Rychlik, E.; Stoś, K.; Charzewska, J. *Normy Żywienia Dla Populacji Polski i ich Zastosowanie*; Narodowy Instytut Zdrowia Publicznego-Państwowy Zakład Higieny: Warsaw, Poland, 2020; pp. 68–437.
55. Bal, L.M.; Meda, V.; Naik, S.N.; Satya, S. Sea buckthorn berries: A potential source of valuable nutrients for nutraceuticals and cosmoceuticals. *Food Res. Int.* **2011**, *44*, 1718–1727. [CrossRef]
56. Glew, R.H.; Ayaz, F.A.; Sanz, C.; VanderJagt, D.J.; Huang, H.S.; Chuang, L.T.; Strnad, M. Changes in sugars, organic acids and amino acids in medlar (*Mespilus germanica* L.) during fruit development and maturation. *Food Chem.* **2003**, *83*, 363–369. [CrossRef]
57. Mazza, G. Compositional and functional properties of saskatoon berry and blueberry. *Int. J. Fruit Sci.* **2005**, *5*, 101–120. [CrossRef]
58. Zhang, Y.; Li, P.; Cheng, L. Developmental changes of carbohydrates, organic acids, amino acids, and phenolic compounds in 'Honeycrisp'apple flesh. *Food Chem.* **2010**, *123*, 1013–1018. [CrossRef]
59. Olszowy, M. What is responsible for antioxidant properties of polyphenolic compounds from plants? *Plant Physiol. Biochem.* **2019**, *144*, 135–143. [CrossRef]
60. Piłat, B. Owoce rokitnika (*Hippophae rhamnoides* L.) jako źródło substancji biologicznie aktywnych. Ph.D. Thesis, Biblioteka UWM, Olsztyn, Poland, 2014. (In Polish).
61. Teleszko, M.; Wojdyło, A.; Rudzinska, M.; Oszmianski, J.; Golis, T. Analysis of lipophilic and hydrophilic bioactive compounds content in sea buckthorn (*Hippophae rhamnoides* L.) berries. *J. Agric. Food Chem.* **2015**, *63*, 4120–4129. [CrossRef] [PubMed]
62. Cho, K.M.; Joo, O.S. Quality and antioxidant charactistics of *Elaeagnus multiflora* wine through the thermal processing of juice. *Korean J. Food Preserv.* **2014**, *21*, 206–214. [CrossRef]
63. Lachowicz, S.; Oszmiański, J.; Kalisz, S. Effects of various polysaccharide clarification agents and reaction time on content of polyphenolic compound, antioxidant activity, turbidity and colour of chokeberry juice. *LWT* **2018**, *92*, 347–360. [CrossRef]
64. Spínola, V.; Pinto, J.; Llorent-Martínez, E.J.; Castilho, P.C. Changes in the phenolic compositions of Elaeagnus umbellata and Sambucus lanceolata after in vitro gastrointestinal digestion and evaluation of their potential anti-diabetic properties. *Food Res. Int.* **2019**, *122*, 283–294. [CrossRef]
65. Meléndez-Martínez, A.J. An overview of carotenoids, apocarotenoids, and vitamin A in agro-food, nutrition, health, and disease. *Mol. Nutr. Food Res.* **2019**, *63*, 1801045. [CrossRef]
66. Meléndez-Martínez, A.J.; Stinco, C.M.; Mapelli-Brahm, P. Skin carotenoids in public health and nutricosmetics: The emerging roles and applications of the UV radiation-absorbing colourless carotenoids phytoene and phytofluene. *Nutrients* **2019**, *11*, 1093. [CrossRef]
67. Ignat, I.; Volf, I.; Popa, V.I. A critical review of methods for characterisation of polyphenolic compounds in fruits and vegetables. *Food Chem.* **2011**, *126*, 1821–1835. [CrossRef] [PubMed]
68. Delgado-Pelayo, R.; Hornero-Méndez, D. Identification and quantitative analysis of carotenoids and their esters from sarsaparilla (*Smilax aspera* L.) berries. *J. Agric. Food Chem.* **2012**, *60*, 8225–8232. [CrossRef] [PubMed]
69. Andersson, S.C.; Olsson, M.E.; Johansson, E.; Rumpunen, K. Carotenoids in sea buckthorn (*Hippophae rhamnoides* L.) berries during ripening and use of pheophytin a as a maturity marker. *J. Agric. Food Chem.* **2009**, *57*, 250–258. [CrossRef] [PubMed]
70. Eggersdorfer, M.; Wyss, A. Carotenoids in human nutrition and health. *Arch. Biochem. Biophys.* **2018**, *652*, 18–26. [CrossRef] [PubMed]
71. Dias, M.G.; Olmedilla-Alonso, B.; Hornero-Méndez, D.; Mercadante, A.Z.; Osorio, C.; Vargas-Murga, L.; Meléndez-Martínez, A.J. Comprehensive database of carotenoid contents in ibero-american foods. A valuable tool in the context of functional foods and the establishment of recommended intakes of bioactives. *J. Agric. Food Chem.* **2018**, *66*, 5055–5107. [CrossRef]
72. Piłat, B.; Zadernowski, R. Fruits of sea buckthorn (*Hippophae rhamnoides* L.)—Rich source of biologically active compounds. *Postępy Fitoterapii.* **2016**, *17*, 298–306.
73. Kallio, H.; Yang, B.; Peippo, P.; Tahvonen, R.; Pan, R. Triacylglycerols, Glycerophospholipids, Tocopherols, and Tocotrienols in Berries and Seeds of Two Subspecies (ssp. sinensis and mongolica) of Sea Buckthorn (*Hippophaë rhamnoides*). *J. Agric. Food Chem.* **2002**, *50*, 3004–3009. [CrossRef]
74. Srinivasan, R.; Aruna, A.; Manigandan, K.; Pugazhendhi, A.; Kim, M.; Shivakumar, M.S.; Natarajan, D. Phytohemical, antioxidant, antimicrobial and antiproliferative potential of *Elaeagnus indica*. *Biocatal. Ana Agric. Biotechnol.* **2019**, *20*, 101265. [CrossRef]
75. Mahomoodally, M.F.; Zengin, G.; Aumeeruddy, M.Z.; Sezgin, M.; Aktumsek, A. Phytochemical profile and antioxidant properties of two *Brassicaceae* species: *Cardaria draba* subsp. *Draba and Descurainia sophia*. *Biocatalysis Agric. Biotechnol.* **2018**, *16*, 453–458. [CrossRef]
76. Gołba, M.; Sokół-Łętowska, A.; Kucharska, A.Z. Health properties and composition of honeysuckle berry *Lonicera caerulea* L. an update on recent studies. *Molecules* **2020**, *25*, 749. [CrossRef]

77. Kolniak-Ostek, J.; Kłopotowska, D.; Rutkowski, K.P.; Skorupińska, A.; Kruczyńska, D.E. Bioactive Compounds and Health-Promoting Properties of Pear (*Pyrus communis* L.) Fruits. *Molecules* **2020**, *25*, 4444. [CrossRef] [PubMed]
78. Gayer, B.A.; Avendano, E.E.; Edelson, E.; Nirmala, A.; Johnson, E.J.; Raman, G. Effects of intake of apples, pears, or their products on cardiometabolic risk factors and clinical outcomes: A systematic review and meta-Analysis. *Curr. Dev. Nutr.* **2019**, *3*, 1–14. [CrossRef] [PubMed]
79. Pawlowska, E.; Szczepanska, J.; Koskela, K.; Kaarniranta, K.; Blasiak, J. Dietary polyphenols in age-related macular degeneration: Protection against oxidative stress and beyond. *Oxid. Med. Cell. Longev.* **2019**, *2019*, 9682318. [CrossRef] [PubMed]
80. Lizardo, R.C.M.; Cho, H.-D.; Won, Y.S.; Seo, K.-I. Fermentation with mono- and mixedcultures of *Lactobacillus plantarum* and *casei* enhances the phytochemical content and biological activities of cherry silverberry (*Elaeagnus multiflora* Thunb.) fruit. *J. Sci. Food Agric.* **2020**, *100*, 3687–3696. [CrossRef]
81. Lizardo, R.C.M.; Cho, H.-D.; Lee, J.-H.; Won, Y.S.; Seo, K.-I. Extracts of *Elaeagnus multiflora* Thunb. Fruit fermented by lactic acid bacteria ihibit SW 480 human colon adenocarcinoma via induction of cell cycle arrest and suppression of metastatic potential. *J. Food Sci. Health Nutr. Food* **2020**, *85*, 2565–2577. [CrossRef]
82. Jung, M.-A.; Jo, A.; Shin, J.; Kang, H.; Kim, Y.; Oh, D.-R.; Choi, C.-Y. Anti-fatigue effects of *Elaeagnus multiflora* fruit extracts in mice. *J. Appl. Biol. Chem.* **2020**, *63*, 69–74. [CrossRef]
83. Jung, M.-A.; Shin, J.; Jo, A.; Kang, H.; Lee, G.; Oh, D.-R.; Yun, H.J.; Im, S.; Bae, D.; Kim, J.; et al. Alleviating effects of the mixture of *Elaeagnus multiflora* and *Cynanchum wilfordii* extracts on testosteronedeficiency syndrome. *J. Appl. Biol. Chem.* **2020**, *63*, 451–455. [CrossRef]
84. Dehghan, M.H.; Soltani, J.; Kalantar, E.; Farnad, M.; Kamalinejad, M.; Khodaii, Z.; Hatami, S.; Natanzi, M.M. Characterization of an Antimicrobial Extract from *Elaeagnus angustifolia*. *Int. J. Enteric. Pathog.* **2014**, *2*, e20157. [CrossRef]
85. Sá, M.B.; Ralph, M.T.; Nascimento, D.C.O.; Ramos, C.S.; Barbosa, I.M.S.; Sá, F.B.; Lima-Filho, J.V. Phytochemistry and Preliminary Assessment of the Antibacterial Activity of Chloroform Extract of Amburana cearensis (Allemão) AC Sm. against Klebsiella pneumoniae Carbapenemase-Producing Strains. *Evid. Based Complementary Altern. Med.* **2014**, *2014*. [CrossRef]
86. Nageeb, A.; Al-Tawashi, A.; Emwas, A.M.; Al-Talla, Z.A.; Al-Rifai, N. Comparison of *Artemisia annua* Bioactivities between Traditional Medicine and Chemical Extracts. *Curr. Bioact. Compd.* **2013**, *9*, 324. [CrossRef]
87. Gurbuz, I.; Ustun, O.; Yesilada, E.; Sezik, E.; Kutsal, O. Anti-ulcerogenic activity of some plants used as folk remedy in Turkey. *J. Ethnopharmacol.* **2003**, *88*, 93–97. [CrossRef]
88. Rawat, S.; Singh, R.; Thakur, P.; Kaur, S.; Semwal, A. Wound healing Agents from Medicinal Plants: A Review. *Asian Pac. J. Trop. Biomed.* **2012**, *2*, S1910–S1917. [CrossRef]
89. Lima-Filho, J.V.; Martins, L.V.; de Oliveira Nascimento, D.C.; Ventura, R.F.; Batista, J.E.C.; Silva, A.F.B.; Taciana Ralpha, M.; ValençaVaza, R.; Boa-Viagem Rabello, C.; da Silvac, I.M.M.; et al. Zoonotic potential of multidrug-resistant extraintestinal pathogenic Escherichia coli obtained from healthy poultry carcasses in Salvador, Brazil. *Braz. J. Infect. Dis.* **2013**, *17*, 54–61. [CrossRef] [PubMed]
90. Cowann, M.M. Plant products as antimicrobial agents. *Clin. Microbiol. Rev.* **1999**, *12*, 564–582. [CrossRef]
91. Uddin, G.; Rauf, A. Phytochemical screening and biological activity of the aerial parts of *Elaeagnus Umbellate*. *Sci. Res. Essays* **2012**, *7*, 3690–3694.
92. Khan, S.U.; Khan, A.U.; Ali Shah, A.U.; Shah, S.M.; Hussain, S.; Ayz, M.; Ayz, S. Heavy metals content, phytochemical composition, antimicrobial and insecticidal evaluation of *Elaeagnus angustifolia*. *Toxicol. Ind. Health* **2013**, *9*, 92. [CrossRef]
93. Okmen, G.; Turkcan, O. The antibacterial activity of *Elaeagnus angustifolia* L. against mastitis pathogens and antioxidant capacity of the leaf methanolic extracts. *J. Anim. Vet. Adv.* **2013**, *12*, 491–496.
94. Okmen, G.; Turkcan, O. A study on antimicrobial, antioxidant and antimutagenic activities of *Elaeagnus angustifolia* L. leaves. *Afr. J. Tradit. Complementary Altern. Med.* **2014**, *11*, 116–120. [CrossRef]
95. Liu, J.P.; Chang, Y.B.; Ya-ling, J.; Xiang, Y.G.; Chao, L. Study on the Antibacterial Activity of *Elaeagnus Macrophylla* Thunb. Leaf Extract. *North. Hortic.* **2001**, *1*, 144–145.
96. Fenjuan, S.; Zhiyong, C.; Ling, X.; Guiqin, Y. Study on Antimicrobial Activity of the Alkaloids from *Elaeagnus Mollis*. *Plant Prot.* **2009**, *1*, 126–128.
97. Merculieff, Z.; Ramnath, S.; Sankoli, S.M.; Venkataramegowda, S.; Murthy, S.G.; Ceballos, R.M. Phytochemical, antioxidant and antibacterial potential of *Elaeagnus kologa* (Schlecht.) leaf. *Asian Pac. J. Trop. Biomed.* **2014**, *4*, 687–691. [CrossRef]
98. Arias, R.M.; Prado, A.; Hernandez-Perez, B.M.; Sanchez Mateo, C.C. Antimicrobial studies on three species of *Hypericum* from the Canary Islands. *J. Ethnopharmacol.* **2002**, *40*, 287–292.
99. Mubasher, S.; Sabir; Dilnawaz, S.A.; Imtiaz, M.H.; Kaleem, M.T. Antibacterial activity of *Elaeagnus umbellata* (Thunb.) a medicinal plant from Pakistan. *Saudi Med. J.* **2007**, *28*, 259.
100. Lee, H.B.; Kim, C.S.; Ahn, Y.J. Anti-helicobacter pylori activity of methanol extracts from Korean native plant species in Jeju Island. *Agric. Chem. Biotechnol.* **2004**, *47*, 91–96.
101. RameshKannan, N.; Nayagam, A.A.J.; Gurunagara, S.; Muthukumar, B.; Ekambaram, N.; Manimaran, A. Photochemical screening from *Elaeagnus indica* activity against human pathogens and cancer cells. *Adv. Biol. Res.* **2013**, *7*, 95–103.
102. Nikolaeva, G.; Krivenchuk, P.E. Prokopenko. *Elaeagnus angustifolia flavonoids*. *Farm.* **1971**, *26*, 56–60.
103. Zargari, A. *Medicinal Plants*; Tehran University Press: Tehran, Iran, 1990; Volume 4, pp. 275–277.

104. Minhas, F.A.; Rehaman, H.; Yasin, A.; Awan, Z.I.; Ahmed, N. Antimicrobial activities of the leaves and roots of *Elaeagnus umbellate* Thunb. *Afr. J. Biotechnol.* **2013**, *12*, 6754–6760.
105. Wang, B.; Lin, L.; Ni, Q.; Lian Su, C. *Hippophae rhamnoides* Linn. For treatment of diabetes mellitus: A review. *J. Med. Plants Res.* **2011**, *5*, 2599–2607.
106. Lavinia, S.; Gabi, D.; Drinceanu, D.; Stef, D.; Daniela, M.; Julean, C.; Tetileanu, R.; Corcionivoschi, N. The effect of medicinal plants and plant extracted oils on broiler duodenum morphology and immunological profile. *Rom. Biotech. Lett.* **2009**, *14*, 4606–4616.
107. Nazir, N.; Zahoor, M.; Nisar, M.; Khan, I.; Karim, N.; Abdel-Halim, H.; Ali, A. Phytochemical analysis and antidiabetic potential of *Elaeagnus umbellate* (Thunb.) in streptozotocin-induced diabetic rats: Pharmacological and computational approach. *BMC Complement. Altern. Med.* **2018**, *18*, 332. [CrossRef]
108. Saltan, F.Z.; Okutucu, B.; Canby, H.S.; Ozel, D. In vitro α-Glucosidase and α-Amylase Enzyme Inhibitory Effects in *Elaeagnus angustifolia* Leaves Extracts. *Eurasian J. Anal. Chem.* **2017**, *12*, 117–126. [CrossRef]
109. Ducep, J.B.; Kastner, P.R.; Marshall, F.N.; Danzin, C. New potent α-glucohydrolase inhibitor MDL 73945 with long duration of action in rats. *Diabetes* **1991**, *40*, 825–830.
110. Fernandes, B.; Sagman, U.; Auger, M.; Demetrio, M.; Dennism, J.W. B-6 branched oligosaccharides as a marker of tumor progression in human breast and clon neoplasia. *Cancer Res.* **1991**, *51*, 718–723. [PubMed]
111. Ogawa, S.; Maruyama, A.; Odagiri, T.; Yuasa, H.; Hashimoto, H. Synthesis and biological evaluation of α-L-fucosidase inhibitors: 5a-carba- α-L-fucopyranosylamine and related compounds. *Eur. J. Org. Chem.* **2001**, 967–974. [CrossRef]
112. Skiepko, N.; Chwastowska-Siwiecka, I.; Kondratowicz, J. Properties of lycopene and utilizing it to produce functional foods. *ŻYWOŚĆ Nauka Technol. Jakość* **2015**, *6*, 20–32.
113. Bramley, P.M. Is lycopene beneficial to human health? *Phytochemistry* **2000**, *54*, 233–236. [CrossRef]
114. Larsson, S.C.; Orsini, N.; Wolk, A. Processed meat consumption and stomach cancer risk: A metaanalysis. *J. Natl. Cancer Inst.* **2006**, *98*, 1078–1087. [CrossRef]
115. Omoni, A.O.; Aluko, R.E. The anti-carcinogenic and anti-atherogenic effects of lycopene: A review. *Trends Food Sci. Technol.* **2005**, *16*, 344–350. [CrossRef]
116. Rao, A.V.; Agarwal, S. Role of antioxidant lycopene in cancer and heart disease. *J. Am. Coll. Nutr.* **2000**, *5*, 563–569. [CrossRef]
117. Yang, T.; Yang, X.; Wang, X.; Wang, Y.; Song, Z. The role of tomato products and lycopene in the prevention of gastric cancer: A meta-analysis of epidemiologic studies. *Med. Hypotheses* **2013**, *80*, 383–388. [CrossRef]
118. De Stefani, E.; Oreggia, F.; Boffetta, P.; Deneo-Pellegrini, H.; Ronco, A.; Mendilaharsu, M. Tomatoes, tomato-rich foods, lycopene and cancer of the upper aerodigestive tract: A case-control in Uruguay. *Oral Oncol.* **2000**, *36*, 47–53. [CrossRef]

Article

Triterpenoids of Three Apple Cultivars—Biosynthesis, Antioxidative and Anti-Inflammatory Properties, and Fate during Processing

Łukasz Woźniak [1,*], Anna Szakiel [2], Agnieszka Głowacka [3], Elżbieta Rozpara [3], Krystian Marszałek [4] and Sylwia Skąpska [4]

[1] Department of Food Safety and Chemical Analysis, Institute of Agricultural and Food Biotechnology—State Research Institute, 36 Rakowiecka Street, 02532 Warsaw, Poland
[2] Department of Plant Biochemistry, Faculty of Biology, University of Warsaw, 1 Miecznikowa Street, 02096 Warsaw, Poland
[3] Department of Pomology, Gene Resources and Nursery, The National Institute of Horticulture Research, 1/3 3 Maja Street, 96100 Skierniewice, Poland
[4] Department of Fruit and Vegetable Product Technology, Institute of Agricultural and Food Biotechnology—State Research Institute, 36 Rakowiecka Street, 02532 Warsaw, Poland
* Correspondence: lukasz.wozniak@ibprs.pl

Abstract: Triterpenoids are a group of secondary plant metabolites, with a remarkable pharmacological potential, occurring in the cuticular waxes of the aerial parts of plants. The aim of this study was to analyze triterpenoid variability in the fruits and leaves of three apple cultivars during the growing season and gain new insights into their health-promoting properties and fate during juice and purée production. The identification and quantification of the compounds of interest were conducted using gas chromatography coupled with mass spectrometry. The waxes of both matrices contained similar analytes; however, their quantitative patterns varied: triterpenic acids prevailed in the fruits, while higher contents of steroids and esterified forms were observed in the leaves. The total triterpenoid content per unit area was stable during the growing season; the percentage of esters increased in the later phases of growth. Antioxidative and anti-inflammatory properties were evaluated with a series of in vitro assays. Triterpenoids were found to be the main anti-inflammatory compounds in the apples, while their impact on antioxidant capacity was minor. The apples were processed on a lab scale to obtain juices and purées. The apple purée and cloudy juice contained only some of the triterpenoids present in the raw fruit, while the clear juices were virtually free of those lipophilic compounds.

Keywords: antioxidants; cuticular waxes; cyclooxygenases; GC-MS; Golden Delicious; Ligol; *Malus domestica*; phytosterols; Redkroft; triterpenes; ursolic acid

1. Introduction

The cuticle is a complex hydrophobic layer covering the non-woody aerial parts of plants, which has been developed through evolution to allow plants to survive in a terrestrial environment and endure its challenges. Consequently, the cuticle's main function is to act as a barrier to transpirational loss of water, although it is also responsible for protecting the plant against pests and pathogens, screening UV-B radiation, and defining organ boundaries during development [1]. The typical composition of cuticles includes a macromolecular scaffold of linked cutin and a variety of lipids that are collectively termed waxes. The chemical composition of the waxes shows great variability between species, but also among the organs of the same plant, development stages, and environmental conditions [2–4]. The primary constituents of waxes are very long chain fatty acids (typically C20-C34) and their derivatives, which are responsible for the mechanical and waterproof properties of cuticles. In addition to these, cuticles often contain secondary metabolites, the most important of which are triterpenoids [1,5].

The term "triterpenoids" is used to describe the group of plant metabolites synthetized from a common intermediate, 2,3-oxidosqualene. Its cyclization and rearrangement lead to the formation of two groups of metabolites: sterols (containing a tetracyclic scaffold and a side chain) and triterpenes (containing a pentacyclic scaffold), while further enzymatic modifications result in the occurrence of more than 20,000 different structures found in nature [6]. Triterpenes from the oleane, ursane, and lupane families are the most common constituents of cuticular waxes [7]. They have important ecological and agronomical significance connected with resistance against pests and pathogens; additionally, their impact on consumer health is considered very promising [3,6]. Ursolic acid, which is widely recognized as being the most abundant of the triterpenoids present in apples, exhibits a wide array of pharmacological activities. The features of ursolic acid include anti-cancer potential, antioxidative and anti-inflammatory properties, protection of internal organs against chemically induced damage, and anti-microbial potential directly connected with the acid's role in plants' defense system [8]. Plant sterols (phytosterols) can also be found in plant cuticles, although, unlike triterpenes, they are ubiquitous in plant cells since they are responsible for the stabilization of cellular membranes. Dietary phytosterols are important agents in the prevention and treatment of hypercholesterolemia, and in a wider context, they have a significant impact on the absorption of fat-soluble diet components [9]. Plant peels can also contain other groups of secondary metabolites, such as anthocyanins and lycopene, which are responsible for coloration, although their presence was not investigated in this study.

Apples are among the most widely produced and consumed fruits worldwide, with approximately 80 million tons harvested each year [10]. They are the most important fruit considering the total revenue of their sale, while the cultivation, trade, and processing of apples provide maintenance to thousands of people around the world [11]. In addition to their nutritional value and economic significance, apples are also a rich source of secondary metabolites with bioactive potential, which are claimed to have a beneficial impact on health [12,13]. Thus far, the majority of the research has been focused on apple polyphenols, which include, inter alia, chlorogenic acid, catechin, epicatechin, phloridzin, and procyanidins [14]. The polyphenol levels are considered the main quality parameter in studies evaluating apples' nutraceutical composition and improvement possibilities [15]. In recent years, an increasing number of publications have reported the content and properties of the lipophilic compounds present in the cuticular waxes of apples. The majority of the data in the literature [16–20] show that the typical triterpenoid pattern of apple peel includes ursolic acid as the dominant compound, followed by oleanolic acid; however, many contradictory results have been published over the years, which justifies further investigation. Additionally, the formation of triterpenoid esters with various compound groups (fatty acids, carbohydrates, phenolic acids) has also been reported [21–24], but these compounds are often overlooked during analysis.

The aim of this work was to analyze the triterpenoid content in cuticular waxes encompassing the fruits and leaves of apples and its changes during the growing season. The experiments included both free and esterified forms of triterpenic acids, neutral triterpenes, and sterols in three apple cultivars from the same origin. This paper broadens the knowledge on the presence of triterpenoids in apples and, according to the authors' best knowledge, is the first work that investigates the content of these compounds in apple leaves. Furthermore, the focus was placed on the biological activities of the triterpenoids. Their antioxidative and anti-inflammatory properties were analyzed in vitro, including their stability during processing.

2. Results

2.1. Identification of Triterpenoids

As stated below, the analyses of triterpenoids were conducted using GC-FID-MS. Three methods of analyte identification were applied simultaneously: use of analytical standards, examination of GC-MS spectra, and comparison of the results to samples of known composition from earlier studies. Meanwhile, a qualitative analysis was performed using an FID. Figure 1 presents the structures of the quantified compounds. Some of the compounds shown were not detected in the samples.

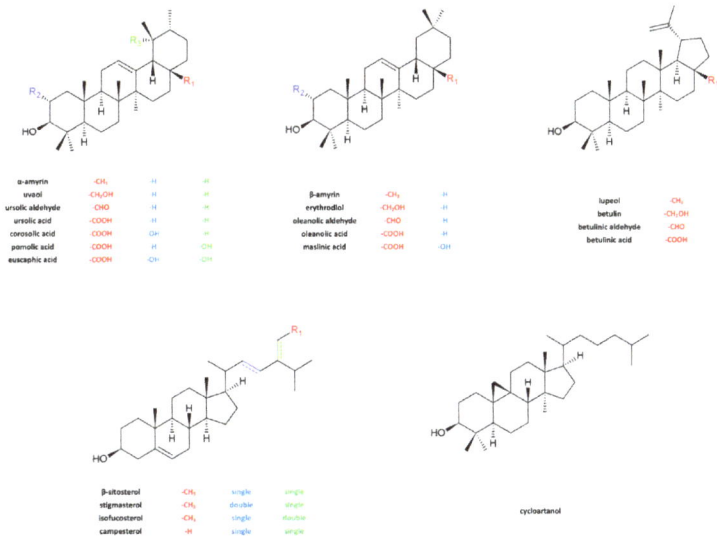

Figure 1. Triterpenoids quantified in apple fruits and leaves.

2.2. Triterpenoids in the Fruit

The cuticular waxes of all cultivars were dominated by ursolic acid and oleanolic acid. The content of minor triterpenic acids varied between the cultivars: 'Redkroft' contained only betulinic acid, 'Ligol' contained small amounts of pomolic acid, and 'Golden Delicious' contained betulinic acid as well as a wide spectrum of hydroxylated acids from the oleane and ursane families. Intermediates of triterpenic acid biosynthesis were detected at lower levels. The triterpenoid content was roughly constant, with slightly higher levels at the beginning of the growing season. Esterified forms of the triterpenes were detected at significant levels on the last two collection dates. Despite the lower content of free forms than ursolic acid, esters of oleanolic acid were more abundant in all cultivars. The phytosterol content was stable during the growing season and did not vary significantly between the cultivars. β-sitosterol was the most abundant compound from this group, with an approximately 90% share. Figure 2 presents the changes in the total triterpenoid content during the growing season, while Table 1 shows numerical data on the content of the most abundant constituents of the waxes in the last stage of development. Additionally, Table S1 in the Supporting Information presents all of the obtained data.

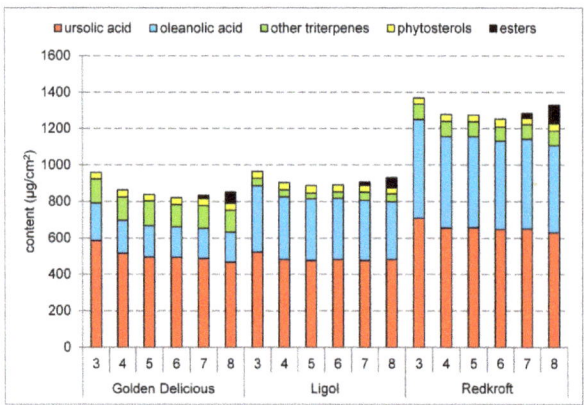

Figure 2. The changes in the triterpenoid content in fruits per unit area during the growth of three apple cultivars. Numbers on the horizontal axis denote the collection dates.

Table 1. The content of selected triterpenoids in apple fruits in the last stage of development. Data in Roman script represent unbounded analytes, while data in italic script correspond to their esterified forms. Letters represent the statistical significance of the differences between cultivars (Tukey test, α = 0.05).

Content (µg cm^{-2})	'Golden Delicious'	'Ligol'	'Redkroft'
Ursolic acid	466.2 ± 24.0 c	581.0 ± 14.2 b	630.4 ± 11.0 a
	13.1 ± 6.4 b	*12.1 ± 2.8 b*	*31.5 ± 3.1 a*
Oleanolic acid	166.1 ± 13.2 c	317.8 ± 19.8 b	476.1 ± 7.7 a
	39.4 ± 2.0 b	*38.1 ± 5.4 b*	*63.1 ± 3.4 a*
Betulinic acid	30.2 ± 4.1 b	nd	52.1 ± 7.1 a
	1.4 ± 0.7 a	*nd*	*4.3 ± 2.0 a*
Pomolic acid	49.4 ± 5.7 a	10.2 ± 2.5 b	nd
	4.1 ± 2.4	*nd*	*nd*
Corosolic acid	10.2 ± 1.5	nd	nd
	1.1 ± 0.4	*nd*	*nd*
α-amyrin	9.0 ± 0.7 a	8.2 ± 0.5 ab	6.3 ± 0.5 b
	nd	*nd*	*nd*
Uvaol	4.1 ± 0.7 a	5.1 ± 0.8 a	5.3 ± 0.6 a
	nd	*nd*	*nd*
β-amyrin	7.1 ± 0.9 a	5.0 ± 1.1 ab	3.3 ± 0.2 b
	nd	*nd*	*nd*
Erythrodiol	3.2 ± 0.5 a	2.1 ± 0.3 a	2.9 ± 0.2 a
	nd	*nd*	*nd*
β-sitosterol	35.2 ± 1.6 a	33.3 ± 1.0 a	37.4 ± 1.9 a
	1.4 ± 0.3 a	*1.7 ± 0.4 a*	*1.4 ± 0.4 a*
Campesterol	1.4 ± 0.2 a	1.0 ± 0.1 a	1.2 ± 0.2 a
	nd	*nd*	*nd*

nd—not detected (≤0.2 µg cm^{-2}).

2.3. Triterpenoids in the Leaves

The cuticular waxes of the apple leaves contained the same analytes as those of the fruits; however, their levels were different. Ursolic acid and oleanolic acid were the most abundant triterpenic acids, although their content was almost an order of magnitude lower.

The content of other triterpenic acids and neutral triterpenes decreased proportionally. The degree of esterification was much higher than in the fruits; additionally, esters were detected during the entire growing season. The phytosterol levels were stable and approximately 25% higher than in the fruits. The total triterpenoid content increased at the beginning of the season and was maintained in the later stages. Figure 3 presents the changes in the total triterpenoid content during the growing season, while Table 2 shows numerical data on the content of the most abundant constituents of the waxes in the last stage of development. Additionally, Table S2 in the Supporting Information presents all of the obtained data.

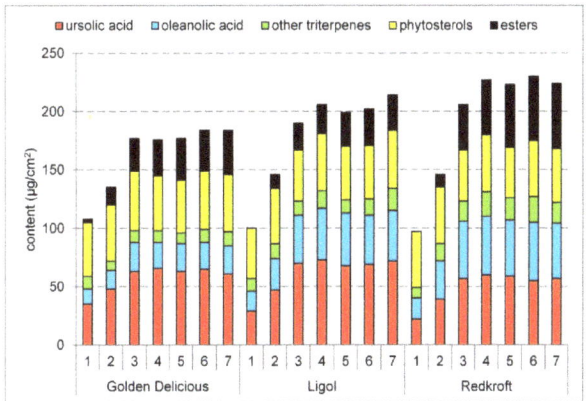

Figure 3. The changes in the triterpenoid content in leaves per unit area during the growth of three apple cultivars. Numbers on the horizontal axis denote the collection dates.

Table 2. The content of selected triterpenoids in apple leaves in the last stage of development. Data in Roman script represent unbounded analytes, while data in italic script correspond to their esterified forms. Letters represent the statistical significance of the differences between cultivars (Tukey test, $\alpha = 0.05$).

Content ($\mu g\ cm^{-2}$)	'Golden Delicious'	'Ligol'	'Redkroft'
Ursolic acid	61.2 ± 4.1 a *12.1 ± 1.4 b*	71.9 ± 5.0 a *13.9 ± 1.2 ab*	57.3 ± 3.2 a *18.3 ± 1.9 a*
Oleanolic acid	24.0 ± 2.0 b *18.4 ± 1.2 b*	43.4 ± 1.6 a *12.1 ± 0.9 c*	46.5 ± 2.4 a *31.5 ± 0.7 a*
Betulinic acid	1.6 ± 0.2 b *nd*	nd *nd*	8.2 ± 1.4 a *0.9 ± 0.4*
Pomolic acid	1.4 ± 0.2 a *0.5 ± 0.2*	0.7 ± 0.0 b *nd*	nd *nd*
Corosolic acid	nd *nd*	nd *nd*	nd *nd*
α-amyrin	4.1 ± 0.6 ab *0.6 ± 0.2 a*	5.2 ± 0.3 a *0.8 ± 0.0 a*	3.7 ± 0.5 b *0.5 ± 0.2 a*
Uvaol	2.1 ± 0.4 a *nd*	1.7 ± 0.3 a *nd*	1.9 ± 0.3 a *nd*
β-amyrin	2.0 ± 0.2 b *nd*	4.9 ± 0.7 a *0.6 ± 0.1*	2.6 ± 0.4 b *nd*
Erythrodiol	1.1 ± 0.3 ab *nd*	1.4 ± 0.2 a *nd*	0.8 ± 0.0 b *nd*
β-sitosterol	45.8 ± 1.6 a *2.6 ± 0.4 a*	48.3 ± 2.1 a *3.4 ± 0.5 a*	44.7 ± 1.3 a *2.5 ± 0.2 a*
Campesterol	1.7 ± 0.4 a *nd*	1.2 ± 0.2 a *nd*	1.0 ± 0.3 a *nd*

nd—not detected ($\leq 0.2\ \mu g\ cm^{-2}$).

2.4. Antioxidative and Anti-Inflammatory Properties of Triterpenoids

Three main triterpenoids of the apples as well as two phenolic compounds characteristic of these fruits were subjected to a series of analyses in order to evaluate their antioxidative and anti-inflammatory properties. Table 3 summarizes the IC_{50} values obtained for the pure compounds. Figure 4 compares the activities exhibited by apple extracts with the theoretical activity of their triterpenoid constituents.

Table 3. Antioxidative and anti-inflammatory properties of the main triterpenoids and phenolic compounds found in the apples. The results are expressed per dry mass. All results are presented as IC_{50} expressed in mg L^{-1}.

Test	Ursolic Acid	Oleanolic Acid	β-Sitosterol	Chlorogenic Acid	Phloridzin	Apple Extract
ABTS•+	163 ± 8	155 ± 7	130 ± 4	23 ± 2	34 ± 4	140 ± 8
DPPH•	94 ± 3	96 ± 5	88 ± 5	21 ± 3	23 ± 3	82 ± 9
COX-1	52 ± 4	104 ± 6	542 ± 6	1047 ± 73	960 ± 61	205 ± 12
COX-2	31 ± 3	73 ± 7	382 ± 4	612 ± 31	644 ± 38	144 ± 10
5-LOX	717 ± 43	641 ± 22	1740 ± 52	>5000	>5000	2084 ± 301

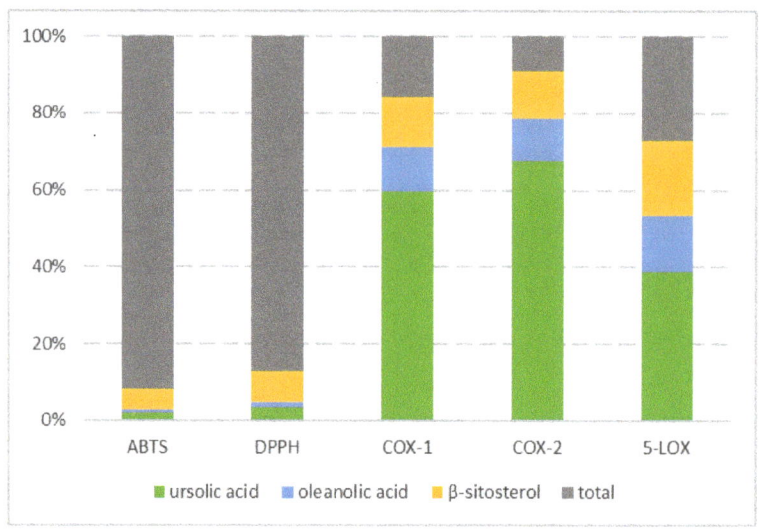

Figure 4. The contribution of the triterpenoids to the total antioxidative and anti-inflammatory activities of apple extracts. The activity of the apple extracts was compared with the theoretical activity of their selected constituents.

2.5. Impact of Processing on Triterpenoid Content

The 'Golden Delicious' apples were subjected to lab-scale processing in order to evaluate the changes in the triterpenoid content during purée and juice production. Additionally, commercial samples of apple-based products were bought and analyzed. A summary of the results is presented in Table 4.

Table 4. Content of the main triterpenoids in apple products from laboratory processing (triplicate production process) and commercial sale (ten distinct products per category). The results are expressed per wet mass. Letters represent the statistical significance of the differences between groups (Tukey test, α = 0.05).

Content (mg L^{-1}) or (mg kg^{-1})	Ursolic Acid	Oleanolic Acid	β-Sitosterol
Apple	56.1 ± 3.1 a	24.1 ± 1.3 a	80.5 ± 3.1 a
Purée (laboratory)	19.6 ± 2.1 b	8.0 ± 1.1 c	53.6 ± 4.2 b
Cloudy juice (laboratory)	6.2 ± 0.9 c	2.8 ± 0.4 d	17.0 ± 2.9 d
Clear juice (laboratory)	<0.1 d	<0.1 e	<0.1 e
Purée (commercial)	22.3 ± 3.9 b	15.2 ± 4.0 b	34.1 ± 11.1 c
Cloudy juice (commercial)	4.1 ± 2.0 c	2.3 ± 0.5 d	14.1 ± 5. d
Clear juice (commercial)	<0.1 d	<0.1 e	<0.1 e

3. Discussion

3.1. Triterpenoids in the Fruit

The literature provides numerous publications dealing with the triterpenoid content in apples, including the analysis of the whole fruit, its selected parts, and processed apples; however, the results are often contradictory and difficult to compare due to the method of their presentation.

The typical triterpenic acid pattern was reported by Andre et al. in their study investigating 109 apple cultivars. They quantified the content of three triterpenic acids and found ursolic acid to be the most prevalent (median content in the peel of 1.32 mg g^{-1}), followed by oleanolic acid (0.45 mg g^{-1}) and betulinic acid (44 µg g^{-1}). All cultivars except one showed a similar acid content profile: ursolic > oleanolic > betulinic, whilst the waxes of the 'Merton Russet' cultivar were dominated by betulinic acid [16]. Subsequent publications by these authors confirmed the findings, but also reported the presence of trans- and cis-caffeates of triterpenic acids in the apple cuticular waxes. These compounds were present in all the fruits tested; however, their content was significantly higher in russetted cultivars [21,25]. Similar results were obtained by other research teams. Butkevičiūte et al. analyzed the triterpenic acid content in six apple cultivars. In the peel, they found ursolic acid at levels of 4.03–6.43 mg g^{-1}, oleanolic acid at 0.95–1.24 mg g^{-1}, corosolic acid at 0.22–0.83 mg g^{-1}, and betulinic acid at 39–83 µg g^{-1} (all expressed per dry weight), while the flesh contained only trace amounts of triterpenic acids [18]. Dashbaldan and colleagues reported the composition of cuticular waxes of the 'Antonovka' cultivar. The observed patterns were dominated by ursolic and oleanolic acid, while the minor triterpenic compounds included betulinic acid, corosolic acid, and oxo derivatives of ursolic and oleanolic acid [26]. Woźniak et al. analyzed the triterpene content in dried apple pomace obtained from an industrial plant and thus containing a mixture of cultivars. They found ursolic acid to be the most abundant (7.13 mg g^{-1}), followed by oleanolic acid (1.59 mg g^{-1}), pomolic acid (0.87 mg g^{-1}), and neutral terpenoids at levels not exceeding 0.1 mg g^{-1} [27]. Considering the findings of Sut et al., the uniformity of the results can be connected with the common heritage of contemporary cultivars. The authors analyzed the phenolics and triterpenic acids in old Italian apple cultivars. Whereas commercial cultivars exhibited typical UA-dominated profiles with a total content in the dried peel of 10–50 mg g^{-1}, the ancient varieties had a higher total content (25–70 mg g^{-1}) and different qualitative and quantitative patterns: UA and OA were minor constituents, while large amounts of pomolic acid, maslinic acid, corosolic acid, and cuneataol were found [28]. The results obtained in our study are consistent with the data in the literature for commercial apple cultivars.

The aforementioned results refer to the mass of the peel, which, since it is mechanically removed, may differ between the studies and, therefore, is not the optimal method for expressing the triterpenoid content. Instead, we decided to express it per amount of area, in line with other authors. The series of papers by Lv and co-workers investigated the impact of cultivation and storage parameters on the content of terpenic compounds in apples. The first paper stated that cold storage does not affect triterpenoid levels [29], and the second reported that the cultivar and sun exposure can affect the triterpenoid content [30], while the last showed the impact of the root stock and harvest time [31]. The ursolic acid content varied slightly between the fruits; however, it was typically in the range of 250–600 µg cm^{-2}, while oleanolic acid was at levels of 50–80 µg cm^{-2}. Ju and Bramladge reported an increase in the content of all constituents of the cuticular waxes after ethylene treatment of 'Delicious' apples; the reported ursolic acid levels were in the range of 60–250 µg cm^{-2} [32]. The ursolic acid levels per area of fruit were also provided by Frighetto et al. in a paper focusing on the isolation of this compound. The authors analyzed four cultivars, obtaining a content of 210–820 µg cm^{-2} [33]. Significantly lower levels of ursolic and oleanolic acid in the 'Florina' and 'Prima' cultivars were reported by Leide et al., with 50 µg cm^{-2} and 5 µg cm^{-2}, respectively [19]. The levels of ursolic acid (466–708 µg cm^{-2}) and oleanolic acid (166–541 µg cm^{-2}) presented in Table 1 are similar to those found in the aforementioned reports. The absence of literature on the levels of minor wax constituents makes a comparison impossible.

The literature includes studies implementing high-resolution mass spectrometry to identify the minor triterpenoid constituents of apple waxes. McGhie and colleagues detected 43 triterpenic acids and their derivatives in the peel of seven apple cultivars. In addition to monohydroxy acids (such as ursolic acid), di- and trihydroxy acids, oxo derivatives, and coumaric acid esters were also found; the results, however, were expressed only in relation to the total peak area [23]. Poirier et al. analyzed the content and partitioning of terpenoids in 'Granny Smith' apple cuticular waxes. In addition to the typical acidic and neutral triterpenes, the authors also found fatty acid esters of ursolic acid, uvaol, and α-amyrin. Remarkably, the first two were located mainly in the wax, while the latter was accumulated in the peel. The authors did not provide quantitative data on the content of the compounds in the samples [24]. Our study included an analysis of over 20 triterpenoids in free and esterified forms. An exact identification of the esters was not possible; however, due to the analytical approach, all bounded forms were quantified regardless of their chemical character. The presence of esterified forms of triterpenic acids and neutral triterpenes in apple waxes was confirmed.

It should be noted that contradictory results also appear in some papers. Bars-Cortina et al. analyzed phytochemicals in the flesh and peel of white- and red-fleshed apples. The peel contained ursolic (132–326 mg kg^{-1}), hydroxyursolic (42–83 mg kg^{-1}), euscaphic (9–127 mg kg^{-1}), maslinic (13–30 mg kg^{-1}), and betulinic (2–29 mg kg^{-1}) acid; however, oleanolic acid was not reported in any of the nine cultivars [17]. Wildner and co-authors reported the ursolic and betulinic acid content in five apple cultivars from southern Brazil, but they did not present any data on oleanolic acid [20]. He and Liu investigated the constituents of the peel of 'Red Delicious' apples. They presented only low levels of neutral triterpenes and terpenic acid esters, while neither ursolic nor oleanolic acid was reported [22]. The discrepancies could be the result of focusing the analyses on particular compounds and therefore neglecting others. Misidentifications are possible as the compounds from the ursane and oleane groups create pairs of corresponding isomers.

The data in the literature on the sterol content in apples typically describe β-sitosterol-dominated patterns. The amount of this compound in fresh apples is in the range of 79–157 µg g^{-1} [34–37], while in pomace, it is 1147 µg g^{-1} [27]. Campesterol is typically the second most abundant sterol, although its levels are more than an order of magnitude lower [35,36]. The pattern of sterols in our samples was similar to the findings in the literature; however, the amounts expressed per weight of sample were much lower. Sterols are ubiquitous in cells, while our protocol only included an extraction of those present in the peel. Poirier and co-workers performed an in-depth analysis of the content and partitioning of steroids in 'Granny Smith' apple cuticular waxes. They found that, in addition to free sterols, acyl esters, glycosides, and acyl ester glycosides are also present. Most of these compounds were located in the peel rather than in the waxes. The authors did not provide quantitative data on the content of the compounds in the samples [24]. The results obtained in our study agree with the findings in the literature.

A few other teams reported changes in the triterpenoid content in other fruits. Dashbaldan and co-workers investigated the triterpenoid content in three phenological stages of the development of four edible berry species (*Vaccinium myrtillus*, *Vaccinium vitis-ideae*, *Arbutus unedo*, and *Lonicera caerulea*). The most abundant triterpenoids in their waxes were ursolic acid, oleanolic acid, α-amyrin, and β-sitosterol. Unfortunately, the results were expressed only in relation to the total mass of the extracted waxes. Nevertheless, the content of the majority of the analytes was constant during their development [38]. Salvador and colleagues focused their work on the phytochemicals of three elderberry (*Sambucus nigra* L.) cultivars. The cuticular waxes of all the cultivars were dominated by ursolic acid, which accounted for approx. 70% of the total content. The reported triterpenoid content decreased during the season; however, the results were only expressed as the dry mass of the fruit and, while the mass/area ratio changed during the growth of the fruit, are hard to compare with other findings [39].

3.2. Triterpenoids in the Leaves

According to our best knowledge, the literature contains only one report on the triterpenoid content in apple leaves. Bringe et al. analyzed adaxial (upper) surfaces and reported that the content of ursolic acid and oleanolic acid was in the range of 177–390 ng cm^{-2} and 20–97 ng cm^{-2}, respectively [40]. The levels observed in our study were a few orders of magnitude higher. A more similar triterpenoid content was reported by Jetter et al., who analyzed the composition of cuticular waxes of *Prunus laurocerasus* leaves. The experiments showed a difference in the triterpenoid content between the abaxial (15 µg cm^{-2}) and adaxial surfaces (5 µg cm^{-2}). The publication reported the limited usability of mechanical methods of collecting surface wax for the analysis of triterpenoids, as they are present in the intracuticular part of the waxes [41].

Two publications by Pensec and colleagues investigated the cuticular waxes of grapes. The first one reported that oleanolic acid was the dominant compound in the fruit waxes of six cultivars. The changes in the composition of the waxes were also analyzed: a decrease in the triterpenoid content per gram of wax was reported; however, no recalculation considering fruit area was conducted [42]. The second publication focused on leaves and, surprisingly, reported a different pattern: the waxes were rich in lupeol and taraxeol, while oleanolic acid was one of the minor constituents [43]. In our study, the observed qualitative patterns were similar, but the levels of particular compounds differed between the fruits and the leaves.

3.3. Antioxidative and Anti-Inflammatory Properties of Triterpenoids

Three main triterpenes as well as two phenolic compounds typically found in apples were subjected to a series of assays to evaluate their potential health-promoting properties. The antioxidative capacity was evaluated via quantification of ABTS$^{\bullet+}$ and DPPH$^{\bullet}$ scavenging in simple chemical tests. Triterpenoids were inferior antioxidants compared to phenolic compounds; the assays showed that 4–7 times higher mass concentrations should be used to obtain a similar scavenging effect. The high antioxidative potential of phenolic compounds is a result of their structure; multiple double bonds as well as oxidation-prone hydroxyl and carbon groups allow them to accept several electrons per molecule [44,45]. The analyzed triterpenoids have a sole double bond; therefore, their potential in such tests is limited. Reports on antioxidative properties observed in vivo can be found in the literature [46–49], although the activity should be attributed to altered cell metabolism rather than a simple chemical reaction.

On the other hand, all three tests evaluating anti-inflammatory features showed a higher potential of triterpenes compared to phenolic compounds. The obtained IC_{50} values are similar to the literature findings [50–52], although their levels vary significantly between enzymes. Inhibition of cyclooxygenases can be considered promising, especially considering the COX-2/COX-1 ratio [53]. The high IC_{50} values for 5-LOX inhibition suggest that it could be impossible to obtain therapeutic concentrations in in vivo models without triggering cytotoxicity.

The contribution of triterpenoids to the overall antioxidative potential was minor. As stated before, phenolic compounds can be considered as the main antioxidants in apples, prevailing in both content and potential. Simultaneously, it was observed that triterpenic acids play a pivotal role in the inhibition of all of the tested pro-inflammatory enzymes, being responsible for up to 90% of the total activity. These findings should be considered as especially important since terpenes are often overlooked by food scientists and nutritionists investigating apples and apple-based products.

3.4. Impact of Processing on Triterpenoid Content

The levels of triterpenoids in the juices and purées were significantly lower than in the raw apples used for their production. In the case of ursolic and oleanolic acid, their content in the purée was approx. 3 times lower, while in the cloudy juice, their content was approx. 10 times lower. The content of β-sitosterol in the above products was 67% and 21% of the content in the apples, respectively. The contents of triterpenoids in the commercial products were similar to their laboratory equivalents; however, distinctions in the raw material and processing led to higher variance in the results for the commercial products. Meanwhile, in the clear juices, the levels of all analytes were below the limit of detection. These results are consistent with the literature; the presence of triterpenic acids is usually attributed to dried fruits and pomace, while their levels in juices are negligible [27,54]. The observed phenomena can be explained considering the low polarity of the triterpenoids. Their mass transfer from initial placement is insignificant; therefore, their content can be connected with the amount of apple tissue in products. Triterpenic acids are present mainly in the peel, the majority of which is discarded during processing; meanwhile, sterols are abundant in all cells, which results in higher retention in products.

The commercial products were subjected to thermal processing as part of their production. During the conventional pasteurization process, samples are kept at the temperature of 90–95 °C for 15–20 min, while newer applications are heading towards decreasing temperatures and process durations to limit the impact of heat on the sensory quality of the product and its bioactive compounds. For comparison, a significant rate of sitosterol oxidation is observed above 150 °C [55], while ursolic and oleanolic acids are even more stable in degradative processes occurring above 200 °C [56]. Therefore, it can be assumed that, in this case, temperature is not a significant factor.

The combination of the data levels of triterpenoids in various apple-based products and their anti-inflammatory properties can be used to draw another conclusion: the health benefits from the consumption of cloudy juices and purées will be superior to those from the consumption of clear juices. This supports earlier works highlighting the impact of fiber components in shaping the health-promoting activity of apple products [57,58].

4. Materials and Methods

4.1. Material

Three apple (*Malus domestica* Borkh.) cultivars varying in fruit peel coloration were selected for the experiment: 'Golden Delicious' (yellow), 'Ligol' (yellow with red blush), and 'Redkroft' (red). The plant material was collected in the experimental orchard of the National Institute of Horticulture Research in Dąbrowice (51°55′ N, 20°06′ E) during the 2017 growing season. The fruits and leaves were picked randomly from one tree per cultivar at regular intervals from April to September (eight times in total); Table 5 presents the dates when the material was collected. At least three fruits and six leaves were collected on each date to ensure that the samples were representative. The material collected was weighed, measured, and frozen at −18 °C prior to extraction. The Supporting Information presents the method of estimating the surface area [59].

Table 5. The dates of sample collection. The pluses and minuses denote whether the material was available on that day.

Term	1	2	3	4	5	6	7	8
Date	21 April	12 May	2 June	27 June	19 July	14 August	8 September	29 September
Fruits	-	-	+	+	+	+	+	+
Leaves	+	+	+	+	+	+	+	-

The 'Golden Delicious' apples used during the processing and in vitro antioxidative and anti-inflammatory tests were bought in a supermarket in Warsaw, Poland. The apples were processed using lab-scale methods. A Miniprimer 9 blender (Braun, Kronberg im Taunus, Germany) was used for purée preparation, and a Robot Coupe J80 Ultra (Robot Coupe, Vincennes, France) was used for juice pressing, while the clear juice was obtained via centrifugation and subsequent filtration. Additionally, ten samples of each product, namely, clear juice, cloudy juice, and purée, were acquired from a local market in Warsaw.

4.2. Chemicals and Standards

The analytical standards of ursolic acid, oleanolic acid, betulinic acid, β-sitosterol, cholesteryl acetate, chlorogenic acid, and phloridzin were acquired from Merck (Darmstadt, Germany), while the standard of ursolic acid methyl ester was obtained from Carl Roth (Karlsruhe, Germany). Analytical-grade pyridine was acquired from Honeywell (Charlotte, NC, USA), while N, O-bis(trimethylsilyl)trifluoroacetamide with trimethylchlorosilane (BSTFA + 1% TMCS), p-anisaldehyde, and toluene were bought from Merck. Potassium hydroxide, acetic acid, sulfuric acid, and HPLC-grade methanol and chloroform were acquired from POCh (Gliwice, Poland). Hydrogen for chromatographic analysis was produced in situ by a Balston Hydrogen Generator (Parker Hannifin, Cleveland, OH, USA); other compressed gases were supplied by a local vendor.

4.3. Analysis of Triterpenoid Content

The multistep analytical procedure was conducted using an approach described by Pensec et al. [42]. The minor modifications that were implemented are described and justified in the descriptions of the individual stages.

The fruits and leaves were submerged in chloroform and stirred gently for 60 s. The amount of chloroform was selected to ensure at least 1 mL of solvent per square centimeter of fruit/leaf area. The extracts obtained were evaporated to dryness under reduced pressure using a Rotavapor R-300 vacuum dryer (Büchi, Flawil, Switzerland).

Part of the extracts obtained was subjected to alkaline hydrolysis using a protocol described by Woźniak et al. [60]. The reaction mixture was prepared by dissolving 0.75 g of potassium hydroxide in 1 mL of water, adding 4 mL of methanol, and dissolving the extract in 5 mL of toluene. The reaction was conducted for 60 min at 90 °C. The organic phase was collected, while the aqueous phase was re-extracted three times with toluene. The organic fractions were merged and evaporated to dryness. The original method included the separation of low polarity esters and their subsequent saponification and analysis [42]. The presence of hydrophobic fatty acid esters in the apple waxes is reported in the data in the literature [24]; however, reports describing the presence of esters of higher polarity, including esters of coumaric and caffeic acid, are also available [21–23]. Therefore, the selected approach allowed for the quantification of a wider array of terpenoid and triterpene derivatives.

The obtained extracts (both raw and hydrolyzed) were fractionated using preparative thin-layer chromatography (TLC). The samples were applied to TLC silica gel 60 glass plates (10 cm × 20 cm) (Merck, Darmstadt, Germany) and developed in a chloroform/methanol (97:3, v/v) mixture. The analytical standards of oleanolic acid and β-sitosterol were used to localize fractions containing triterpenic acids (R_F 0.2–0.3) and steroids and neutral triterpenes (RF 0.3–0.9), respectively. Due to safety concerns, the standards were visualized by spraying them with an anisaldehyde sulfuric acid reagent (recipe reported in the Supporting Information) and heating [61], instead of the use of 50% sulfuric acid described in the original method. Zones of the plate coating containing the desired fractions were scraped, and, subsequently, the analytes were rinsed with chloroform/methanol (2:1, v/v).

The fractions containing triterpenic acids were subjected to derivatization prior to the chromatographic analysis. The silylation protocol presented by Sánchez Ávila et al. [62] was used, instead of the original derivatization method utilizing diazomethane, which can be hazardous. An amount of 1 mL of the sample was placed in a 1.5 mL reaction tube and evaporated under a gentle stream of nitrogen. Afterwards, 600 µL of pyridine and 300 µL of BSTFA + 1% TMCS were added, and the tube was heated at 80 °C for 2 h.

Quantitative analyses of the samples were conducted using a Varian 430-GC gas chromatograph with a built-in flame ionization detector (FID) and CP-8400 autosampler (Varian, Palo Alto, CA, USA). The separations were conducted using a DB-5, 30 m × 0.25 mm, 0.25 µm column (Agilent Technologies, Palo Alto, CA, USA) that was eluted with helium at a flowrate of 1.0 mL min^{-1}. Samples (2.5 µL) were injected using a 1:10 split ratio and an injector temperature of 280 °C. Silylated triterpenic acids were analyzed under isothermal conditions at 280 °C for 60 min, while steroids and neutral triterpenes were analyzed using a temperature program: initial temperature of 160 °C for 2 min, which was increased at 5 °C min^{-1} to a final temperature of 280 °C that was held for a further 44 min. The detector was kept at a temperature of 300 °C, while the gas flows were 25 mL min^{-1} of N_2, 30 mL min^{-1} of H_2, and 300 mL min^{-1} of air. An internal standard (cholesteryl acetate) was added to each sample prior to the analysis.

The identification of the analytes was performed based on a comparison of their retention times with the standards. The five-point calibration curves showed good linearity in the tested range; the LOQ value for all analytes was set to 0.2 µg cm^{-2}. Additionally, selected samples were subjected to GC-MS analysis, and the obtained mass spectra were compared with the data in the literature. The GC-MS analyses were conducted using an Agilent 7890A gas chromatograph and 5975C mass spectrometric detector (both Agilent Technologies). The separation parameters were the same as those described above for the GC-FID analysis; the mass spectrometer worked with an energy of ionization of 70 eV and an m/z range of 33–500.

4.4. Antioxidative and Anti-Inflammatory Properties

The selected peel extracts as well as the standards of triterpenic compounds were analyzed for their antioxidative and anti-inflammatory properties. Due to their hydrophobic character, the samples were dissolved in dimethyl sulfoxide (DMSO). The content of DMSO in the reaction media in all the tests described below was 1% for all the tested samples as well as the controls.

The scavenging activity against ABTS$^{\bullet+}$ was measured with a protocol described by Re et al. [63]. A solution of 7 mM of ABTS and 2.45 mM of potassium persulfate was used for the generation of radical cations. After overnight incubation in darkness, the solution was diluted with ethanol to obtain an absorbance of approx. 0.7 at 734 nm. Then, 25 µL samples were mixed with 2.5 mL of ABTS$^{\bullet+}$ solution and incubated for 6 min. The scavenging of ABTS$^{\bullet+}$ molecules was measured spectrophotometrically at 734 nm. Trolox was chosen as a reference antioxidant.

A method described by Yen and Chen was used to determine the activity against DPPH• [64]. Briefly, 0.1 mL samples were mixed with 2 mL of DPPH• solution (1 mM) and incubated at room temperature for 30 min. The scavenging of DPPH• radicals was quantified spectrophotometrically at 517 nm. Trolox was chosen as a reference antioxidant.

The anti-inflammatory activity was evaluated using commercial kits for three enzymes of biological activity: cyclooxygenases 1 and 2 (COX-1 and COX-2) and 5-lipoxygenase (5-LOX), according to the instructions of the producers. Inhibition of the cyclooxygenases was measured using kits from Cayman Chemicals (Ann Arbor, MI, USA; items 701070 and 701080, respectively), while 5-LOX inhibition was measured using a kit from abcam (Cambridge, UK; item no. ab284521).

4.5. Statistics

Three independent samples were analyzed for each of the matrices. The data were analyzed using Statistica 7.1 software (StatSoft, Tulsa, OK, USA). ANOVA with a post hoc Tukey test at $\alpha = 0.05$ was used to determine the statistical significance of the differences.

Supplementary Materials: The following Supporting Information can be downloaded at: https://www.mdpi.com/article/10.3390/molecules28062584/s1. Table S1. Content of triterpenoids in apple fruit. Only compounds detected in at least one sample were listed. All data in $\mu g/cm^2$. nd—not detected ($\leq 0.2\ \mu g/cm^2$) Table S2. Content of triterpenoids in apple leaves. Only compounds detected in at least one sample were listed. All data in $\mu g/cm2$. nd—not detected ($\leq 0.2\ \mu g/cm^2$).

Author Contributions: Conceptualization, Ł.W.; methodology, Ł.W. and A.S.; formal analysis, Ł.W.; investigation, Ł.W.; resources, A.S., A.G. and E.R.; writing—original draft preparation, Ł.W.; writing—review and editing, A.S., A.G., E.R., K.M. and S.S.; supervision, K.M. and S.S.; funding acquisition, Ł.W. All authors have read and agreed to the published version of the manuscript.

Funding: This research was funded by the National Science Center (Narodowe Centrum Nauki), grant number 2016/21/N/NZ9/03442.

Institutional Review Board Statement: Not applicable.

Informed Consent Statement: Not applicable.

Data Availability Statement: Data is contained within the article or supplementary material.

Conflicts of Interest: The authors declare no conflict of interest.

References

1. Yeats, T.H.; Rose, J.K.C. The formation and function of plant cuticles. *Plant Physiol.* **2013**, *163*, 5–20.
2. Buschhaus, C.; Jetter, R. Composition differences between epicuticular and intracuticular wax substructures: How do plants seal their epidermal surfaces? *J. Exp. Bot.* **2011**, *62*, 841–853. [CrossRef]
3. Szakiel, A.; Pączkowski, C.; Pensec, F.; Bertsch, C. Fruit cuticular waxes as a source of biologically active triterpenoids. *Phytochem. Rev.* **2012**, *11*, 263–284.
4. Li, D.; Zaman, W.; Lu, J.; Niu, Q.; Zhang, X.; Ayaz, A.; Saqib, S.; Yang, B.; Zhang, J.; Zhao, H.; et al. Natural lupeol level variation among castor accessions and the upregulation of lupeol synthesis in response to light. *Ind. Crops Prod.* **2023**, *192*, 116090.
5. Kunst, L.; Samuels, A.L. Biosynthesis and secretion of plant cuticular waxes. *Prog. Lipid Res.* **2003**, *42*, 51–80. [CrossRef]
6. Thimmappa, R.; Geisler, K.; Louveau, T.; O'Maille, P.; Osbourn, A. Triterpene biosynthesis in plants. *Annu. Rev. Plant Bio.* **2014**, *65*, 225–257.
7. Lara, I.; Beige, B.; Goulao, L.F. A focus on the biosynthesis and composition of cuticle in fruits. *J. Agric. Food Chem.* **2015**, *63*, 4005–4019.
8. Woźniak, Ł.; Skąpska, S.; Marszałek, K. Ursolic acid—A pentacyclic triterpenoid with a wide spectrum of pharmacological activities. *Molecules* **2015**, *20*, 20614–20641.
9. Moreau, R.A.; Nyström, L.; Whitaker, B.D.; Winkler-Moser, J.K.; Baer, D.J.; Gebauer, S.K.; Hicks, K.B. Phytosterols and their derivatives: Structural diversity, distribution, metabolism, analysis, and health-promoting uses. *Prog. Lipid Res.* **2018**, *70*, 35–61.
10. FAOSTAT. Available online: http://faostat.fao.org (accessed on 10 October 2021).
11. O'Rourke, D. Economic Importance of the World Apple Industry. In *The Apple Genome*; Korban, S.S., Ed.; Springer: Cham, Switzerland, 2021; pp. 1–18.
12. Nezbedova, L.; McGhie, T.; Christensen, M.; Heyes, J.; Nasef, N.A.; Mehta, S. Onco-Preventive and Chemo-Protective Effects of Apple Bioactive Compounds. *Nutrients* **2021**, *13*, 4025.

13. Oyehini, A.B.; Belay, Z.A.; Mditshwa, A.; Caleb, O.J. "An apple a day keeps the doctor away": The potentials of apple bioactive constituents for chronic disease prevention. *J. Food Sci.* **2022**, *87*, 2291–2309. [CrossRef]
14. Kalinowska, M.; Bielawska, A.; Lewandowska-Siwkiewicz, H.; Priebe, W.; Lewandowski, W. Apples: Content of phenolic compounds vs. variety, part of apple and cultivation model, extraction of phenolic compounds, biological properties. *Plant Physiol. Biochem.* **2014**, *84*, 169–188. [CrossRef] [PubMed]
15. Farneti, B.; Masuero, D.; Costa, F.; Magnago, P.; Malnoy, M.; Costa, G.; Vrhovsek, U.; Mattivi, F. Is There Room for Improving the Nutraceutical Composition of Apple? *J. Agric. Food Chem.* **2015**, *63*, 2750–2759. [CrossRef] [PubMed]
16. Andre, C.M.; Greenwood, J.M.; Walker, E.G.; Rassam, M.; Sullivan, M.; Evers, D.; Perry, N.B.; Laing, W.A. Anti-inflammatory procyanidins and triterepenes in 109 apple varieties. *J. Agric. Food Chem.* **2012**, *60*, 10546–10554. [CrossRef]
17. Bars-Cortina, D.; Macià, A.; Iglesias, I.; Paz Romero, M.; Motilva, M.J. Phytochemical profiles of new red-fleshed apple varieties compared with traditional and new white-fleshed varieties. *J. Agric. Food Chem.* **2017**, *65*, 1684–1696. [CrossRef]
18. Butkevičiūte, A.; Liaudanskas, M.; Kviklys, D.; Zymonė, K.; Raudonis, J.V.; Uselis, N.; Janulis, V. Detection and analysis of triterpenic compounds in apple extracts. *Int. J. Food Prop.* **2018**, *21*, 1716–1727. [CrossRef]
19. Leide, J.; Xavier de Souza, A.; Papp, I.; Riederer, M. Specific characteristics of the apple fruit cuticle: Investigation of early and late season cultivars 'Prima' and 'Florina' (*Malus domestica* Borkh.). *Sci. Hortic.* **2018**, *229*, 137–147. [CrossRef]
20. Wildner, A.C.; Ferreira, P.L.; Oliveira, S.S.; Gnoatto, S.B.; Bergold, A.M. Variation of ursolic acid and betulinic acid in five *Malus domestica* clones form Southern Brazil. *J. Appl. Pharm. Sci.* **2018**, *8*, 158–165.
21. Andre, C.M.; Larsen, L.; Burgess, E.J.; Jensen, D.J.; Cooney, J.M.; Evers, D.; Zhang, J.; Perry, N.B.; Laing, W.A. Unusual immuno-modulatory triterpene-caffeates in the skins of russeted varieties of apples and pears. *J. Agric. Food Chem.* **2013**, *61*, 2273–2279. [CrossRef]
22. He, X.; Liu, R.H. Phytochemicals in apple peels: Isolation, structure elucidation, and their antiproliferative and antioxidant activities. *J. Agric. Food Chem.* **2008**, *56*, 9905–9910. [CrossRef] [PubMed]
23. McGhie, T.K.; Hudault, S.; Lunken, R.C.M.; Christeller, J.T. Apple peels, from seven cultivars, have lipase-inhibitory activity and contain numerous ursenoic acids as identified by LC-ESI-QTOF-HRMS. *J. Agric. Food Chem.* **2012**, *60*, 482–491. [CrossRef]
24. Poirier, B.C.; Buchanan, D.A.; Rudell, D.R.; Mattheis, J.P. Differential partitioning of triterpenes and triterpene esters in apple peel. *J. Agric. Food Chem.* **2018**, *66*, 1800–1806. [CrossRef]
25. Andre, C.M.; Legay, S.; Deleruelle, A.; Nieuwenhuizen, N.; Punter, M.; Brendolise, C.; Cooney, J.M.; Lateur, M.; Hausman, J.F.; Larondelle, Y.; et al. Multifunctional oxidosqualene cyclases and cytochrome P450 involved in the biosynthesis of apple fruit triterpenic acids. *New Phytol.* **2016**, *211*, 1279–1294. [CrossRef]
26. Dashbaldan, S.; Pączkowski, C.; Szakiel, A. Variations in triterpenoid deposition in cuticular waxes during development and maturation of selected fruits of Rosaceae family. *Int. J. Mol. Sci.* **2020**, *21*, 9762. [CrossRef]
27. Woźniak, Ł.; Szakiel, A.; Pączkowski, C.; Marszałek, K.; Skąpska, S.; Kowalska, H.; Jędrzejczak, R. Extraction of triterpenic acids and phytosterols from apple pomace with supercritical carbon dioxide: Impact of process parameters, modelling of kinetics, and scaling-up study. *Molecules* **2018**, *23*, 2790. [CrossRef]
28. Sut, S.; Zengin, G.; Maggi, F.; Malagoli, M.; Dall'Acqua, S. Triterpene acids and phenolics from ancient apples of Friuli Venezia Giulia as Nutraceutical Ingredients: LC-MS study and in vitro activities. *Molecules* **2019**, *24*, 1109. [CrossRef]
29. Lv, Y.; Tahir, I.; Olsson, M. Changes in triterpene content during storage of three apple cultivars. *Acta Hortic.* **2015**, *1071*, 365–368. [CrossRef]
30. Lv, Y.; Tahir, I.; Olsson, M. Factors affecting the content of the ursolic acid and oleanolic acid in apple peel: Influence of cultivars, sun exposure, storage conditions, bruising and *Penicillium expanum* infections. *J. Sci. Food Agric.* **2015**, *96*, 2161–2169. [CrossRef]
31. Lv, Y.; Tahir, I.; Olsson, M. Influence of rootstock, harvest time, and storage conditions on triterpene content of apple peel. *Acta Hortic.* **2016**, *1120*, 405–408. [CrossRef]
32. Ju, Z.; Bramlage, W.J. Developmental changes of cuticular constituents and their association with ethylene during fruit ripening in 'Delicious' apples. *Postharvest Biol. Technol.* **2001**, *21*, 257–263. [CrossRef]
33. Frighetto, R.T.S.; Welendorf, R.M.; Nigro, E.N.; Frighetto, N.; Siani, A.C. Isolation of ursolic acid from apple peels by high speed counter-current chromatography. *Food Chem.* **2008**, *106*, 767–771. [CrossRef]
34. Bartley, I.M. Lipid metabolism of ripening apples. *Phytochemistry* **1985**, *24*, 2857–2859. [CrossRef]
35. Han, J.H.; Yang, Y.X.; Feng, M.Y. Content of phytosterols in vegetables and fruits commonly consumed in China. *Biomed. Environ. Sci.* **2008**, *21*, 449–453. [CrossRef]
36. Piironen, V.; Toivo, J.; Puupponen-Pimiä, R.; Lampi, A.M. Plant sterols in vegetables, fruits, and berries. *J. Sci. Food Agric.* **2003**, *83*, 330–337. [CrossRef]
37. Normén, L.; Johnsson, M.; Andersson, H.; van Gameren, Y.; Dutta, P. Plant sterols in vegetables and fruits commonly consumed in Sweden. *Eur. J. Nutr.* **1999**, *38*, 84–89. [CrossRef] [PubMed]
38. Dashbaldan, S.; Becker, R.; Pączkowski, C.; Szakiel, A. Various patterns of composition and accumulation of steroids and triterpenoids in cuticular waxes from selected *Ericaceae* and *Caprifoliaceae* berries during fruit development. *Molecules* **2019**, *24*, 3826. [CrossRef]
39. Salvador, Â.C.; Rocha, S.M.; Silvestre, A.J.D. Lipophilic phytochemicals from elderberries (*Sambucus nigra* L.): Influence of ripening, cultivar and season. *Ind. Crops Prod.* **2015**, *71*, 15–23. [CrossRef]
40. Bringe, K.; Schumacher, C.F.A.; Schmitz-Eiberger, M.; Steriner, U.; Oerke, E.C. Ontogenic variation in chemical and physical characteristics of adaxial apple leaf surfaces. *Phytochemistry* **2006**, *67*, 161–170. [CrossRef]

41. Jetter, R.; Schäffer, S.; Riederer, M. Leaf cuticular waxes are arranged in chemically and mechanically distinct layers: Evidence from *Prunus laurocerasus* L. *Plant Cell Environ.* **2000**, *23*, 619–628. [CrossRef]
42. Pensec, F.; Pączkowski, C.; Grabarczyk, M.; Woźniak, A.; Bénard-Gellon, M.; Bertsch, C.; Chong, J.; Szakiel, A. Changes in the triterpenoid content of cuticular waxes during fruit ripening of eight grape (*Vitis vinifera*) cultivars grown in the Upper Rhine Valley. *J. Agric. Food Chem.* **2014**, *62*, 7998–8007. [CrossRef]
43. Pensec, F.; Szakiel, A.; Pączkowski, C.; Woźniak, A.; Grabarczyk, M.; Bertsch, C.; Fischer, M.J.C.; Chong, J. Characterization of triterpenoid profiles and triterpene synthase expression in the leaves of eight *Vitis vinifera* cultivars grown in the Upper Rhine Valley. *J. Plant Res.* **2016**, *129*, 499–512. [CrossRef] [PubMed]
44. Li, X.; Chen, B.; Xie, H.; He, Y.; Zhong, D.; Chen, D. Antioxidant structure-activity relationship analysis of five dihydrochalcones. *Molecules* **2018**, *23*, 1162. [CrossRef] [PubMed]
45. Spagnol, C.M.; Assis, R.P.; Brunetti, I.L.; Isaac, V.L.B.; Salgado, H.R.N.; Corrêa, M.A. In vitro methods to determine the antioxidant activity of caffeic acid. *Spectrochim. Acta A Mol. Biomol. Spectrosc.* **2019**, *219*, 358–366. [CrossRef] [PubMed]
46. Balanehru, S.; Nagarajan, B. Protective effect of oleanolic and ursolic acid against lipid peroxidation. *Biochem. Int.* **1991**, *24*, 981–990.
47. Ramachandran, S.; Rajendra Prasad, N.; Pugalendi, K.V.; Menon, V.P. Modulation of UVB-induced oxidative stress by ursolic acid in human blood lymphocytes. *Asian J. Biochem.* **2008**, *3*, 11–18. [CrossRef]
48. Ramos, A.A.; Pereira-Wilson, C.; Collins, A.R. Protective effects of ursolic acid and luteolin against oxidative DNA damage include enhancement of DNA repair in Caco-2 cells. *Mutat. Res.* **2010**, *692*, 6–11. [CrossRef] [PubMed]
49. Ramachandran, S.; Rajendra Prasad, N. Effect of ursolic acid, a triterpenoid antioxidant, on ultraviolet-B radiation-induced cytotoxicity, lipid peroxidation and DNA damage in human lymphocytes. *Chem. Biol. Interact.* **2008**, *176*, 99–107. [CrossRef]
50. Ringbom, T.; Segura, L.; Noreen, Y.; Perera, P.; Bohlin, L. Ursolic acid from *Plantago major*, a selective inhibitor of cyclooxygenase-2 catalyzed prostaglandin biosynthesis. *J. Nat. Prod.* **1998**, *61*, 1212–1215. [CrossRef]
51. Bowen-Forbes, C.S.; Mulabagal, V.; Liu, Y.; Nair, M.G. Ursolic acid analogues: Non-phenolic functional food components in Jamaican raspberry fruits. *Food Chem.* **2009**, *116*, 633–637. [CrossRef]
52. Lončarić, M.; Strelec, I.; Moslovac, T.; Šubarić, D.; Pavić, V.; Molnar, M. Lipoxygenase inhibition by plant extracts. *Biomolecules* **2021**, *11*, 152. [CrossRef]
53. Brooks, P.; Emery, P.; Evans, J.F.; Fenner, H.; Hawkey, C.J.; Patrono, C.; Smolen, J.; Breedveld, F.; Day, R.; Dougados, M.; et al. Interpreting the clinical significance of the differential inhibition of cyclooxygenase-1 and cyclooxygenase-2. *Rheumathology* **1999**, *38*, 779–788. [CrossRef]
54. Zhang, F.; Daimaru, E.; Ohnishi, M.; Kinoshita, M.; Tokuji, Y. Oleanolic acid and ursolic acid in commercial dried fruits. *Food Sci. Technol. Res.* **2013**, *19*, 113–116. [CrossRef]
55. Zhang, X.; Julien-David, D.; Miesch, M.; Geoffroy, P.; Raul, F.; Roussi, S.; Aoudé-Werner, D.; Marchioni, E. Identification and quantitative analysis of β-sitosterol oxides in vegetable oils by capillary gas chromatography–mass spectrometry. *Steroids* **2005**, *70*, 896–906. [CrossRef]
56. Fuliaş, A.; Ledeţi, I.; Vlase, G.; Vlase, T.; Şoica, C.; Dehelean, C.; Oprean, C.; Bojin, F.; Şuta, L.M.; Bercean, V.; et al. Thermal degradation, kinetic analysis, and apoptosis induction in human melanoma for oleanolic and ursolic acids. *J. Therm. Anal. Calorim.* **2016**, *125*, 759–768. [CrossRef]
57. Ravn-Haren, G.; Dragsted, L.O.; Buch-Andersen, T.; Jensen, E.N.; Jensen, R.I.; Németh-Balogh, M.; Paulovicsová, B.; Bergström, A.; Wilcks, A.; Licht, T.R.; et al. Intake of whole apples or clear apple juice has contrasting effects on plasma lipids in healthy volunteers. *Eur. J. Nutr.* **2013**, *52*, 1875–1889. [CrossRef]
58. Marcotte, B.V.; Verheyde, M.; Pomerleau, S.; Doyen, A.; Couillard, C. Health Benefits of Apple Juice Consumption: A Review of Interventional Trials on Humans. *Nutrients* **2022**, *14*, 821. [CrossRef]
59. Clayton, M.; Amos, N.D.; Banks, N.H.; Morton, R.H. Estimation of apple fruit surface area. *New Zeal. J. Crop Hort. Sci.* **1995**, *23*, 345–349. [CrossRef]
60. Woźniak, Ł.; Marszałek, K.; Skąpska, S.; Jędrzejczak, R. Novel method for HPLC analysis of triterpenic acids using 9-anthryldiazomethane derivatization and fluorescence detection. *Chromatographia* **2017**, *80*, 1527–1533. [CrossRef]
61. Janicsák, G.; Veres, K.; Kállai, M.; Máthé, I. Gas chromatographic method for routine determination of oleanolic and ursolic acids in medicinal plants. *Chromatographia* **2003**, *58*, 295–299.
62. Sánchez Ávila, N.; Priego Capote, F.; Luque de Castro, M.D. Ultrasound-assisted extraction and silylation prior to gas chromatography-mass spectrometry for the characterization of the triterpenic fraction in olive leaves. *J. Chromatogr. A* **2007**, *1665*, 158–165. [CrossRef] [PubMed]
63. Re, R.; Pellegrini, N.; Proteggente, A.; Pannala, A.; Yang, M.; Rice-Evans, C. Antioxidant activity applying an improved ABTS radical cation decolorization assay. *Free Radic. Biol. Med.* **1999**, *26*, 1231–1237. [CrossRef] [PubMed]
64. Yen, G.C.; Chen, H.Y. Antioxidant activity of various tea extracts in relation to their antimutagenicity. *J. Agric. Food Chem.* **1995**, *43*, 27–32. [CrossRef]

Disclaimer/Publisher's Note: The statements, opinions and data contained in all publications are solely those of the individual author(s) and contributor(s) and not of MDPI and/or the editor(s). MDPI and/or the editor(s) disclaim responsibility for any injury to people or property resulting from any ideas, methods, instructions or products referred to in the content.

Article

Accumulation of Antioxidative Phenolics and Carotenoids Using Thermal Processing in Different Stages of *Momordica charantia* Fruit

Ji Hye Kim [†], You Jin Lim [†], Shucheng Duan, Tae Jung Park and Seok Hyun Eom *

Department of Smart Farm Science, College of Life Sciences, Kyung Hee University, Yongin 17104, Republic of Korea
* Correspondence: se43@khu.ac.kr; Tel.: +82-31-201-3860
[†] These authors contributed equally to this work.

Abstract: The bitter taste of *M. charantia* fruit limits its consumption, although the health benefits are well known. The thermal drying process is considered as an alternative method to reduce the bitterness. However, processing studies have rarely investigated physiochemical changes in fruit stages. The antioxidant activities and physiochemical properties of various fruit stages were investigated using different thermal treatments. The color of the thermally treated fruit varied depending on the temperature. When heat-treated for 3 days, the samples from the 30 °C and 90 °C treatments turned brown, while the color of the 60 °C sample did not change significantly. The antioxidant activities were increased in the thermally processed samples in a temperature-dependent manner, with an increase in phenolic compounds. In the 90 °C samples, the 2,2-diphenyl-1-picrylhydrazyl radical scavenging activity presented a 6.8-fold higher level than that of nonthermal treatment in mature yellow fruit (S3), whereas the activity showed about a 3.1-fold higher level in immature green (S1) and mature green (S2) fruits. Regardless of the stages, the carotenoid content tended to decrease with increasing temperature. In terms of antioxidant activities, these results suggested that mature yellow fruit is better for consumption using thermal processing.

Keywords: antioxidant; bitter gourd; carotenoid; fruit stage; polyphenol; thermal processing

Citation: Kim, J.H.; Lim, Y.J.; Duan, S.; Park, T.J.; Eom, S.H. Accumulation of Antioxidative Phenolics and Carotenoids Using Thermal Processing in Different Stages of *Momordica charantia* Fruit. *Molecules* **2023**, *28*, 1500. https://doi.org/10.3390/molecules28031500

Academic Editor: Sabina Lachowicz-Wiśniewska

Received: 20 January 2023
Revised: 31 January 2023
Accepted: 1 February 2023
Published: 3 February 2023

Copyright: © 2023 by the authors. Licensee MDPI, Basel, Switzerland. This article is an open access article distributed under the terms and conditions of the Creative Commons Attribution (CC BY) license (https://creativecommons.org/licenses/by/4.0/).

1. Introduction

Momordica charantia, commonly known as bitter gourd, belongs to the Cucurbitaceae family and contains pharmaceuticals that are employed in traditional Asian medicines. The fruit of *M. charantia* is widely cultivated in tropical and subtropical climates, such as India, China, and Thailand for vegetables and medicinal usage [1]. The fruit are widely used with not only fresh salad and juice but also pre-boiling, drying, stir-frying, and frying to reduce their bitter flavor [2]. In addition, *M. charantia* has been used as a source of medicine to treat cough-, liver heating-, anthelmintic-, and diabetes-related diseases [3–5]. In particular, its fruit contains health-beneficial bioactive compounds such as charantin and it is particularly attractive for use in food and pharmaceuticals. In addition, it contains plenty of vitamin C, phenolic acids, and carotenoids, which have been considered antioxidants in food ingredients [6].

Despite these health advantages, its consumption has been limited due to a strong bitter taste, especially in raw fruit. Therefore, numerous attempts are being performed to reduce the taste while maintaining the health benefits. Heat treatments such as baking, roasting, and pressure cooking are known to reduce bitterness [7]. It is known that saponins, including momodicoside F, which contribute to the bitter taste in *M. charantia*, are also reduced by heat treatment [8]. Furthermore, hot-air drying is one of the most commonly employed heat treatments that alters not only the physical properties (such as hardness) but also the chemical properties (such as polyphenol decomposition) of food materials.

Heat treatment increases taste preferences by promoting nonenzymatic reactions in food or inducing a change in flavor components [6–13]. Heat treatment can cause an increase in phenolic compounds and antioxidant activity in various fruit crops such as citrus, persimmon peel, and eggplant fruit [10,14,15]. On the other hands, phenolic compounds and antioxidant activity were decreased by heating in some crops such as olive, persimmon flesh, and plum [10,16,17]. According to Choi et al. [18], *M. charantia* fruit roasted at 200 °C for 15 min showed similar antioxidant activity compared to the unroasted sample, but the flavan-3-ol and phenolic acid contents were about 1.4 times higher. Furthermore, Ng and Kuppusamy [19] found that heat treatments such as microwave heating and boiling were effective in increasing the antioxidant activity of *M. charantia* extract. Several studies have focused on the biological activities and variations in bioactive compounds during different thermal processing methods; however, current scientific information does not explain how these changes in the fruit occur at different hot-air drying temperatures.

M. charantia changes morphological and physicochemical characteristics depending on the growth stage. The fruit turns yellow as it ripens and the bitterness of the fruit decreases and the sweetness increases at this stage [20,21]. However, the fruit shows signs of decay and splitting, including cracking or bursting, causing it to be impossible for consumption as fresh fruit. Therefore, *M. charantia* is usually used in its mature green skin stage. Phenolic compounds can be increased or decreased depending on the crops, maturity, and processing methods. It was reported that the phenolic content decreased as the olive fruit matured, while it increased as soybean seed matured [9,22]. Moreover, the total phenolic content decreased during thermal processing in the flesh of persimmon fruit with the decrease in antioxidant activity, while it increased in the peel [10]. In a previous study, it was discovered that the polyphenolic compounds and antioxidant capacities of *M. charantia* were altered at various maturity stages [23,24]. It is reported that several phenolic compounds increased as *M. charantia* fruit matures [20,24]. Moreover, certain studies have shown the comparison between mature and immature fruits or leaves in different cultivars [25,26]; however, thermal processing at the different stages has not been studied in terms of the processing and antioxidant effects.

Therefore, the present study aimed to investigate the changes in antioxidant activities, phenolic compounds, carotenoids, and chlorophylls according to thermal treatments in the growth stage of *M. charantia* fruit. Furthermore, we have analyzed the major antioxidants using Pearson correlation coefficient analysis between antioxidant activities and the active compounds.

2. Results
2.1. Changes of Color during Thermal Process

Figure 1 shows the morphological characteristics of *M. charantia* with different maturity stages, presenting stage 1 (S1) as immature green fruit about 15 days after fertilization (DAF), stage 2 (S2) as mature green fruit at about 25 DAF, and stage 3 (S3) as mature yellow fruit at about 30 DAF. The treatments for each temperature at different stages were heat treated for 3 days and the freeze-dried samples were used as control. In the nonthermal process, the S1 and S2 fruits presented a light green color, while the fruits turned greenish yellow after 90 °C temperature treatment. S3 was yellow in nonthermal conditions and turned dark brown after 90 °C temperature treatment. The fruits dried at 60 °C in all stages showed a lighter color than those in the 30 °C and 90 °C temperature treatments. Figure 1C presents the CIE-Lab color values of these samples. The L^* values of *M. charantia* in all stages were similar at temperatures below 60 °C. Furthermore, the L^* value of S3 fruit decreased and was lower than that of S1 and S2 at 90 °C. The increasing a^* value representing the color varied from greenness to redness. The increased a^* values were observed as the drying temperature was enhanced regardless of the growth stages. The values of S3 ranged from 7.2 to 19.1 and were higher than those of S1 and S2 at all drying temperatures. The increasing b^* value representing the color varied from blueness to yellowness. The b^* values were maintained regardless of drying temperatures in S1 and S2.

However, without heat treatment, the value in S3 was higher than that in S1 and S2 but decreased as the drying temperature increased.

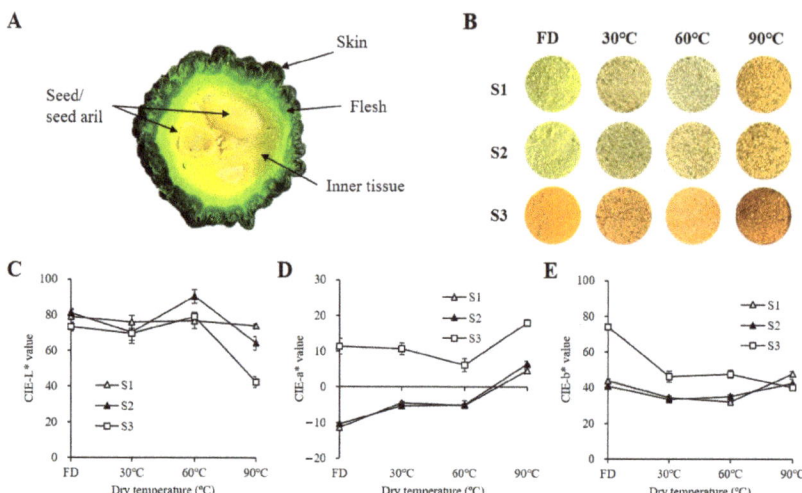

Figure 1. Morphological characteristics of *M. charantia* (**A**) cross section at the central part of *M. charantia*, (**B**) visual color, and (**C**) their CIE L*, (**D**) a*, and (**E**) b* values of *M. charantia* (ground powder) at different stages after thermal processing for 3 days. FD indicates freeze dry. S1, S2, and S3 indicate immature green fruit about 15 days after fertilization (DAF), mature green fruit about 25 DAF, and mature yellow fruit about 30 DAF.

2.2. Radical Scavenging Activities and Reducing Power

In Figure 2, the changes in antioxidant activity in *M. charantia* dried at different temperatures were evaluated using 2,2-diphenyl-1-picrylhydrazyl (DPPH) and 2,2′-azino-bis (3-ethylbenzothiazoline-6-sulfonic acid (ABTS) radical scavenging activities and ferric-reducing antioxidant power (FRAP). In all growth stages of *M. charantia*, the radical scavenging activities and reducing power increased as the drying temperature increased and were significantly higher in the 90 °C heat-treated fruit than other treatments.

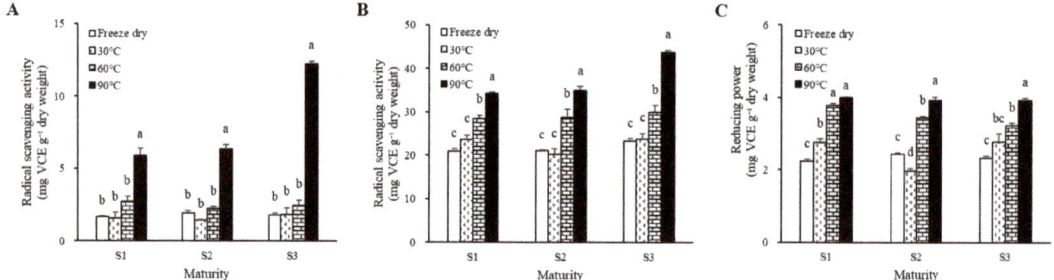

Figure 2. Radical scavenging activity of (**A**) DPPH and (**B**) ABTS and (**C**) reducing power evaluated using FRAP assay according to drying temperature of different maturities of *M. charantia* fruit. Different letters within each maturity stage indicate significant differences according to Tukey's studentized test at $p < 0.05$.

Similar variation patterns in the DPPH radical scavenging activity were observed between S1 and S2 after thermal treatments, whereas S3 showed distinct differences compared to the earlier stages. The DPPH radical scavenging activity of the 90 °C heat-treated sample in S3 was 6.8-fold higher than that of the freeze-dried (FD) sample, whereas it was 2.7- and 2.9-fold higher in S1 and S2, respectively. The changes of antioxidant activities of *M. charantia* at different stages showed a positive temperature-dependent increase in the ABTS radical scavenging activity and FRAP. Similar to DPPH radical scavenging activity, the 90 °C heat-treated S3 exhibited higher ABTS radical scavenging activity than the early stages. However, the FRAP did not show a clear difference between each maturity stage according to the temperature, in contrast to the radical scavenging assays.

2.3. Phenolic Contents

Figure 3A shows the effects of thermal processing on total phenolic content (TPC) in *M. charantia*. In nonthermal treatment, the TPC was 4.1, 4.9, and 5.6 mg gallic acid equivalent (GE)·g^{-1} dry weight (DW) in S1, S2, and S3, respectively. The TPC increased as the drying temperature increased, regardless of the stages after thermal treatment. In comparison to FD, the TPC increased with temperature in 60 °C and 90 °C heat-treated fruit, regardless of their stages. The higher total phenolic contents were exhibited at 90 °C heat-treated fruit in each stage, with 11.8 mg GE·g^{-1} DW in S1, 13.7 mg GE·g^{-1} DW in S2, and 25.3 mg GE·g^{-1} DW in S3.

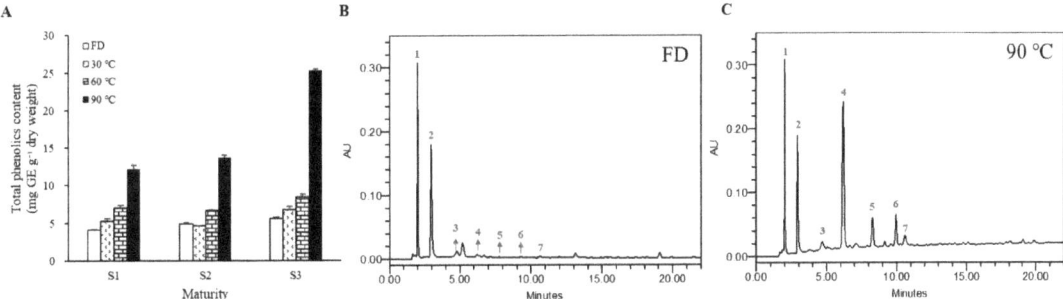

Figure 3. Changes of total phenolic content (**A**) in thermal processing of *M. charantia*. (**B**) Chromatograms of phenolic compounds in (**B**) freeze-dried (FD) and (**C**) 90 °C heat-treated S3 detected at 280 nm using HPLC analysis.

In the HPLC chromatogram, seven distinct peaks, which are potential phenolic compounds, in high temperature-treated *M. charantia* were detected at 280 nm. The relative content of each compound among the thermal treatments was calculated on the bases of the total peak area in FD of S1 (Table 1). Furthermore, regardless of maturity, peaks 1 and 2 were major substances in FD. These contents increased significantly after high-temperature treatments. In the S3 samples, the contents of peaks 1 and 2 increased up to 30 °C and 60 °C and then decreased at higher temperatures. Interestingly, the content of peak 3 showed variations in patterns at the different growth stages. In samples of S1 and S3, the content slightly decreased at 30 °C and then increased with elevation in the drying temperature. The peak 3 contents of dried fruit at 90 °C were about twice as high as those of FD. Peaks 4–7 were not observed in FD *M. charantia* regardless of stages. However, the content of each compound continuously increased in *M. charantia* as the drying temperature increased, resulting in from 5 to 120 times higher content in the 90 °C heat-treated fruit compared to that in the FD at each stage.

Table 1. The change in the content (%) of seven candidate phenolic compounds according to the drying temperature of fruits.

Peak	RT (min)	λmax	S1				S2				S3				LSD
			FD	30 °C	60 °C	90 °C	FD	30 °C	60 °C	90 °C	FD	30 °C	60 °C	90 °C	
1	1.99	217.7/273.1	38.86 f	48.09 de	43.54 ef	72.87 a	44.84 def	23.64 g	61.01 bc	64.13 ab	47.99 de	67.98 a	59.07 bc	51.06 cd	1.88
2	2.9	274.3	50.13 cd	49.15 cd	41.80 d	80.37 a	46.15 cd	21.18 e	70.28 b	68.03 b	54.93 c	73.14 ab	69.52 b	44.25 d	2.35
3	4.69	260.1	8.21 d	7.90 de	11.20 c	16.12 a	7.81 ef	1.80 g	0.00 g	13.45 b	5.81 ef	3.82 f	10.35 c	16.83 a	0.44
4	6.18	296.1	1.88 de	0.45 e	4.69 de	46.24 b	1.03 e	0.60 e	2.42 de	29.58 c	2.12 de	1.53 de	6.57 d	92.01 a	1.25
5	8.26	263.7	0.12 e	0.32 de	1.14 d	9.10 b	0.27 de	0.14 e	0.36 de	6.57 c	0.13 e	0.20 de	0.75 de	21.26 a	0.22
6	9.94	289.7	0.25 e	0.35 e	0.33 e	9.97 b	0.28 e	0.37 e	0.21 e	8.40 c	0.15 e	0.74 d	0.58 d	17.78 a	0.06
7	10.58	213.0/257.7	0.54 f	1.19 e	2.15 d	6.68 b	1.50 de	1.72 de	1.11 e	5.95 c	1.46 de	1.20 e	1.46 de	8.00 a	0.16
	Total		100.00	107.45	104.82	241.35	101.88	49.45	135.38	196.09	112.60	148.60	148.30	251.19	

The relative content of each compound among the thermal treatments was calculated in the bases of total peak area in FD of S1. Alphabetical letters within a row indicate significant difference in Tukey's studentized test at $p < 0.05$. Peak number corresponds to the peak number of Figure 3B chromatograms.

2.4. Changes of Carotenoids and Chlorophylls

Table 2 shows the changes in the carotenoid and chlorophyll content of different maturities of *M. charantia* during thermal processing. Here, nine carotenoids and two chlorophylls were determined. In the nonthermal-treated fruit, the total content of the carotenoids increased as the fruit stage progressed. After thermal treatment, the total content of all stages of the fruits decreased continuously with the increasing drying temperatures. Three patterns were roughly observed after heat treatment for each carotenoid content variation. In all stages, the contents of carotenoid ester 1 and lutein decreased constantly with the increasing drying temperature in all stages, except carotenoid ester 1 did not exist in S3. The content of carotenoid ester 2 decreased as the growth stage progressed but was maintained with relatively small variation during thermal treatment. Regardless of the thermal treatment, carotenoids esters 3–7 and β-carotene were rarely or not detected in S1 and S2. Notably, these compounds were found in relatively large amounts in the FD of S3. However, at temperatures above 60 °C, the patterns demonstrated that each compound decomposed similarly. In the FD-treated fruit, the total chlorophyll content decreased with increasing fruit maturity stages. Moreover, no chlorophyll was found at the mature yellow stage in the fruit. After thermal treatment, the total chlorophyll content in S1 and S2 significantly decreased as the drying temperature increased. The content variation of chlorophyll a and b are responsible for the observed results.

2.5. Correlation between Antioxidant Activities and Physiochemicals

Figure 4 shows the correlation between antioxidant activities and phenolic compounds, carotenoids, and chlorophylls. Overall, the antioxidant activities showed positive correlations with phenolics, whereas the antioxidant activities were negatively correlated with either carotenoids or chlorophylls. Furthermore, DPPH, ABTS, and FRAP activities showed a high correlation with TPC, exhibiting high correlation values (r) of 0.77, 0.72, and 0.70, respectively. These activities showed a significant correlation ($r > 0.6$, $p < 0.05$) with candidate phenolics 4–6, which significantly increased in the 90 °C heat-treated fruits.

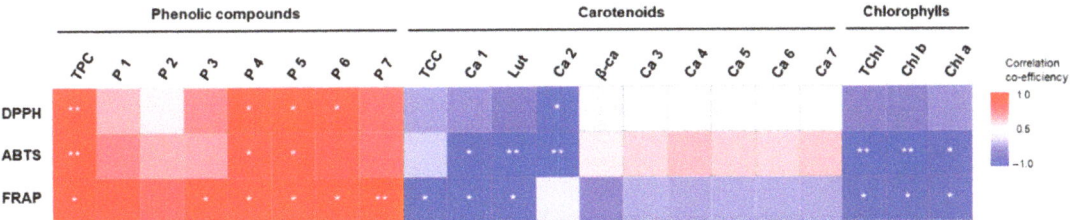

Figure 4. Correlation coefficients between antioxidant activities and phytochemicals in heat-processed *M. charantia*. TPC: total phenolic content; P1–P7: candidate phenolic compound 1–7; TCC: total carotenoid content; Ca1–Ca7: carotenoid ester 1–7; Lut: lutein; β-ca: β-carotene; TChl: total chlorophylls; Chl b: chlorophyll b; and Chl a: chlorophyll a. Asterisks indicate significance (* $p < 0.05$; ** $p < 0.01$) using Pearson's correlation analysis.

Table 2. Carotenoid and chlorophyll content (%) by drying temperature of *M. charantia* at different maturity stages.

				S1				S2				S3			
Peak	Compounds	Rt (min)	λmax	FD	30 °C	60 °C	90 °C	FD	30 °C	60 °C	90 °C	FD	30 °C	60 °C	90 °C	LSD
Carotenoid																
1	Carotenoid ester 1	8.59	412.8/439.4	3.63 a	2.81 c	n.d.	n.d.	3.06 b	1.80 d	n.d.	n.d.	n.d.	n.d.	n.d.	n.d.	0.22
2	Lutein	12.4	462.4/487.9	47.15 a	30.65 b	11.35 d	4.35 e	32.34 b	27.61 c	3.24 e	4.10 e	3.35 e	n.d.	n.d.	n.d.	2.99
3	Carotenoid ester 2	13.5	418.8	45.88 a	42.33 abc	39.81 bcd	39.89 bcd	37.37 cde	38.87 bcd	41.34 ab	39.38 bcd	35.71 de	35.13 de	35.70 de	33.73 e	5.43
4	β-carotene	19.7	427.3/451.5	3.34 b	n.d.	n.d.	n.d.	n.d.	n.d.	n.d.	n.d.	11.25 a	2.07 c	n.d.	n.d.	0.78
5	Carotenoid ester 3	26.1	445.1/481.8	n.d.	n.d.	n.d.	n.d.	n.d.	n.d.	n.d.	n.d.	34.56 a	3.53 b	n.d.	n.d.	1.09
6	Carotenoid ester 4	29.03	447.8/487.9	n.d.	n.d.	n.d.	n.d.	n.d.	n.d.	n.d.	n.d.	13.88 a	4.66 b	n.d.	n.d.	0.85
7	Carotenoid ester 5	30.2	446.6/473.3	n.d.	n.d.	n.d.	n.d.	n.d.	n.d.	n.d.	n.d.	10.12 a	1.66 b	n.d.	n.d.	0.59
8	Carotenoid ester 6	31.0	453.9/478.1	n.d.	n.d.	n.d.	n.d.	n.d.	n.d.	n.d.	n.d.	36.30 a	4.59 b	n.d.	n.d.	1.95
9	Carotenoid ester 7	34.2	452.7/483	n.d.	n.d.	n.d.	n.d.	n.d.	n.d.	n.d.	n.d.	27.22 a	6.62 b	n.d.	n.d.	1.10
	Total carotenoid			100.00	75.79	51.16	44.24	72.77	68.29	44.58	43.48	172.38	58.26	37.91	33.73	
Chlorophyll																
10	Chlorophyll b	14.1	457.5/643.2	88.17 a	68.55 b	23.52 d	2.56 f	66.31 b	51.81 c	6.28 e	n.d.	n.d.	n.d.	n.d.	n.d.	2.92
11	Chlorophyll a	15	429.6/660.4	11.83 a	5.77 b	1.63 d	n.d.	5.39 b	4.34 c	n.d.	n.d.	n.d.	n.d.	n.d.	n.d.	0.95
	Total chlorophyll			100.00	74.32	25.15	2.56	71.69	56.15	6.28						

The relative content of each compound among the thermal treatments was calculated in the bases of total peak areas in FD of S1. Alphabetical letters within a row indicate significant difference in Tukey's studentized test at $p < 0.05$. n.d. indicates not detected.

3. Discussion

Color is an important indicator for judging fruit quality [27,28]. During ripening, the color of *M. charantia* is determined by the quality and quantity of natural pigments, such as the greenish contribution of chlorophylls and the yellowish contribution of carotenoids. According to current studies, the decrease in chlorophylls and the increases in carotenoids in *M. charantia* fruit during maturity strongly support the color variation patterns with low lightless and high redness with yellowness in mature yellow fruit than early stages. Color has also become an index to evaluate the quality of the processing, as pigment compounds can be decomposed or oxidized. The pigment changes during thermal processing are significantly affected by many factors, such as methods, time, and temperature [27,29]. In our results, the color of the hot-air-dried fruit was obviously different from that of the freeze-dried (Figure 1). The degradation of pigment compounds after thermal processing in *M. charantia*, regardless of the stages, should be responsible for these phenomena. On the other hand, it was found that browning occurred in all stages of the fruit treated at 90 °C, accompanied by mature yellow fruit treated at 30 °C. Similar results were reported in persimmon by Lim and Eom [10]. These are tentatively assumed to be the results of enzymatic and nonenzymatic browning. At a relatively low temperature (30 °C), the enzymatic browning occurs actively, causing discoloration. The moisture of the fruits dried fast and the enzyme was destructed at a relatively high temperature (90 °C). Nonenzymatic browning such as caramelization and the Maillard reaction probably happened due to the presence of amino acids and sugars in fruit at high temperatures [10]. Thus, it is important to determine the appropriated drying temperature to ensure the color quality of the fruit.

The significantly Increased antioxidant activities were shown in *M. charantia* after thermal processing, regardless of the fruit stages. The greater antioxidant activities were exhibited at higher drying temperatures (Figure 2). Interestingly, the changes in antioxidant activity of *M. charantia* at different growth stages showed different responses to heat treatment. Furthermore, for radical scavenging assays, more increased antioxidant activities were observed in high-temperature-treated fully matured fruit than in the earlier stages. The DPPH radical scavenging activity of the 90 °C heat-treated S3 was found to be 6.8-fold higher than that of the FD sample, while it was 2.7- and 2.9-fold higher in S1 and S2, respectively (Figure 2). However, the reducing power of the fruit did not show a clear difference between each maturity stage according to the temperature. Here, it is important to point out the characteristics of different antioxidant reactions among the colorimetric antioxidant assays. The DPPH assay tends to react with the hydrophobic compounds, the ABTS assay tends to screen both lipophilic and hydrophilic compounds, and the FRAP assay presents nonspecific properties [30,31]. The difference in antioxidant activity variation patterns among the three assays suggested that thermal processing induced the release of hydrophobic compounds and an increase in lipophilic compounds more than hydrophilic compounds.

The increase in antioxidant activities might be due to the variation of polyphenols in the matrix after thermal processing. The polyphenols are important antioxidant contributors [10–13] and the significantly increased total phenolic content in the thermally treated *M. charantia* can support the increased antioxidant activities, which were evaluated using three different assays (Figures 2 and 3). A chromatogram at 280 nm is widely used to study the polyphenols because the absorption at this wavelength is suitable for detecting a large number of such compounds [32]. Based on the lambda max of the peaks at 280 nm in our data and the phenolic acid profiling of *M. charantia* from a previous report [24], the peaks we detected are considered to be phenolic compounds. The change of each phenolic compound from the heat treatment showed a significantly different pattern (Table 1). Our results demonstrated that peaks 4–6 were significantly increased by heat at all maturities, contributing to antioxidant activity. Although these peaks were increased by heat regardless of maturity, the increased levels were highest in S3, showing a strong correlation with the antioxidant activity. The quantitative increase may result from the thermal conversion of insoluble phenolic compounds into soluble form [33]. According to Horax et al. [24], several phenolic compounds were increased after the 60 °C oven-drying of *M. charantia* fruits, where the increased phenolics differed depending on the part of the fruit; mainly, the increased phenolics were gallic acid and catechin in the flesh, while they were gentisic acid and epicatechin in the inner tissue. It has been reported that food processing, such as heat treatment, can lead to polyphenol degradation from cellular structures, increasing phenolic compounds [33,34]. Thermal processing can also result in an enhanced extract ratio of the polyphenols. High-temperature treatment may have caused the decomposition of high-molecular phenolics into low-molecular phenolic compounds such as gallic acid and epicatechin [33,34]. Thus, our results connote that mature yellow fruits contain relatively more high-molecular phenolics or that low molecular phenolics are easily degraded in mature fruit.

Carotenoids are also known to exhibit antioxidant activity [35,36]. Previous studies found that carotenoid variation accumulates during *M. charantia* maturation [37,38] and that several carotenoids only existed in the mature yellow stage [39]. These results are consistent with our finding that five carotenoids are generated only in S3 of *M. charantia* (Table 2). Regardless of the fruit growth stage, the content of carotenoids decreases with higher processing temperatures. The destruction of these carotenoids is similar to the results of previous studies showing that carotenoids are decomposed by the oxidation of oxygen at high temperatures [23]. Interestingly, although carotenoids are known to be heat-sensitive compounds, carotenoid ester 2, which maintained a relatively high content despite heat treatment, was observed during heat treatment. This may be due to the different thermal stabilities according to the structure of the carotenoid [40,41].

The correlation coefficiency analysis suggested a high positive correlation between the antioxidant activities and phenolics but no or negative correlation between the antioxidant activities and carotenoids or chlorophylls. For individual candidate phenolic compounsd detected at 280 nm, peaks 4 and 5 showed a significant positive correlation with three antioxidant activity assays. Similarly, phenolic compounds in *M. charantia* have previously been studied as potential antioxidants [42,43]. Although thermal treatment has a negative effect on carotenoid preservation during *M. charantia* fruit processing, our results suggest that the maturity of the fruit and heating temperature are critical factors enhancing phenolic compounds and antioxidant activity.

4. Materials and Methods

4.1. Plant Materials

The *M. charantia* plants were cultivated in the greenhouse of Kyung Hee university (Yongin si, Republic of Korea). These plants were planted in horticultural soil (Baroker, Seoulbio Co., Eumseong, Republic of Korea) mixed with perlite (GFC. Co., Ltd. Hongseong, Republic of Korea) on 10 May 2021. The fruits of the *M. charantia* were harvested at three different maturity stages: S1, immature green with undeveloped seeds (15.52 ± 0.61 cm of length); S2, mature green with fully developed seeds (23.14 ± 1.10 cm of length); S3, mature yellow with red ripe seeds (26.54 ± 0.64 cm of length). We randomly harvested 20 individuals in the growth stage from 15 plants three times in July and August 2021. The experiment was carried out with a total of four repetitions with 5 objects as one repetition.

The fruits were washed with distilled water and dried using a paper towel. The fruit samples were cross sectioned at a thickness of 1 cm. The skin, flesh, and inner tissues were obtained and the seeds were removed (Figure 1A). The separated fruits were immediately dried after each harvest without storage. Batches of the samples were freeze-dried in a freeze dryer as a control group. For thermal processing, the samples were hot-air dried using a dry machine (Koencon Co., Ltd., Hanam, Republic of Korea) until the water content was <10% at 30, 60, and 90 °C for 3 days, respectively. The dried samples were pulverized using a commercial mixer and sieved with a 100-mesh size.

4.2. Color Measurement

The colors of the ground *M. charantia* samples were obtained by using a color analyzer (Lutron Electronics, Inc., Coopersburg, PA, USA). The acquired RGB values were converted into L* (lightness), a* (greenness to redness), and b* (blueness to yellowness) using OpenRGB v. 2.30.10125 software.

4.3. Sample Extraction

4.3.1. Preparation of Extract for the Determination of Antioxidant Activities, Total Phenolic Content, and Content of Individual Phenolic Compound

The extraction method was performed as previously described methods [24,44] with some modifications. The dried samples (50 mg) were immersed in 1 mL of 80% ethanol and placed in a shaking incubator at 24 °C for 8 h after 1 h sonication. Later, the mixture was centrifuged at $12,000 \times g$ for 10 min and the supernatant was collected for further analysis.

4.3.2. Preparation of Extract for Carotenoid and Chlorophyll Analysis

The extraction method was modified slightly from that described by Lim and Eom [10]. The ground sample was extracted with 350 µL of methanol and, later, 700 µL of chloroform was added to it. After vortexing, 350 µL of 10% sodium chloride (NaCl) was added to the mixture, which was then centrifuged at $8000 \times g$ for 5 min. The chloroform phase was separated from the mixture in fresh tubes. Potassium hydroxide (350 µL, 1 N) was added to the chloroform phase and the mixture was heated in the dark for 30 min. The mixture was centrifuged after adding 10% NaCl and the chloroform phase was collected. The collected phase was washed with additional 10% NaCl (700 µL) to remove the KOH. The chloroform phase (500 µL) was centrifuged and 800 µL of ethyl acetate was added to the collected phase. The final mixture was filtered using a 0.45 µm syringe filter (Futecs Co., Ltd., Daejeon, Republic of Korea) and the filtrate was used for carotenoid and chlorophyll content analysis.

4.4. Colormetric Assays of Antioxidant Activities

The antioxidant assays of the *M. charantia* extract following thermal processing were measured using the 2,2-diphenyl-1-picrylhydrazyl (DPPH) and 2,2′-azino-bis (3-ethylbenzothiazoline-6-sulfonic acid (ABTS) radical scavenging activity assays. The antioxidant assays were performed as described by the method of Lim and Eom [10] with some modifications. For DPPH radical scavenging activity, the sample or standard (17 µL) was mixed with 983 µL of DPPH solution. The absorbance of the mixture was measured at 517 nm after the reaction in the dark for 30 min. The DPPH solution was adjusted by 0.65 ± 0.02 in the absorbance value with 80% methanol at 517 nm. For ABTS radical scavenging activity, 10 mM ABTS dissolved in DMSO was mixed in a 1:4 ratio with 8 mM of 2,2′-Azobis (2-amidinopropane) dihydrochloride dissolved in 1X phosphate-buffered saline (PBS). The mixture was heated at 70 °C for 40 min. The ABTS solution was filtered using a 0.45 µm syringe filter and adjusted to a 0.65 ± 0.02 in absorbance value with 1X PBS at 734 nm. The solution was added to the sample (20 µL) and measured at 734 nm after a 10 min incubation at room temperature. The DPPH and ABTS radical scavenging activities were expressed as milligrams of vitamin C equivalents (VCE) per gram of DW.

The ferric-reducing antioxidant power assay was performed as described by the method of Lim and Eom [10] with minor modifications. A solution of 300 mM acetate buffer (pH 3.6) was prepared by dissolving 3.1 g sodium acetate trihydrate and 16 mL acetic acid in 1 L of distilled water. The 10 mM 2,4,6-tripyridyl-s-triazine (TPTZ) in 40 mM hydrochloric acid were and 20 mM ferric chloride ($FeCl_3 \cdot 6H_2O$) solution prepared. The FRAP solution was prepared by mixing acetate buffer, TPTZ solution, and $FeCl_3 \cdot 6H_2O$ in a 10:1:1 ratio. The 950 µL of FRAP solution was added to 50 µL of sample extract and reacted in the dark for 30 min. The absorbance of the mixture was measured at 593 nm. The FRAP was expressed as milligrams VCE per gram DW.

4.5. Measurement of Total Phenolic Content

The total phenolic content was determined in accordance with the Folin–Ciocalteu method of Lim et al. [45] with some modifications. The sample extract (50 µL) was added to 650 µL of distilled water. The 50 µL of Folin–Ciocalteu phenol reagent was immediately added and mixed. After 6 min of incubation, 500 µL of 7% sodium carbonate (Na_2CO_3) was added and reacted at room temperature for 90 min.

4.6. Determination of Individual Phenolic Compound

The 0.2 mL of the sample extract was diluted with 0.8 mL of 80% ethanol. The diluted extract was filtered through a 0.45 µm syringe filter. The filtrate was analyzed using reverse-phase HPLC (Waters 2695 Alliance HPLC; Bischoff, Leonberg, Germany) with a prontosil column (120-5-C18-SH, 5 µm, 150 × 4.6 mm; Bischoff, Leonberg, Germany) as previously described methods [18,24]. The mobile phase consisted of (A) water with 0.1% formic acid and (B) acetonitrile. The gradient elution was as follows: 0–23 min, 1–20% B; 23–45 min, 20–60% B; 45–46 min, 60% B; 46–47 min, 60–1% B; and 47–49 min, 1% B. The flow rate of the mobile phase was 1.0 mL·min^{-1} and the injection volume of the sample was 10 µL. The peaks were detected at 280 nm using the Waters 996 photodiode array detector (Waters Inc., Milford, MA, USA).

4.7. Determination of Carotenoids

The saponified sample extract was used for the HPLC analysis. The analysis was performed using Waters 2695 Alliance HPLC as in the previously described method [10]. The column used prontosil 120-5-C18-SH 5.0 µm (4.6 × 250 mm, Bischoff, Leonberg, Germany). Mobile phase A consisted of 90% acetonitrile with 0.1% formic acid and mobile phase B consisted of ethyl acetate with 0.1% formic acid. The gradient was as follows: 0–10 min, 0–60% B; 10–25 min, 60% B; 25–26 min, 60–0% B; and 26–27 min, 100% A. The flow rate was 1.0 mL·min^{-1}. The injection volume was 10 µL. The peaks were detected at 445 nm using a Waters 996 photodiode array detector (Waters Inc., Milford, MA, USA). The quantitative data of the carotenoid was expressed as relative content based on the total carotenoid of the freeze-dried S1 sample. The chlorophyll data were expressed as relative content of the total chlorophyll of freeze-dried S1 sample.

4.8. Statistical Analysis

All the samples were performed in three replicates and expressed as the mean and standard error. The data were analyzed using SAS software (Enterprise Guide 7.1 version; SAS Institute Inc., Cary, NC, USA). A one-way analysis of variance was performed to assess the differences between the mean values using a Fisher's least significant difference (LSD) test. The significant differences among the experimental treatments were evaluated using Tukey's studentized test (HSD) at $p < 0.05$. The relationship between the antioxidant activity and component contents under each treatment was analyzed using Pearson's correlation coefficients.

5. Conclusions

This study investigated the effect of thermal processing on the color, antioxidant activities, phenolics, carotenoids, and chlorophylls in the growth stage of *M. charantia* fruit. Thermal treatment led to distinct color changes in different maturities of *M. charantia*. The changes from green to yellow in greenish fruits were caused by the degradation of chlorophylls, whereas the changes from yellow to brown in mature fruit were caused by two different reactions, indicating enzymatic browning by low-temperature dry processing and nonenzymatic browning by high-temperature dry processing. After thermal treatment, antioxidant activities increased in fruit with increasing drying temperatures, regardless of the maturity stage of the fruit. The maturity of *M. charantia* affected the changes in radical scavenging activity after heat treatment, whereas it had no effect on reducing power. Due to the different mechanisms of each assay, these results are tentatively explained by the release of hydrophobic compounds and the increase in lipophilic compounds. The total phenolic contents were also significantly increased in each maturity of *M. charantia* after thermal treatment. Particularly, the highest content was observed in the 90 °C heat-treated S3 fruit, which is extremely higher than the 90 °C heat-treated S1 and S2. These results may be due to the significantly increased content of the candidate phenolic compounds 4, 5, and 6, which were detected at 280 nm using HPLC. However, after thermal processing, both carotenoids and chlorophylls were significantly decreased. The correlation coefficiency test between antioxidant activity and bioactive compounds suggested that the antioxidant activities of *M. charantia* were mainly contributed to by phenolic compounds. Although thermal processing induced the decrease in the carotenoid of *M. charantia* fruit, our results suggest that the maturity of fruit and the processing temperature are the critical factors enhancing phenolic compounds and antioxidant activity and that mature yellow fruit is better for consumption after using thermal processing. Overall, these results suggest that thermal processing at a high temperature can be usefully applied in industries of health supplements and nutraceuticals of *M. charantia* fruit and provide an optimized harvesting time and processing method for the development of functional foods.

Author Contributions: Conceptualization, S.H.E.; methodology, J.H.K., Y.J.L., and S.D.; software, J.H.K. and Y.J.L.; validation, J.H.K. and S.H.E.; formal analysis, J.H.K., Y.J.L., T.J.P., and S.D.; investigation, J.H.K. and S.H.E.; resources, J.H.K.; data curation, J.H.K., Y.J.L., and S.D., and S.H.E.; writing—original draft preparation, J.H.K., Y.J.L., and S.D.; writing—review and editing, S.H.E.; visualization, J.H.K.; supervision, S.H.E.; funding acquisition, S.H.E. All authors have read and agreed to the published version of the manuscript.

Funding: This research was supported by a National Research Foundation of Korea (NRF) grant, which was funded by the Ministry of Science and ICT (MSIT) of Korea government (NRF-2022R1A2C 100769511).

Institutional Review Board Statement: Not applicable.

Informed Consent Statement: Not applicable.

Data Availability Statement: All of the data is contained within the article.

Conflicts of Interest: The authors declare no conflict of interest.

Sample Availability: Samples are not available from the authors.

References

1. Subratty, A.; Gurib-Fakim, A.; Mahomoodally, F. Bitter melon: An exotic vegetable with medicinal values. *Nutr. Food Sci.* **2005**, *35*, 143–147. [CrossRef]
2. Wang, L.; Clardy, A.; Hui, D.; Gao, A.; Wu, Y. Antioxidant and antidiabetic properties of Chinese and Indian bitter melons (*Momordica charantia* L.). *Food Biosci.* **2019**, *29*, 73–80. [CrossRef]
3. Boy, H.I.A.; Rutilla, A.J.H.; Santos, K.A.; Ty, A.M.T.; Yu, A.I.; Mahboob, T.; Tangpoong, J.; Nissapatorn, V. Recommended Medicinal Plants as Source of Natural Products: A Review. *DCM* **2018**, *1*, 131–142. [CrossRef]
4. Leung, L.; Birtwhistle, R.; Kotecha, J.; Hannah, S.; Cuthbertson, S. Anti-diabetic and hypoglycaemic effects of *Momordica charantia* (bitter melon): A mini review. *Br. J. Nutr.* **2009**, *102*, 1703–1708. [CrossRef] [PubMed]
5. Saeed, F.; Afzaal, M.; Niaz, B.; Arshad, M.U.; Tufail, T.; Hussain, M.B.; Javed, A. Bitter melon (*Momordica charantia*): A natural healthy vegetable. *Int. J. Food Prop.* **2018**, *21*, 1270–1290. [CrossRef]
6. Hsieh, H.; Lin, J.; Chen, K.; Cheng, K.; Hsieh, C. Thermal treatment enhances the α-glucosidase inhibitory activity of bitter melon (*Momordica charantia*) by increasing the free form of phenolic compounds and the contents of Maillard reaction products. *J. Food Sci.* **2021**, *86*, 3109–3121. [CrossRef]
7. Goldberg, E.; Grant, J.; Aliani, M.; Eskin, M.N.A. Methods for Removing Bitterness in Functional Foods and Nutraceuticals. In *Bitterness: Perception, Chemistry and Food Processing*; John Wiley & Sons, Inc.: New York, NY, USA, 2017; pp. 209–237. [CrossRef]
8. Liu, Y.-J.; Lai, Y.-J.; Wang, R.; Lo, Y.-C.; Chiu, C.-H. The Effect of Thermal Processing on the Saponin Profiles of *Momordica charantia* L. *J. Food Qual.* **2020**, *2000*, 8862020. [CrossRef]
9. Qu, S.; Kwon, S.J.; Duan, S.; Lim, Y.J.; Eom, S.H. Isoflavone changes in immature and mature soybeans by thermal processing. *Molecules* **2021**, *26*, 7471. [CrossRef]
10. Lim, Y.J.; Eom, S.H. The different contributors to antioxidant activity in thermally dried flesh and peel of astringent persimmon fruit. *Antioxidants* **2022**, *11*, 597. [CrossRef]

11. Duan, S.C.; Kwon, S.J.; Eom, S.H. Effect of thermal processing on color, phenolic compounds, and antioxidant activity of faba bean (*Vicia faba* L.) leaves and seeds. *Antioxidants* **2021**, *10*, 1207. [CrossRef]
12. Duan, S.; Kwon, S.J.; Gil, C.S.; Eom, S.H. Improving the antioxidant activity and flavor of faba (*Vicia faba* L.) leaves by domestic cooking methods. *Antioxidants* **2022**, *11*, 931. [CrossRef] [PubMed]
13. Eom, S.H.; Park, H.J.; Seo, D.W.; Kim, W.W.; Cho, D.H. Stimulating effects of far-infrared ray radiation on the release of antioxidative phenolics in grape berries. *Food Sci. Biotechnol.* **2009**, *18*, 362–366.
14. Xu, G.; Ye, X.; Chen, J.; Liu, D. Effect of Heat Treatment on the Phenolic Compounds and Antioxidant Capacity of Citrus Peel Extract. *J. Agric. Food Chem.* **2007**, *55*, 330–335. [CrossRef] [PubMed]
15. Chumyam, A.; Whangchai, K.; Jungklang, J.; Faiyue, B.; Saengnil, K. Effects of heat treatments on antioxidant capacity and total phenolic content of four cultivars of purple skin eggplants. *Scienceasia* **2013**, *39*, 246–251. [CrossRef]
16. Pérez-Nevado, F.; Cabrera-Bañegil, M.; Repilado, E.; Martillanes, S.; Martín-Vertedor, D. Effect of different baking treat-ments on the acrylamide formation and phenolic compounds in Californian-style black olives. *Food Control* **2018**, *94*, 22–29. [CrossRef]
17. Ghafoor, K.; Ahmed, I.A.M.; Doğu, S.; Uslu, N.; Fadimu, G.J.; Al Juhaimi, F.; Babiker, E.E.; Özcan, M.M. The Effect of Heating Temperature on Total Phenolic Content, Antioxidant Activity, and Phenolic Compounds of Plum and Mahaleb Fruits. *Int. J. Food Eng.* **2019**, *15*, 11–12. [CrossRef]
18. Choi, J.S.; Kim, H.Y.; Seo, W.T.; Lee, J.H.; Cho, K.M. Roasting enhances antioxidant effect of bitter melon (*Momordica charantia* L.) increasing in flavan-3-ol and phenolic acid contents. *Food Sci. Biotechnol.* **2012**, *21*, 19–26. [CrossRef]
19. Ng, Z.X.; Kuppusamy, U.R. Effects of different heat treatments on the antioxidant activity and ascorbic acid content of bitter melon, *Momordica charantia*. *Braz. J. Food Technol.* **2019**, *22*, e2018283. [CrossRef]
20. Aminah, A.; Anna, P.K. Influence of ripening stages on physicochemical characteristics and antioxidant properties of bitter gourd (*Momordica charantia*). *Int. Food Res. J.* **2011**, *18*, 895–900.
21. Zong, R.-J.; Morris, L.; Cantwell, M. Postharvest physiology and quality of bitter melon (*Momordica charantia* L.). *Postharvest Biol. Technol.* **1995**, *6*, 65–72. [CrossRef]
22. Franco, M.N.; Galeano-Díaz, T.; López, Ó.; Fernández-Bolaños, J.G.; Sánchez, J.; De Miguel, C.; Gil, M.V.; Martín-Vertedor, D. Phenolic compounds and antioxidant capacity of virgin olive oil. *Food Chem.* **2014**, *163*, 289–298. [CrossRef]
23. Zhang, M.; Hettiarachchy, N.S.; Horax, R.; Chen, P.; Over, K.F. Effect of Maturity Stages and Drying Methods on the Retention of Selected Nutrients and Phytochemicals in Bitter Melon (*Momordica charantia*) Leaf. *J. Food Sci.* **2009**, *74*, C441–C448. [CrossRef] [PubMed]
24. Horax, R.; Hettiarachchy, N.; Chen, P. Extraction, quantification, and antioxidant activities of phenolics from pericarp and seeds of bitter melons (*Momordica charantia*) harvested at three maturity stages (immature, mature, and ripe). *J. Agric. Food Chem.* **2010**, *58*, 4428–4433. [CrossRef] [PubMed]
25. Valyaie, A.; Azizi, M.; Kashi, A.; Sathasivam, R.; Park, S.U.; Sugiyama, A.; Motobayashi, T.; Fujii, Y. Evaluation of growth, yield, and biochemical attributes of bitter gourd (*Momordica charantia* L.) cultivars under Karaj conditions in Iran. *Plants* **2021**, *10*, 1370. [CrossRef]
26. Lee, S.H.; Jeong, Y.S.; Song, J.; Hwang, K.A.; Noh, G.M.; Hwang, I.G. Phenolic acid, carotenoid composition, and antioxidant activity of bitter melon (*Momordica charantia* L.) at different maturation stages. *Int. J. Food Prop.* **2017**, *20*, 3078–3087. [CrossRef]
27. Kulapichitr, F.; Borompichaichartkul, C.; Fang, M.; Suppavorasatit, I.; Cadwallader, K.R. Effect of post-harvest drying process on chlorogenic acids, antioxidant activities and CIE-Lab color of Thai Arabica green coffee beans. *Food Chem.* **2022**, *366*, 130504. [CrossRef] [PubMed]
28. Lim, Y.J.; Eom, S.H. Kiwifruit cultivar 'Halla gold' functional component changes during preharvest fruit maturation and postharvest storage. *Sci. Hortic.* **2018**, *234*, 134–139. [CrossRef]
29. Barrett, D.M.; Beaulieu, J.; Shewfelt, R. Color, Flavor, Texture, and Nutritional Quality of Fresh-Cut Fruits and Vegetables: Desirable Levels, Instrumental and Sensory Measurement, and the Effects of Processing. *Crit. Rev. Food Sci. Nutr.* **2010**, *50*, 369–389. [CrossRef]
30. Floegel, A.; Kim, D.-O.; Chung, S.-J.; Koo, S.I.; Chun, O.K. Comparison of ABTS/DPPH assays to measure antioxidant capacity in popular antioxidant-rich US foods. *J. Food Compos. Anal.* **2011**, *24*, 1043–1048. [CrossRef]
31. Shah, P.; Modi, H.A. Comparative study of DPPH, ABTS and FRAP assays for determination of antioxidant activity. *Int. J. Res. Appl. Sci. Eng. Technol.* **2015**, *3*, 636–641.
32. Duan, S.; Liu, J.R.; Wang, X.; Sun, X.M.; Gong, H.S.; Jin, C.W.; Eom, S.H. Thermal Control Using Far-Infrared Irradiation for Producing Deglycosylated Bioactive Compounds from Korean Ginseng Leaves. *Molecules* **2022**, *27*, 4782. [CrossRef] [PubMed]
33. Faller, A.; Fialho, E. The antioxidant capacity and polyphenol content of organic and conventional retail vegetables after domestic cooking. *Food Res. Int.* **2009**, *42*, 210–215. [CrossRef]
34. Xie, P.-J.; Huang, L.-X.; Zhang, C.-H.; Zhang, Y.-L. Phenolic compositions, and antioxidant performance of olive leaf and fruit (*Olea europaea* L.) extracts and their structure–activity relationships. *J. Funct. Foods* **2015**, *16*, 460–471. [CrossRef]
35. Stahl, W.; Sies, H. Antioxidant activity of carotenoids. *Mol. Asp. Med.* **2003**, *24*, 345–351. [CrossRef] [PubMed]
36. Paiva, S.A.R.; Russell, R.M. β-Carotene and Other Carotenoids as Antioxidants. *J. Am. Coll. Nutr.* **1999**, *18*, 426–433. [CrossRef]
37. Rodriguez, D.B.; Raymundo, L.C.; Lee, T.-C.; Simpson, K.L.; Chichester, C.O. Carotenoid Pigment Changes in Ripening Momordica charantia Fruits. *Ann. Bot.* **1976**, *40*, 615–624. [CrossRef]

38. Tuan, P.A.; Kim, J.K.; Park, N.I.; Lee, S.Y.; Park, S.U. Carotenoid content and expression of phytoene synthase and phytoene desaturase genes in bitter melon (*Momordica charantia*). *Food Chem.* **2011**, *126*, 1686–1692. [CrossRef]
39. Cuong, D.M.; Jeon, J.; Morgan, A.M.A.; Kim, C.; Kim, J.K.; Lee, S.Y.; Park, S.U. Accumulation of Charantin and Expression of Triterpenoid Biosynthesis Genes in Bitter Melon (*Momordica charantia*). *J. Agric. Food Chem.* **2017**, *65*, 7240–7249. [CrossRef]
40. D'Evoli, L.; Lombardi-Boccia, G.; Lucarini, M. Influence of Heat Treatments on Carotenoid Content of Cherry Tomatoes. *Foods* **2013**, *2*, 352–363. [CrossRef]
41. Achir, N.; Randrianatoandro, V.A.; Bohuon, P.; Laffargue, A.; Avallone, S. Kinetic study of β-carotene and lutein degradation in oils during heat treatment. *Eur. J. Lipid Sci. Technol.* **2010**, *112*, 349–361. [CrossRef]
42. Tan, S.P.; Kha, T.C.; Parks, S.E.; Roach, P.D. Bitter melon (*Momordica charantia* L.) bioactive composition and health benefits: A review. *Food Rev. Int.* **2016**, *32*, 181–202. [CrossRef]
43. Kubola, J.; Siriamornpun, S. Phenolic contents and antioxidant activities of bitter gourd (*Momordica charantia* L.) leaf, stem and fruit fraction extracts in vitro. *Food Chem.* **2008**, *110*, 881–890. [PubMed]
44. Tan, S.P.; Stathopoulos, C.; Parks, S.; Roach, P. An Optimised Aqueous Extract of Phenolic Compounds from Bitter Melon with High Antioxidant Capacity. *Antioxidants* **2014**, *3*, 814–829. [CrossRef]
45. Lim, Y.; Kwon, S.-J.; Qu, S.; Kim, D.-G.; Eom, S. Antioxidant Contributors in Seed, Seed Coat, and Cotyledon of γ-ray-Induced Soybean Mutant Lines with Different Seed Coat Colors. *Antioxidants* **2021**, *10*, 353. [CrossRef] [PubMed]

Disclaimer/Publisher's Note: The statements, opinions and data contained in all publications are solely those of the individual author(s) and contributor(s) and not of MDPI and/or the editor(s). MDPI and/or the editor(s) disclaim responsibility for any injury to people or property resulting from any ideas, methods, instructions or products referred to in the content.

Article

Untargeted Metabolomics Reveals New Markers of Food Processing for Strawberry and Apple Purees

Gabriela Salazar-Orbea [1], Rocío García-Villalba [1], Luis M. Sánchez-Siles [2,3], Francisco A. Tomás-Barberán [1,*] and Carlos J. García [1,*]

[1] Quality, Safety and Bioactivity of Plant-Derived Foods, Centro de Edafología y Biología Aplicada del Segura-Consejo Superior de Investigaciones Científicas (CEBAS-CSIC), 30100 Murcia, Spain
[2] Research and Nutrition Department, Hero Group, 30820 Alcantarilla, Spain
[3] Institute for Research and Nutrition, Hero Group, 5600 Lenzburg, Switzerland
* Correspondence: fatomas@cebas.csic.es (F.A.T.-B.); cjgarcia@cebas.csic.es (C.J.G.)

Abstract: In general, food processing and its conditions affect nutrients, bioactive compounds, and sensory characteristics of food products. This research aims to use a non-targeted metabolomics approach based on UPLC-ESI-QTOF-MS to determine how fruit processing can affect the metabolic profile of fruits and, through a comprehensive metabolic analysis, identify possible markers to assess their degree of processing. The present study uses a real case from the food industry to evaluate markers of the processing of strawberry and apple purees industrially elaborated with different processing techniques and conditions. The results from the multivariate analysis revealed that samples were grouped according to the type of processing, evidencing changes in their metabolic profiles and an apparent temperature-dependent effect. These metabolic profiles showed changes according to the relevance of thermal conditions but also according to the exclusively cold treatment, in the case of strawberry puree, and the pressure treatment, in the case of apple puree. After data analysis, seven metabolites were identified and proposed as processing markers: pyroglutamic acid, pteroyl-D-glutamic acid, 2-hydroxy-5-methoxy benzoic acid, and 2-hydroxybenzoic acid β-D-glucoside in strawberry and di-hydroxycinnamic acid glucuronide, caffeic acid and lysoPE(18:3(9Z,12Z,15Z)/0:0) in apple purees. The use of these markers may potentially help to objectively measure the degree of food processing and help to clarify the controversial narrative on ultra-processed foods.

Keywords: untargeted metabolomics; markers; food processing; processing degree

Citation: Salazar-Orbea, G.; García-Villalba, R.; Sánchez-Siles, L.M.; Tomás-Barberán, F.A.; García, C.J. Untargeted Metabolomics Reveals New Markers of Food Processing for Strawberry and Apple Purees. *Molecules* 2022, 27, 7275. https://doi.org/10.3390/molecules27217275

Academic Editor: Sabina Lachowicz-Wiśniewska

Received: 28 September 2022
Accepted: 23 October 2022
Published: 26 October 2022

Publisher's Note: MDPI stays neutral with regard to jurisdictional claims in published maps and institutional affiliations.

Copyright: © 2022 by the authors. Licensee MDPI, Basel, Switzerland. This article is an open access article distributed under the terms and conditions of the Creative Commons Attribution (CC BY) license (https://creativecommons.org/licenses/by/4.0/).

1. Introduction

Food processing refers to different unit operations that modify foodstuff chemically and physically to extend its shelf life and ensure food quality and safety [1]. Some traditional unit operations used in the fruit industry include peeling, chopping, crushing, sieving, and thermal treatments [2]. Many studies have determined how different processing techniques and conditions affect some sensitive nutrients (e.g., vitamins) and bioactive compounds, as well as the quality characteristics of fruit products [3,4].

Industries must ensure the supply of fresh fruits to manufacture fruit products throughout the year. However, sometimes fruits are processed in advance to be used as raw materials due to the difficulty of local production or fruits (fresh) not available in all seasons. Overall, thermal treatment is the most used processing technology to produce fruit products such as purees, jams, juices, concentrates, and baby foods [1]. Nevertheless, in light of an increasing demand of consumers towards minimally processed and healthier products [5], non-thermal technologies such as high-pressure processing (HPP), cold atmospheric plasma, and pulsed electric fields, among others, have been developed [6]. It has been shown that the degree of processing in thermal and non-thermal processing techniques affects the stability of polyphenols in fruit products [3]. However, the degree of processing has not been fully assessed yet, as many authors categorize the degree of processing according to the number of ingredients in a food product or to the extent and purpose of food processing, and not based on the process itself [7–9].

Therefore, defining the degree of processing of the fruit products will help assess the preservation level of the phenolic compounds and the quality characteristics of the final product (e.g., fruit juices, fruit purees, jams). This represents a significant challenge for industries and consumers. A potential strategy would be to study the overall metabolic profile of the final product, which could be affected by the degree of processing. Metabolomics-based approaches have been demonstrated to be a powerful tool for identifying changes in food metabolic profiles due to processing [10]. In particular, untargeted metabolomics has been used to study modifications after minimal processing and storage of meat, fruit, and vegetable products [11–13]. A previous study examining how processing (thermal pasteurization, pulsed electric field, and high-pressure processing) influences the quality attributes of cloudy apple juice using a targeted and non-targeted metabolomics approach revealed that heat treatment resulted in brighter color together with increased stability of cloudiness in the juice, mainly as a result of inactivation of polyphenol oxidase, peroxidase and pectin methylesterase. However, this heat treatment reduced most of the volatile compounds, principally esters, and induced the formation of off-odor compounds [14]. In addition, Utpott et al.'s [10] review recently reported changes in the metabolic profile due to thermal treatments, drying technologies, fermentation, and chemical and enzymatic treatments. However, the elucidation of potential markers is challenging [15]. Therefore, there is a need for research that examines the influence of thermal and non-thermal processing techniques on the metabolites of fruit products to determine common processing markers to provide helpful information to consumers, industries, and governmental food regulators.

This study will focus on strawberries and apples, two of the most used fruits in the food industry, not only because of their pleasant flavor and sensory characteristics but also due to the presence of bioactive compounds to which their potential health benefits are attributed [16]. The aim is to determine potential markers of food processing in strawberries and apples subjected to different industrial processing techniques, as previously described [17], to produce purees using an untargeted metabolomics approach.

2. Results

2.1. Multivariate Model Analysis

2.1.1. Multivariate Model Analysis of Strawberry

The pre-processing operations gave a data matrix from the full dataset based on 4554 entities (ions that present the necessary features to be a metabolite) for negative mode. In the PCA analysis, the first two principal components (PC1 and PC2) explained 65.3% of the total variability in the negative mode (Figure 1a). Samples were grouped according to the processing technique (Fresh strawberries (FS), No heat treatment (NT), mild heat treatment (MT), standard thermal treatment, vacuum concentration (VC)), evidencing changes in their metabolic profiles. The calculated PLS-DA model, based on 15 samples, described 96.8% of the variance ($R^2 = 0.968$) according to the cross-validation prediction of $Q2 = 0.612$ (Figure 1b) in negative polarity. The discrimination models showed significant differences between FS and VC, but not very high discrimination across the cold-crushing processing samples NT and MT, which showed similar metabolic profiles (Figure 1b). The summarized variance explained by component 1 classified the variables' thermal treatment effect, whereas component 2 classified the variables' cold crushing effect. These results showed how the temperature variation affected the discrimination of samples. Samples were classified along component 1 according to the relevance of the thermal treatment (FS-NT/MT-ST-VC). On the other hand, the samples subjected to cold crushing processing (NT and MT) were classified together in the model and separated by component 2 from fresh samples (FS) and the hot crushing processing samples (ST and VC). The results show the sample metabolome variability according to the temperature used during the processing stages.

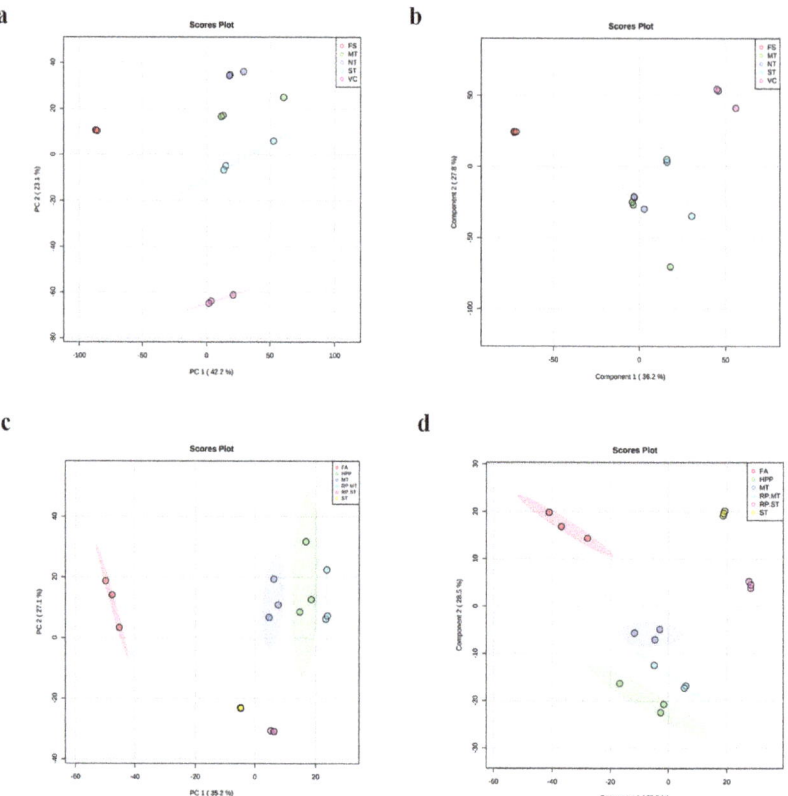

Figure 1. PCA and PLS-DA model plots. Strawberry: (**a**) PCA plot, (**b**) PLS-DA plot. Apple: (**c**) PCA plot, (**d**) PLS-DA plot. Strawberry samples: Fresh strawberries (FS), No heat treatment (NT), mild heat treatment (MT), standard thermal treatment (ST), vacuum concentration (VC). Apple samples: Fresh apple (FA), high-pressure processing (HPP), mild heat treatment (MT), standard thermal treatment (ST), reprocessed mild heat treatment (RP.MT), reprocessed standard thermal treatment (RP.ST).

2.1.2. Multivariate Model Analysis of Apple

For negative mode, the pre-processing procedures gave a data matrix from the full dataset based on 2551. PC1 and PC2 explained 62.3% of the variance in the negative mode (Figure 1c). The global trend of the data variation in the PCA plots describes a clear sample grouping affected by processing treatment [Fresh apple (FA), High-pressure processing (HPP), mild heat treatment (MT), standard thermal treatment (ST), reprocessed mild heat treatment (RP.MT), reprocessed standard thermal treatment (RP.ST), with no significant outliers. The PCA model showed accurate technical reproducibility. The calculated PLS-DA model based on 18 samples described 98.8% of the variation ($R^2 = 0.988$) according to the cross-validation prediction of Q2 = 0.911 for negative polarity (Figure 1d). The summarized variance explained by component 1 exclusively classified the variables' thermal treatment effect, whereas that component 2 classified the variables' cold crushing effect. These results showed how the cold-crushing treatment could differentiate the metabolome of these samples, similar to HPP, MT, and RP.MT treatments.

In summary, the variance explained by component 1 discriminated the fresh product, in both study cases, from the treatments mainly affected by the temperature observed in apple puree, while component 2 grouped the metabolomes affected by cold crushing.

2.2. Metabolites Trend and Markers Identification

After evaluating the multivariate model and the application of univariate operations, a list of candidate markers was obtained.

Strawberry. The filtering and statistics layers resulted in 2902 entities in the negative mode. According to the highest VIP value score (VIP > 1), the accuracy of the tentative database identification, and the p-value, a final list of 1444 entities was obtained. Finally, four entities were identified using accurate mass and MS/MS fragmentation pattern and, in some cases, confirmed with authentic standards (Table 1): Pyroglutamic acid (**1**), Pteroyl-D-glutamic acid (**2**), 2-hydroxy- 5-methoxy benzoic acid (**3**) and 2-hydroxybenzoic acid β-D-glucoside (**4**).

Table 1. Metabolites identified and confirmed by MS/MS in strawberry and apple samples.

ID	m/z	Name	Formula	RT	Polarity	Regulation	MS/MS Fragments
1	128.0344	Pyroglutamic acid *	$C_5H_7NO_3$	1.37	NEG	UP	128.0338; 85.0287; 82.0285; 72.0091
2	472.1577	Pteroyl-D-glutamic acid [a]	$C_{20}H_{23}N_7O_7$	3.70	NEG	DOWN	Unclear fragments
3	167.0339	2-hydroxy-5-methoxy benzoic acid	$C_8H_8O_4$	7.52	NEG	UP	108.0217;109.0243;152.0109;123.0019; 167.0360
4	299.0766	2-hydroxybenzoic acid beta-D-glucoside	$C_{13}H_{16}O_8$	2.16	NEG	DOWN	137.0246; 179.0437; 299.0761
5	355.0666	Dihydroxycinnamic acid glucuronide	$C_{15}H_{16}O_{10}$	2.88	NEG	UP	207.0297; 265.0358; 247.0250;193.0609;191.0555; 135.0488
6	179.0345	Caffeic acid *	$C_9H_8O_4$	5.78	NEG	UP	135.0455; 134.0369
7	474.2621	LysoPE(18:3(9Z,12Z,15Z)/0:0)	$C_{23}H_{42}NO_7P$	25.01	NEG	UP	474.2626; 277.2177; 214.0487; 152.9955

[a] Tentative identification; * Confirmed by an authentic standard; ID 1-4: detected in strawberry; ID 5-7: detected in apple. MS/MS fragments compared with Metlin database, MassBank of North America (MoNA), and calculated by CFM-ID spectrum prediction.

Only pteroyl-D-glutamic acid was not confirmed by MS/MS analysis due to the low intensity of the fragments generated. However, the presence of pyroglutamic acid, the only metabolite confirmed with an authentic standard, in the processing samples supports the occurrence of compound **2** in fresh strawberries. We initially hypothesized that compound **3** with m/z 167.0339 and a fragmentation pattern corresponding to a hydroxy-methoxy benzoic acid could be vanillic acid (4-hydroxy-3-methoxy benzoic acid), a metabolite naturally found in fruits such as banana, mango, blueberry, blackberry and strawberry and in processed products such as juice, wine, beer and cider [18]. However, its retention time did not match that of an authentic standard. Among the other isomers, considering its fragmentation pattern, this compound was tentatively identified as 2-hydroxy- 5-methoxy benzoic acid (5-methoxysalicylic acid).

The area of these potential markers was compared among the different processing treatments (Figure 2). Pyroglutamic acid (Figure 2a) showed an increasing trend mainly correlated with thermal processing. It was practically absent in the fresh fruits, and its presence was higher when increasing the temperature (thermal treatment) in the processing (FS < NT < MT < ST < VC). On the contrary, pteroyl-D-glutamic acid (Figure 2b) showed the opposite trend, a decrease in the area correlated with the thermal treatment applied to the samples. 2-Hydroxy- 5-methoxy benzoic acid (5-Methoxysalicylic acid) (Figure 2c) showed a similar trend to pyroglutamic acid, increasing with the processing intensity. Still, in this case, the temperature and cold-crushing (NT) could affect. 2-Hydroxybenzoic acid β-D-glucoside (Figure 2d) was a downregulated marker showing a decreasing trend correlated with the thermal processing intensity. However, other processing parameters applied during vacuum concentration (VC) apart from temperature could affect it.

Apple. After applying the pre-configured process layer to the apple products (see above), a total of 1723 entities were found in negative mode. According to the highest VIP value, p-value, and accuracy of the tentative database identification, a final list of 767 candidates was obtained. Finally, three entities were identified in databases using their accurate mass and MS/MS fragmentation pattern, and they were proposed as processing markers: di hydroxycinnamic acid glucuronide (**5**), caffeic acid (**6**), and lysoPE(18:3(9Z,12Z,15Z)/0:0) (**7**) (Table 1). Caffeic acid was also confirmed with an authentic standard. Compound **5** showed an MS/MS profile that matched the predicted spectrum of a hydroxycinnamic acid glucuronide.

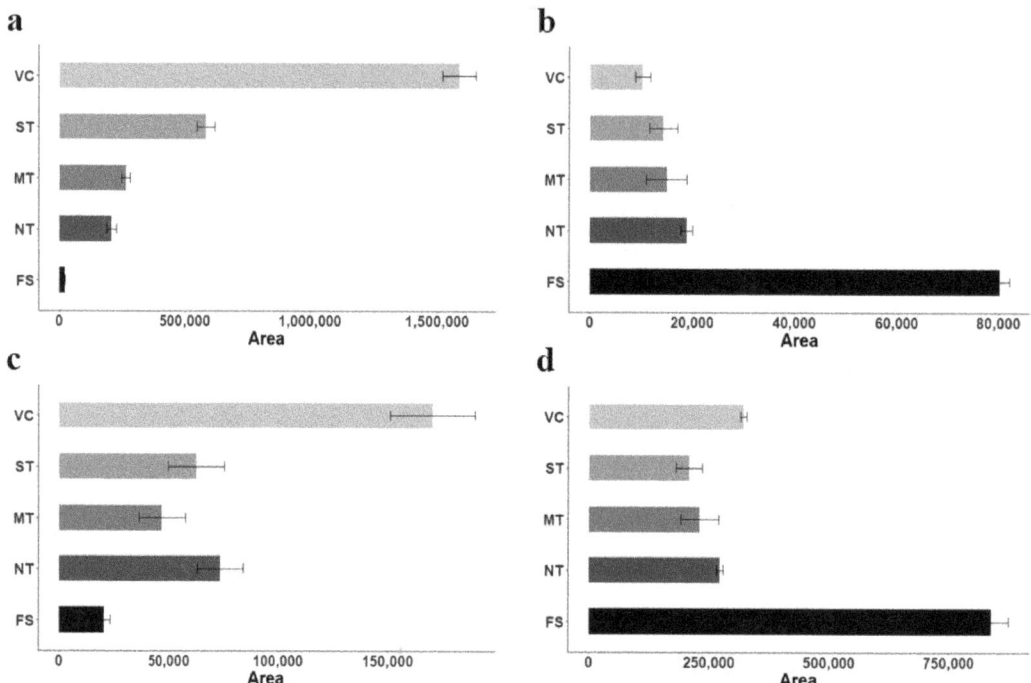

Figure 2. Bar plot of the metabolites identified in strawberry puree across the processing sample groups. X-axis: total abundance of the metabolite; Y-axis: treatment. (**a**) Pyroglutamic acid; (**b**) Pteroyl-D-glutamic acid; (**c**) 4-hydroxy-3-methoxy benzoic acid; (**d**) 2-hydroxybenzoic acid β-D-glucoside. Strawberry samples: Fresh strawberries (FS), No heat treatment (NT), mild heat treatment (MT), standard thermal treatment, vacuum concentration (VC).

After a first look, it seemed that it could be caffeic acid glucuronide. Still, its experimental fragmentation from 5 to 40 V produced the fragment at m/z 135 with low intensity, while in the theoretical MS/MS spectra of caffeic acid glucuronide, this was the major fragment. Accordingly, based on this evidence, it was impossible to specify the dihydroxycinnamic acid type or the glucuronide residue's position. When the trend of these markers in the different processed samples was evaluated, it was observed that all of them (di-hydroxycinnamic acid glucuronide, caffeic acid, and lysoPE(18:3(9Z,12Z,15Z)/0:0)) were not detected in FA samples and were upregulated after processing. They showed an increasing trend correlated with processing, mainly with the thermal treatment applied to the samples (FA < MT < ST < RP.MT < RP.ST) (Figure 3). An exception was observed with the HPP treatment, which showed higher intensity than MT and was similar to ST but was subjected to lower thermal treatment. This behavior could be due to the high pressure applied during treatment, which could also affect these compounds' presence.

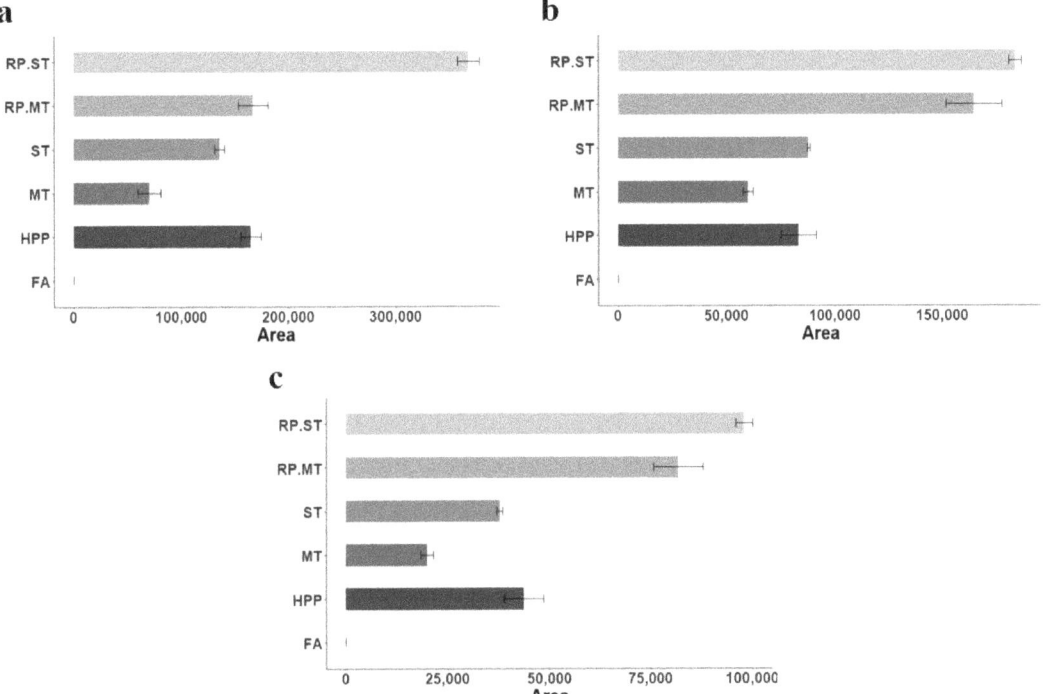

Figure 3. Bar plot of the metabolites identified in apple puree across the processing sample groups. X-axis: total abundance of the metabolite; Y-axis: treatment; (**a**) hydroxycinnamic acid glucuronide; (**b**) Caffeic acid; (**c**) LysoPE(18:3(9Z,12Z,15Z)/0:0). Apple samples: Fresh apple (FA), High-pressure processing (HPP), mild heat treatment (MT), standard thermal treatment (ST), reprocessed mild heat treatment (RP.MT), reprocessed standard thermal treatment (RP.ST).

3. Discussion

In the present study, untargeted metabolomics was successfully applied to identify changes in the metabolic profiles of strawberry and apple purees subjected to different thermal and non-thermal processing techniques [17]. Notably, multivariate analyses (PCA and PLS-DA) evidenced a clear separation between the other fruits' treatments, highlighting a temperature-dependent effect. It was possible to discriminate each treatment based on its metabolomic profiling. It became evident that the presence or absence of temperature used during the crushing process is crucial in the change of the metabolomic profile of the final product. Thus, in the case of strawberries and apples, purees produced with cold crushing clustered together, while those extracted with heat treatment also formed a separate cluster. The effect of thermal processing on the overall metabolome of different foods has been previously reported using untargeted metabolomics approaches [10]. Previous studies have reported changes in the composition of peanuts due to dry roasting [19], in tiger nut milk caused by ultra-high temperature [20], and in black raspberry powder thermally treated at 95 °C [13]. In another study, thermal treatment influenced the final metabolite composition of vegetable purees differently depending on the blending and heating conditions applied [4]. Different molecules were identified as variables to discriminate between processed and fresh samples. Metabolites that increased or decreased along processing treatment, mainly affected by the temperature, were found.

In the case of strawberries, we found that pteroyl-D-glutamic acid could be a marker of non-processed fruits. Pteroyl-D-glutamic acid, the active metabolite of folic acid [21], is essential to prevent congenital malformations such as neural tube defects [22]. Folic acid and its derivatives are present in meat products (chicken liver), leafy green vegetables (spinach, parsley), grains (green beans, peas, chickpeas), and fruits (apple, orange, strawberry), with concentrations ranging from 1 to 580 µg of folate/100 g fresh weight [23,24]. Due to processing, a significant decrease in pteroyl-D-glutamic acid was observed in strawberries. These folates have been reported to be sensitive to oxygen exposure, light intensity, and thermal treatments [23]. Depending on the intensity of the processing conditions, folates could be transformed into their less available derivatives, resulting in losses ranging from 10% to 80% [25]. The decrease in pteroyl D-glutamic acid came with an increase in pyroglutamic acid, which was identified as an upregulated marker of processing in strawberry puree (Figure 2a). These results suggest that the pteroyl-D-glutamic acid in fresh strawberries is degraded with the temperature, releasing pteroic acid and glutamic acid. Then, glutamic acid is converted into pyroglutamic acid due to the loss of a water molecule and internal cyclization (Figure 4). Both metabolites, pteroic acid and glutamic, were not detected in samples, most likely due to the further degradation of pteroic acid into other molecules and the total conversion of glutamic acid to pyroglutamic acid. Transformation of glutamic acid or glutamine to pyroglutamic acid has been previously reported through enzymatic or non-enzymatic reactions [26,27]. Pyroglutamic acid formation after enzymatic reactions have been registered in dairy, meat, and fermented products, whereas that produced after non-enzymatic reactions, has been described for tomato puree and beer [28–30]. Prior research has also found that levels of pyroglutamic acid in foods influence various sensory attributes, such as metallic aroma and bitter taste [29,31]. In addition, evidence shows that L-pyroglutamic acid induces the formation of lower-molecular-weight colored products through non-enzymatic browning reactions [32]. Accordingly, accurately identifying this compound with the authentic standard supports pyroglutamic acid as a good marker of strawberry processing.

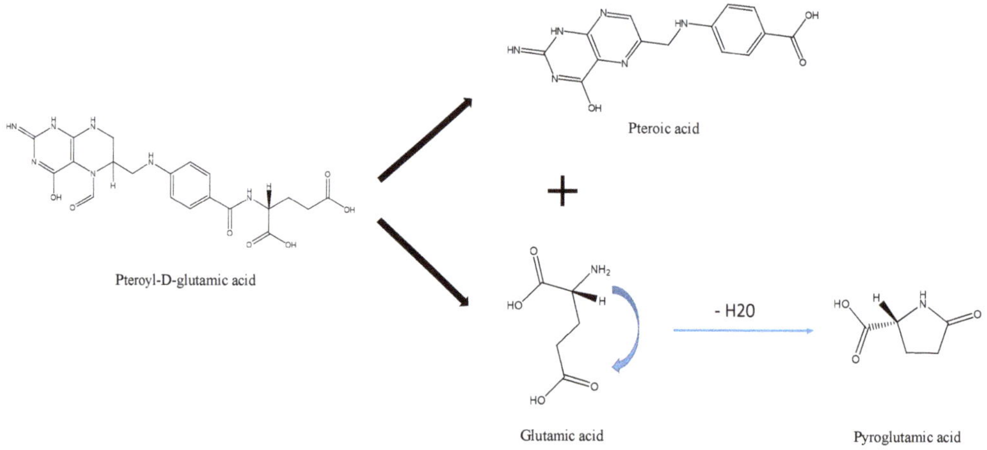

Figure 4. Diagram of pyroglutamic acid production.

Conversely, 2-hydroxy-5-methoxy benzoic acid increased its intensity with thermal processing. It was identified as an upregulated marker of processing in strawberry purees according to the relevance of the thermal treatment, except for NT treatment containing the cold treatment exclusively, which produced an increase. Our results suggest that the thermal treatment may result in the release of 2-hydroxybenzoic (salicylic acid) from the β-D-glucoside conjugate. Then this compound could undergo oxidation and methylation reactions under thermal conditions. 2-Hydroxy-5-methoxy benzoic could also come from the methylation of 2,5-dihydroxybenzoic acid, but this is unlikely as we did not detect this metabolite in fresh strawberries. The 2-hydroxybenzoic acid β-D-glucoside (also known as glucosyl salicylate) was recognized as a downregulated processing marker according to the relevance of the thermal treatment in strawberry purees (Figure 2d), except for VC treatment where the vacuum applied increased its production specifically. Salicylic acid (SA) and its derivatives (usually called salicylates) are naturally present in plants, where they play an essential role in pathogen defense and the regulation of stress response [33]. Although there is no actual data about the presence of salicylic acid glucoside in fresh foods and vegetables, the content of salicylates in different food products has been reported in prior studies. For example, salicylic acid was present in free and conjugated forms in fruits (2–3140 μg/100 g dry weight) and vegetables (1–2693 μg/100 g dry weight), as well as in beverages, meat, dairy, and cereal products (2–1226 μg/100 g dry weight) [34]. Depending on the processing methods, different salicylate levels were observed between fresh and processed products [34]. Cooking particularly impacted the salicylate content, with vegetables boiled in water containing less salicylate than raw vegetables (beans, broccoli, cauliflower).

In the case of apple puree, caffeic acid, di-hydroxycinnamic acid glucuronide, and lysoPE(18:3(9Z,12Z,15Z)/0:0) were identified as upregulated markers of processing (Figure 3e–g) according to the relevance of the thermal treatment, except for the HPP treatment, which increased the metabolite profiles. The presence of caffeic acid, a hydroxycinnamic acid commonly found in fruits, vegetables, and processed products such as coffee, wine, and beer [32], was confirmed with an authentic standard. Caffeic acid, with numerous potential biological activities, is the main contributor to the diet among the hydroxycinnamic acids, with intakes ranging from 188 to 626 mg/day [32]. A negative correlation between the chlorogenic acid degradation and the progressive increments in caffeic acid, as a result of temperature, has been previously reported [33,34]. The increase in caffeic acid with processing can most likely be explained by two factors: the release from bound caffeic acid derivatives from plant structures due to the thermal processing and/or the thermal inactivation of polyphenol oxidase that degrades caffeic acid in fresh apples immediately after processing when thermal treatments are not applied.

Regarding di-hydroxycinnamic acid glucuronide, it has not been described in fresh products or as a result of any processing technique. However, other fruits have reported other glucuronides, such as kaempferol glucuronide and quercetin glucuronide [35]. In these cases, however, the glucuronides were naturally present in the fruits and not induced by processing (they are probably released better after thermal treatments). As mentioned earlier, thermal processing may release caffeic acid or other di-hydroxycinnamic acids from the tissue.

LysoPE(18:3(9Z,12Z,15Z)/0:0) (lysophosphatidylethanolamine) is a naturally present lipid with regulatory effects in senescence and ripening, found in the extraplastidial membranes of all plants and has been identified as a polar lipid in apple tissue and apple callus [36,37]. The temperature may act as an abiotic stress that triggers lipid-dependent signaling cascades [38]. Lysophospholipids are released from membrane phospholipids after the damage of the plant tissues (wounding, cutting, and probably also by non-thermal processing) through phospholipase activity. They are detected in the tissues immediately after cutting, as in fresh-cut lettuce tissues [39]. These lysophospholipids are then quickly metabolized by other enzymes of the jasmonate pathway (9-lipoxygenase, allene-oxide synthase, and other enzymes) to produce jasmonic acid and trigger phenolic compound biosynthesis to provide substrates for the development of tissue browning and tissue wound repairing [39]. In apple tissues, LysoPE can be released by cutting and other damages during processing and its enzymatic conversion to other metabolites of the jasmonic acid pathway can be prevented by the thermal treatments. This is why the LysoPE content correlates with the processing intensity as thermal processing can inactivate the jasmonate pathway enzymes and supports the occurrence of higher levels of lysophospholipids with thermal treatments.

The compounds identified in this study represent a good starting point as metabolites to be examined by future research. The variation in the metabolites' intensity identified in this study is correlated with the degree of processing, in terms of thermal processing, except for the exclusively cold treatment, in the case of strawberry puree, and the pressure treatment, in the case of apple puree. This might contribute to detecting new processing markers, likely to help various stakeholders. In particular, the food industry could use them to optimize the degree of processing of food products. In contrast, academia or governments may use the markers to classify foods based on the degree of processing, which is currently misclassified following attributes not directly related to processing (e.g., NOVA classification system [40]). Moreover, such insights could be used as criteria to measure the degree of food naturalness, which is increasingly demanded by consumers [5,41].

4. Materials and Methods

4.1. Chemicals

Authentic standards of caffeic acid (purity > 96%), pyroglutamic acid (purity > 96%), vanillic acid (purity > 96%), salicylic acid (purity > 96%) and gentisic acid (purity > 96%) were purchased from Sigma-Aldrich (St. Louis, MO, USA). Methanol, acetic acid, acetonitrile, and water 0.1% (v/v) formic acid were from J.T. Baker (Deventer, The Netherlands), and formic acid was obtained from Panreac (Barcelona, Spain). Ultrapure water was obtained through the Milli-Q system (Millipore Corp., Bedford, MA, USA).

4.2. Processing of Strawberry and Apple Purees

This research involves the study of two fruit products, strawberry, and apple purees, processed through different industrial processing techniques, as reported previously [17] (Figure 5). A total of 15 tons of strawberries (*Fragaria* × *ananassa*) of cultivar Primoris (Huelva, Spain) and 10 tons of apples (*Malus domestica*) of cultivar Golden Delicious (Zaragoza, Spain) were harvested ripe and transported under refrigeration conditions to the processing facility. The total amount of fruits was equally distributed to the different industrial processing technologies used to obtain purees from strawberries and apples [17].

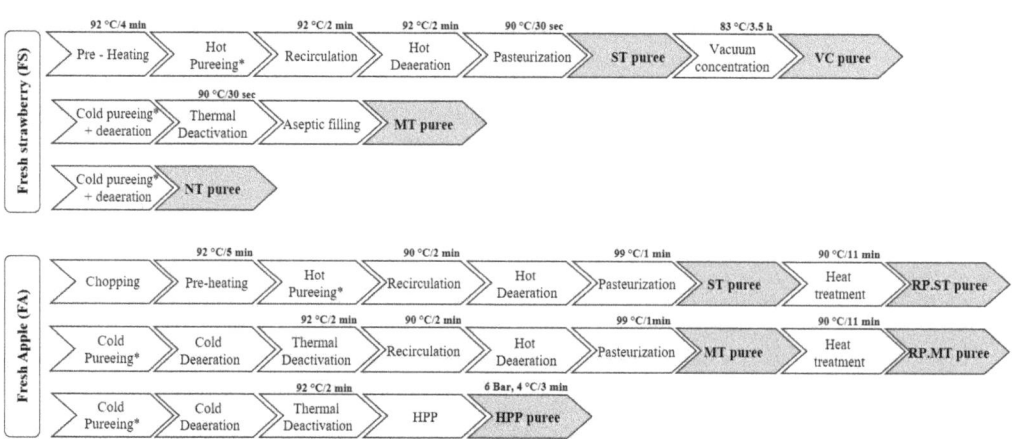

Figure 5. Diagram of the industrial processing techniques employed to obtain strawberry (Fresh strawberries (FS), No heat treatment (NT), mild heat treatment (MT), standard thermal treatment, vacuum concentration (VC)) and apple (Fresh apple (FA), High-pressure processing (HPP), mild heat treatment (MT), standard thermal treatment (ST), reprocessed mild heat treatment (RP.MT), reprocessed standard thermal treatment (RP.ST)) purees. * Removal of peel and seeds.

In general, two extraction techniques were used to obtain the purees, cold crushing, and hot crushing. Cold crushing was performed in a cold extraction line, including deaeration and enzymatic deactivation. Hot crushing was performed in a processing line with a turbo extractor, hot deaerator, and pasteurizer. In the case of cold crushing, fruits, seeds, stems, and skin were separated from the mash after the crushing. For strawberries, the puree extracted by cold crushing was divided into two. One part was subjected to a mild heat treatment of 90 °C/30 s, whereas the other did not receive any additional heat treatment. However, for the puree extracted by hot crushing, entire strawberries were preheated (92 °C/4 min) and then crushed. The hot mash was deaerated (92 °C/2 min) and pasteurized at 90 °C for 30 s resulting in a standard treated puree. Finally, to produce vacuum-concentrated puree, ST puree was subjected to a vacuum concentration (0.3–0.4 Bar) at 83 °C for 3.5 h. In the case of apples, the puree obtained after the cold crushing was thermally deactivated (92 °C/2 min) and divided into two portions. One part was packed and treated by high-pressure processing (6 bar, 4 °C/ 3 min), whereas the other part was hot deaerated and pasteurized at 99 °C for 1 min, obtaining mildly treated puree. To obtain the purée by standard thermal treatment, apples were chopped, pre-heated at 92 °C for 5 min, and hot crushed, to finally being refined, separating the skin and seeds from the puree. The obtained puree was deaerated and pasteurized at 99 °C for 1 min. To evaluate the effects of re-processing, samples of mild heat-treated puree and standard thermal-treated apple puree stored at 24 °C for six months were subjected to an additional thermal treatment of 90 °C for 11 min re-processed apple purees. In both studies, fresh fruits were used as a control. After the different processing, samples of strawberry and apple purees were taken. Their control samples were lyophilized to remove the moisture and ground into powder using a dry bean blender to homogenize the sample.

4.3. Sample Preparation

Extractions for strawberry and apple samples were carried out as previously described by Buendía et al., 2010 [42] and Jakobek et al., 2013 [43], respectively, with some modifications focused on the extraction of as many compounds as possible for the untargeted analysis. A total of 50 mg of lyophilized samples was extracted with 1 mL of methanol/water/acetic acid (70:29:1, $v/v/v$) for strawberries and methanol/water (70:30, v/v) for apples. The samples were homogenized in a vortex for one minute and then sonicated for 30 min at room temperature. Subsequently, these were centrifuged for 15 min at $20,627 \times g$ at 12 °C. The resultant supernatant was filtered through a 0.22 μm PVDF filter before UPLC-MS analysis. Three replicates for each condition were extracted and analyzed.

4.4. UPLC-ESI-QTOF-MS Analysis

Samples were analyzed using an Agilent 1290 Infinity LC system coupled to the 6550 Accurate-Mass Quadrupole time-of-flight (QTOF) (Agilent Technologies, Waldbronn, Germany) using an electrospray interface (Jet Stream Technology). Chromatographic separation was carried out on reversed-phase C18 column (3.0 × 100 mm, 2.1 μm particle size) (ACE Excel, Scotland) at 30 °C, using as mobile phases water + 0.1% formic acid (Phase A) and acetonitrile + 0.1% formic acid (Phase B) with a flow rate of 0.5 mL/min. The following gradient was used: 0–7 min, 5–18% B; 7–17 min, 18–28% B; 17–22 min, 28–50% B, 22–27 min, 50–90% B, 27–29 min, whereafter the gradient comes back to the initial conditions (5% B), which are maintained for 6 min. The injection volume was 5 μL. The optimal conditions of the electrospray interface were as follows: gas temperature 280 °C, drying gas 11 L/min, nebulizer 45 psi, sheath gas temperature 400 °C, and sheath gas flow 12 L/min. Spectra were acquired in the m/z range 100–1100 in negative and positive mode, and fragmentor voltage was 100 V. MS/MS product ion spectra were collected at an m/z range of 50–1000 using a retention time window of 1 min, collision energy of 10 and 20 eV and an acquisition rate of 4 spectra/s.

4.5. Untargeted Metabolomics Data Treatment

The data generated by UPLC-ESI-QTOF-MS metabolomics system were acquired in profile mode. The raw data were exported to Profinder software (Agilent Technologies) for pre-processing procedures and to build the data matrix for further processing and data treatment. Independent data matrixes were created to process and analyze strawberry and apple samples separately for each negative and positive polarity. The data matrixes were imported in parallel to the Metaboanalyst online platform and Mass Profiler Professional (MPP, Agilent Technologies). Data processing was performed before univariate and multivariate analysis, including data log transformation and Pareto scaling [44].

Regarding multivariate analysis, the principal component analysis (PCA) and partial least squares discriminant analysis (PLS-DA) models of the final data matrix were created using the Metaboanalyst platform to describe the total variance of the full data set and figure out the discriminations groups under data matrix criteria. The VIP (variable importance in projection) score value (VIP > 1) obtained by the discriminant analysis was used for candidate selection. After the multivariate analysis evaluation, univariate operations were performed in MPP software. Data treatment through MPP software included filters by frequency of the data matrix to reduce the sample variability within each study group and the ANOVA statistics analysis (corrected p-value cut-off: 0.05; p-value computation: Asymptotic; Multiple Testing Correction: Benjamini–Hochberg). The candidates must be significant, at least between the extreme samples group (FS and VC in the case of strawberry; FA and RP.MT or RP.ST in the case of apple puree). The VIP > 1 score and p-value were used to create the candidate list for evaluating the processing treatment.

After the selection of the candidates, the authentic standards and the MS/MS spectra data of those ions were used for metabolite confirmation. Metlin and MassBank of Noth America (MoNA) databases were used for checking the tentative identification. In addition to the databases, the competitive fragmentation modeling for metabolite identification (CFM-ID) software was complementary to confirm the metabolites. The positive polarity confirmed no metabolites by MS/MS spectra fragmentation or authentic standards. This may be due to the limitation of the method for achieving good ionization results and the unavailability of authentic standards. Despite this, the data acquired in positive polarity were also uploaded to the data repository, and the experimental data in positive polarity could be used in future investigations. The metabolomics data were deposited in the Metabolights database (https://www.ebi.ac.uk/metabolights/reviewerf89c976a-d48f-4218-9e57-64aa1cce52dd) (accessed on 5 May 2022).

5. Conclusions

Processing is a relevant and pervasive practice in the food industry that can affect the composition and quality of foods. Untargeted metabolomics has been demonstrated to be a valuable tool to visualize changes in the metabolic profile of strawberry and apple purees subjected to different processing techniques in a real industry case. The study supposes a promising source of candidates to be confirmed in further investigation and applied in new treatments of the food industry in the future.

Several metabolites showed changes according to the relevance of thermal conditions but also according to the exclusively cold treatment, in the case of strawberry puree, and the pressure treatment, in the case of apple puree. These findings suggest the possibility of studying the isolated impact of these variables. Seven of these compounds were identified and were proposed as potentially powerful markers to evaluate the processing degree of strawberry and apple puree products. The pyroglutamic acid, pteroyl-D-glutamic acid, 2-hydroxy-5-methoxybenzoic acid, and 2-hydroxybenzoic acid β-D-glucoside were identified in strawberry and di-hydroxycinnamic acid glucuronide, caffeic acid and lysoPE(18:3(9Z,12Z,15Z)/0:0) in apple purees. The metabolites confirmed as pyroglutamic acid in the case of strawberry puree and caffeic acid in the apple puree are potential candidates to be validated in a specific protocol by the food industry due to the availability of authentic standards. This study opens a new field for applying untargeted metabolomics to find markers of processing produced in the food industry.

Author Contributions: C.J.G., F.A.T.-B., L.M.S.-S. and R.G.-V. were involved in the conception and design of the study. G.S.-O. was involved in sample preparation. G.S.-O. and R.G.-V. were involved in the UPLC-ESI-QTOF-MS methodology development. C.J.G. and G.S.-O. were involved in the curation, processing and analysis of the data. C.J.G. and R.G.-V. were involved in the interpretation of the data. F.A.T.-B. and L.M.S.-S. were involved in the discussion and applicability of the results. L.M.S.-S. was involved in processing design, sampling, and funding acquisition. All authors have read and agreed to the published version of the manuscript.

Funding: This research was funded by HERO Group under the FOODPRINT project: Understanding and reframing minimal processing.

Institutional Review Board Statement: Not applicable.

Informed Consent Statement: Not applicable.

Data Availability Statement: The raw metabolomics data was deposited in the Metabolights database (https://www.ebi.ac.uk/metabolights/reviewerf89c976a-d48f-4218-9e57-64aa1cce52dd) (accessed on 5 May 2022).

Conflicts of Interest: L.M.S.-S. is a member of the Research and Nutrition Department of Hero Group, a Swiss international food manufacturer.

References

1. van Boekel, M.; Fogliano, V.; Pellegrini, N.; Stanton, C.; Scholz, G.; Lalljie, S.; Somoza, V.; Knorr, D.; Jasti, P.R.; Eisenbrand, G. A review on the beneficial aspects of food processing. *Mol. Nutr. Food Res.* **2010**, *54*, 1215–1247. [CrossRef]
2. Scholz, R.; Herrmann, M.; Kittelmann, A.; von Schledorn, M.; van Donkersgoed, G.; Graven, C.; van der Velde-Koerts, T.; Anagnostopoulos, C.; Bempelou, E.; Michalski, B. Compendium of Representative Processing Techniques investigated in regulatory studies for pesticides. *EFSA Support. Publ.* **2018**, *15*, 1508E. [CrossRef]
3. Salazar-Orbea, G.L.; García-Villalba, R.; Tomás, F.A.; Sánchez-Siles, L.M. Review: High–Pressure Processing vs. Thermal Treatment: Effect on the Stability of Polyphenols in Strawberry and Apple Products. *Foods* **2021**, *10*, 2919. [CrossRef]
4. Lopez-Sanchez, P.; De Vos, R.C.H.; Jonker, H.H.; Mumm, R.; Hall, R.D.; Bialek, L.; Leenman, R.; Strassburg, K.; Vreeken, R.; Hankemeier, T.; et al. Comprehensive metabolomics to evaluate the impact of industrial processing on the phytochemical composition of vegetable purees. *Food Chem.* **2015**, *168*, 348–355. [CrossRef]
5. Román, S.; Sánchez-Siles, L.M.; Siegrist, M. The importance of food naturalness for consumers: Results of a systematic review. *Trends Food Sci. Technol.* **2017**, *67*, 44–57. [CrossRef]
6. Knorr, D.; Augustin, M.A. Food processing need, advantages and misconceptions. *Trends Food Sci. Technol.* **2021**, *108*, 103–110. [CrossRef]
7. Monteiro, C.A.; Cannon, G.; Levy, R.; Moubarac, J.-C.; Jaime, P.; Martins, A.P.; Canella, D.; Louzada, M.; Parra, D. NOVA. The star shines bright. *World Nutr.* **2016**, *7*, 28–38.
8. Monteiro, C.A. Nutrition and health. The issue is not food, nor nutrients, so much as processing. *Public Health Nutr.* **2009**, *12*, 729–731. [CrossRef]
9. Scrinis, G.; Monteiro, C.A. Ultra-processed foods and the limits of product reformulation. *Public Health Nutr.* **2018**, *21*, 247–252. [CrossRef]
10. Utpott, M.; Rodrigues, E.; de Rios, A.O.; Mercali, G.D.; Flôres, S.H. Metabolomics: An analytical technique for food processing evaluation. *Food Chem.* **2022**, *366*, 130685. [CrossRef]
11. Garcia, C.J.; García-Villalba, R.; Garrido, Y.; Gil, M.I.; Tomás-Barberán, F.A. Untargeted metabolomics approach using UPLC-ESI-QTOF-MS to explore the metabolome of fresh-cut iceberg lettuce. *Metabolomics* **2016**, *12*, 138. [CrossRef]
12. Zhao, X.; Chen, L.; Yang, H. Effect of vacuum impregnated fish gelatin and grape seed extract on metabolite profiles of tilapia (*Oreochromis niloticus*) fillets during storage. *Food Chem.* **2019**, *293*, 418–428. [CrossRef]
13. Teegarden, M.D.; Schwartz, S.J.; Cooperstone, J.L. Profiling the impact of thermal processing on black raspberry phytochemicals using untargeted metabolomics. *Food Chem.* **2019**, *274*, 782–788. [CrossRef]
14. Wibowo, S.; Aba, E.; De Man, S.; Bernaert, N.; Van Droogenbroeck, B.; Grauwet, T.; Van Loey, A.; Hendrickx, M. Comparing the impact of high pressure, pulsed electric field and thermal pasteurization on quality attributes of cloudy apple juice using targeted and untargeted analyses. *Innov. Food Sci. Emerg. Technol.* **2019**, *54*, 64–77. [CrossRef]
15. Lacalle-Bergeron, L.; Izquierdo-Sandoval, D.; Sancho, J.V.; Francisco, L.; Portolés, T. Chromatography hyphenated to high resolution mass spectrometry in untargeted metabolomics for investigation of food (bio) markers. *Trends Anal. Chem.* **2021**, *135*, 116161. [CrossRef]
16. Bradbury, K.E.; Appleby, P.N.; Key, T.J. Fruit, vegetable, and fiber intake in relation to cancer risk: Findings from the European Prospective Investigation into Cancer and Nutrition (EPIC). *Am. J. Clin. Nutr.* **2014**, *100*, 394S–398S. [CrossRef]

17. Salazar-Orbea, G.L.; García-Villalba, R.; Bernal, M.J.; Hernández, A.; Tomás-Barberán, F.A.; Sánchez-Siles, L.M. Stability of phenolic compounds in Apple and strawberry: Effect of different processing techniques in industrial set up. *Food Chem.* **2023**, *401*, 134009. [CrossRef] [PubMed]
18. Kiokias, S.; Proestos, C.; Oreopoulou, V. Phenolic acids of plant origin—A review on their antioxidant activity in vitro (O/W emulsion systems) along with their in vivo health biochemical properties. *Foods* **2020**, *9*, 534. [CrossRef] [PubMed]
19. Klevorn, C.M.; Dean, L.L. A metabolomics-based approach identifies changes in the small molecular weight compound composition of the peanut as a result of dry-roasting. *Food Chem.* **2018**, *240*, 1193–1200. [CrossRef] [PubMed]
20. Rubert, J.; Monforte, A.; Hurkova, K.; Pérez-Martínez, G.; Blesa, J.; Navarro, J.L.; Stranka, M.; Miguel, J.; Hajslova, J. Untargeted metabolomics of fresh and heat treatment Tiger nut (*Cyperus esculentus* L.) milks reveals further insight into food quality and nutrition. *J. Chromatogr. A* **2017**, *1514*, 80–87. [CrossRef]
21. Wishart, D.S.; Feunang, Y.D.; Marcu, A.; Guo, A.C.; Liang, K.; Vázquez-Fresno, R.; Sajed, T.; Johnson, D.; Li, C.; Karu, N.; et al. HMDB 4.0: The human metabolome database for 2018. *Nucleic Acids Res.* **2018**, *46*, D608–D617. [CrossRef]
22. Chitayat, D.; Matsui, D.; Amitai, Y.; Kennedy, D.; Vohra, S.; Rieder, M.; Koren, G. Folic acid supplementation for pregnant women and those planning pregnancy: 2015 update. *J. Clin. Pharmacol.* **2016**, *56*, 170–175. [CrossRef]
23. Scott, J.; Rébeillé, F.; Fletcher, J. Folic acid and folates: The feasibility for nutritional enhancement in plant foods. *J. Sci. Food Agric.* **2000**, *80*, 795–824. [CrossRef]
24. Cieślik, E.; Cieślik, I. Occurrence and significance of folic acid. *Pteridines* **2018**, *29*, 187–195. [CrossRef]
25. USDA. FoodData Central. Available online: https://fdc.nal.usda.gov/ (accessed on 31 December 2018).
26. Wegener, S.; Kaufmann, M.; Kroh, L.W. Influence of L-pyroglutamic acid on the color formation process of non-enzymatic browning reactions. *Food Chem.* **2017**, *232*, 450–454. [CrossRef]
27. Rigo, B.; Cauliez, P.; Fasscur, D.; Sauvage, F. Some aspects on the chemistry of pyroglutamic acid and related compounds. *Trends Heterocycl. Chem.* **1991**, *2*, 155–203.
28. Hartmann, J.; Brand, C.; Dose, K. Formation of specific amino acid sequences during thermal polymerization of aminoacids. *Biosystems* **1981**, *13*, 141–147. [CrossRef]
29. Pfeiffer, P.; König, H. *Biology of Microorganisms on Grapes, in Must and in Wine*; Springer: Berlin, Germany; Heidelberg, Germany, 2009; ISBN 9783540854623.
30. Qiu, J.; Vuist, J.; Boom, R.; Schutyser, M. Change in taste of tomato during thermal processing and drying: Quantitative analyses on ascorbic acid and pyroglutamic acid. In Proceedings of the 6th European Drying Conference, Liege, Belgium, 19–21 June 2017; pp. 1–8.
31. Gazme, B.; Boachie, R.T.; Tsopmo, A.; Udenigwe, C.C. Occurrence, properties and biological significance of pyroglutamyl peptides derived from different food sources. *Food Sci. Hum. Wellness* **2019**, *8*, 268–274. [CrossRef]
32. Vinholes, J.; Silva, B.M.; Silva, L.R. Hydroxycinnamic acids (HCAS): Structure, Biological properties and health effects. *Adv. Med. Biol.* **2019**, *88*, 1–33.
33. Rodríguez-Salinas, P.A.; Muy-Rangel, D.; Urías-Orona, V.; Zavala-García, F.; Suárez-Jacobo, Á.; Heredia, J.B.; Rubio-Carrasco, W.; Niño-Medina, G. Thermal processing effects on the microbiological, physicochemical, mineral, and nutraceutical properties of a roasted purple maize beverage. *Farmacia* **2019**, *67*, 587–595. [CrossRef]
34. Szwajgier, D.; Halinowski, T.; Helman, E.; Tylus, K.; Tymcio, A. Influence of different heat treatments on the content of phenolic acids and their derivatives in selected fruits. *Fruits* **2014**, *69*, 167–178. [CrossRef]
35. Enomoto, H. Mass Spectrometry Imaging of Flavonols and Ellagic Acid Glycosides in Ripe Strawberry Fruit. *Molecules* **2020**, *25*, 4600. [CrossRef]
36. Amaro, A.L.; Almeida, D.P.F. Lysophosphatidylethanolamine effects on horticultural commodities: A review. *Postharvest Biol. Technol.* **2013**, *78*, 92–102. [CrossRef]
37. Prabha, T.N.; Raina, P.L.; Patwardhan, M.V. Lipid profile of cultured cells of apple (*Malus sylvestris*) and apple tissue. *J. Biosci.* **1988**, *13*, 33–38. [CrossRef]
38. Hou, Q.; Ufer, G.; Bartels, D. Lipid signalling in plant responses to abiotic stress. *Plant Cell Environ.* **2016**, *39*, 1029–1048. [CrossRef]
39. Garcia, C.J.; García-Villalba, R.; Gil, M.I.; Tomas-Barberan, F.A. LC-MS Untargeted Metabolomics To Explain the Signal Metabolites Inducing Browning in Fresh-Cut Lettuce. *J. Agric. Food Chem.* **2017**, *65*, 4526–4535. [CrossRef] [PubMed]
40. Levy, R.B.; Louzada, M.L.C.; Moubarac, J.-C.; Jaime, P.C.; Monteiro, C.A.; Cannon, G. The UN Decade of Nutrition, the NOVA food classification and the trouble with ultra-processing. *Public Health Nutr.* **2017**, *21*, 5–17. [CrossRef]
41. Sanchez-Siles, L.M.; Michel, F.; Román, S.; Bernal, M.J.; Philipsen, B.; Haro, J.F.; Bodenstab, S.; Siegrist, M. The Food Naturalness Index (FNI): An integrative tool to measure the degree of food naturalness. *Trends Food Sci. Technol.* **2019**, *91*, 681–690. [CrossRef]
42. Buendía, B.; Gil, M.I.; Tudela, J.A.; Gady, A.L.; Medina, J.J.; Soria, C.; López, J.M.; Tomás-Barberán, F.A. HPLC-MS analysis of proanthocyanidin oligomers and other phenolics in 15 strawberry cultivars. *J. Agric. Food Chem.* **2010**, *58*, 3916–3926. [CrossRef] [PubMed]
43. Jakobek, L.; García-Villalba, R.; Tomás-Barberán, F.A. Polyphenolic characterisation of old local apple varieties from Southeastern European region. *J. Food Compos. Anal.* **2013**, *31*, 199–211. [CrossRef]
44. van den Berg, R.A.; Hoefsloot, H.C.J.; Westerhuis, J.A.; Smilde, A.K.; van der Werf, M.J. Centering, scaling, and transformations: Improving the biological information content of metabolomics data. *BMC Genom.* **2006**, *7*, 142. [CrossRef]

Article

Ultrasound, Acetic Acid, and Peracetic Acid as Alternatives Sanitizers to Chlorine Compounds for Fresh-Cut Kale Decontamination

Maria Clara de Moraes Motta Machado [1], Bárbara Morandi Lepaus [1], Patrícia Campos Bernardes [2] and Jackline Freitas Brilhante de São José [3,*]

1. Postgraduate Program in Nutrition and Health, Federal University of Espírito Santo, Marechal Campos Avenue, Vitoria 29047-105, ES, Brazil
2. Department of Food Engineering, Center of Agrarian Sciences and Engineering, Federal University of Espírito Santo, Alegre 29500-000, ES, Brazil
3. Department of Integrated Health Education, Federal University of Espírito Santo, Marechal Campos Avenue, Vitoria 29047-105, ES, Brazil
* Correspondence: jackline.jose@ufes.br; Tel.: +55-27-3335-7223

Citation: de Moraes Motta Machado, M.C.; Lepaus, B.M.; Bernardes, P.C.; de São José, J.F.B. Ultrasound, Acetic Acid, and Peracetic Acid as Alternatives Sanitizers to Chlorine Compounds for Fresh-Cut Kale Decontamination. *Molecules* 2022, 27, 7019. https://doi.org/10.3390/molecules27207019

Academic Editor: Sabina Lachowicz-Wiśniewska

Received: 20 August 2022
Accepted: 28 September 2022
Published: 18 October 2022

Publisher's Note: MDPI stays neutral with regard to jurisdictional claims in published maps and institutional affiliations.

Copyright: © 2022 by the authors. Licensee MDPI, Basel, Switzerland. This article is an open access article distributed under the terms and conditions of the Creative Commons Attribution (CC BY) license (https://creativecommons.org/licenses/by/4.0/).

Abstract: Chlorinated compounds are usually applied in vegetable sanitization, but there are concerns about their application. Thus, this study aimed to evaluate ultrasound (50 kHz), acetic acid (1000; 2000 mg/L), and peracetic acid (20 mg/L) and their combination as alternative treatments to 200 mg/L sodium dichloroisocyanurate. The overall microbial, physicochemical, and nutritional quality of kale stored at 7 °C were assessed. The impact on *Salmonella enterica* Typhimurium was verified by plate-counting and scanning electron microscopy. Ultrasound combined with peracetic acid exhibited higher reductions in aerobic mesophiles, molds and yeasts, and coliforms at 35 °C (2.6; 2.4; 2.6 log CFU/g, respectively). Microbial counts remained stable during storage. The highest reduction in *Salmonella* occurred with the combination of ultrasound and acetic acid at 1000 mg/L and acetic acid at 2000 mg/L (2.8; 3.8 log CFU/g, respectively). No synergistic effect was observed with the combination of treatments. The cellular morphology of the pathogen altered after combinations of ultrasound and acetic acid at 2000 mg/L and peracetic acid. No changes in titratable total acidity, mass loss, vitamin C, or total phenolic compounds occurred. Alternative treatments presented equal to or greater efficacies than chlorinated compounds, so they could potentially be used for the decontamination of kale.

Keywords: ultrasound; acetic acid; peracetic acid; disinfection; vegetables; ready-to-eat; food safety; non-chlorine sanitizer; emerging technologies

1. Introduction

Fresh-cut or minimally processed vegetables are characterized as small portions that are ready to eat (RTE) or to cook [1]. Consuming fresh leafy vegetables, such as lettuce, spinach, cabbage, and kale, has certain benefits for human health [2,3], and the interest in healthy and well-balanced diets has increased. This interest includes a higher demand for foods such as RTE vegetables [1,4]. Kale is a leafy vegetable in the *Brassica* group that contains many important nutrients, such as calcium, iron, and vitamins A, C, K, and B5, and can be commercialized as an RTE vegetable. Vegetables are usually consumed as an excellent source of phenolic compounds that are known for their antioxidant activity [5].

The main limitations of RTE products are related to the fact that they are much more expensive than their conventional counterparts. In contrast, as an advantage, they often do not require an additional preparation step before consumption. However, the lack of a formal cooking or preparation step can cause concerns for consumers regarding the microbiological, sensorial, and nutritional qualities of the produce [6,7]. Additionally, the cases of foodborne disease outbreaks associated with the consumption of RTE leafy vegetables contaminated with pathogens like *Escherichia coli*, *Salmonella* spp., and *Listeria* spp. have increased over the years [8].

For that reason, cleaning and sanitization are crucial steps to guaranteeing the microbiological safety of RTE vegetables. Several compounds have been widely studied as alternatives to chlorine-based cleaning products, which are the most used chemicals for the sanitization of vegetables by the food industry [9]. Among the newly explored compounds are organic acids, such as acetic acid (AA) and lactic acid (LA), and peracetic acid (PAA), which have advantages over-chlorinated compounds when used as antimicrobial agents [10–18]. Because sanitization with organic molecules does not generate toxic or carcinogenic residues, AA and PAA can be used as alternative chlorinated compounds in the ready-to-eat produce industry [11,16].

Some technologies, such as ultrasound (US), can be combined with chemical cleaners or by themselves for fruit and vegetable decontamination post-harvest [3,12,17]. US is described as sound waves above the human hearing limit that could be applied in several industrial functions, such as food processing [19,20]. US is classified as a physical and nonthermal technology that can prolong shelf life and reduce the microbial counts of heat-sensitive foods to preserve the nutritional and sensorial characteristics of products [1,2,4]. The mechanisms of inactivation of microorganisms by US have been attributed to cavitation, reducing bacterial viability by weakening cell walls and facilitating access to sanitizers [3,17].

Considering the hurdle theory, we hypothesize that US combined with organic acids can generate a synergistic reduction in microorganisms and still preserve food quality. Some studies have investigated the effect of sanitization with organic acids and US, isolated or combined, on the microbiological and physicochemical quality of vegetables [10,11,13,18]. However, studies about the possible synergistic effects of US and organic acids compared to chlorine compounds on the overall quality of RTE kale are still scarce.

Therefore, the present study investigated the microbiological and physicochemical quality of fresh-cut kale subjected to sodium dichloroisocyanurate and alternative sanitization procedures. We aimed to verify four hypotheses in this study: (i) AA and PAA would show similar or better effectiveness in the disinfection of fresh-cut kale than the chlorine-based sanitizer; (ii) the combination of US with AA and PAA would improve the chemical sanitizer's efficiency; (iii) the treatments would contribute to the preservation of physicochemical properties, and; (iv) the treatments proposed in this study would not have a detrimental effect on bioactive compounds.

2. Material and Methods

2.1. Experimental Design, Obtaining Samples, and Sanitizing Procedure

The experiment was conducted in a completely randomized design, and three experimental repetitions, with duplicates, were conducted for each sanitizing treatment. The treatment conditions chosen were based on previous studies and literature.

Kale (Brassica oleracea var. acephala) was obtained from a local market, stored at 7 °C for a maximum of 24 h, and processed according to Figure 1. Any leaves that were damaged and/or yellowish were discarded, and the rest were washed in tap water to remove soil adhered to the leaf surface. The sanitization step consisted of the immersion of 400 g of sliced kale in 4 L of sanitizing solution for 5 min at room temperature. The effects of 1000 mg/L and 2000 mg/L acetic acid (Neon®, São Paulo, Brazil), 20 mg/L peracetic acid (Nippon Chemical®, São Paulo, Brazil), and ultrasound (bath-type, 50 kHz, 150 W, model NI 1207, Nova Instruments®, São Paulo, Brazil) were evaluated. Moreover, the combination of organic acids sanitizers with ultrasound was tested. The reference treatment was a 200 mg/L sodium dichloroisocyanurate solution (Nippon Chemical®, São Paulo, Brazil). The control vegetable sample was only washed in tap water and not subjected to any sanitization procedure.

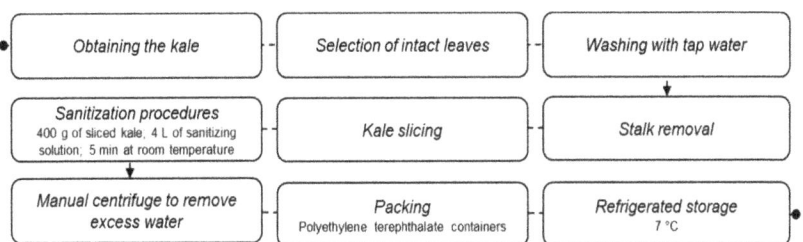

Figure 1. Flowchart of obtaining and processing minimally processed (ready-to-eat) kale used in this study.

2.2. Storage Time Study

After the sanitization procedures were carried out, the kale samples were packed in polyethylene terephthalate containers and stored at 7 ± 1 °C. The assessment of sanitizers, ultrasound, and their combination effects on natural microbiota and physicochemical quality were evaluated immediately after sanitization and after 6 days of storage at 7 ± 1 °C, as described in the following paragraphs. Salmonella populations intentionally inoculated on kale surfaces were analyzed only after sanitizing treatments (first day).

2.3. Evaluation of the Efficiency of Sanitization in the Natural Microbiota

This step was performed according to the methodology of the American Public Health Association (APHA) [21]. Kale samples (25 g), previously subjected to sanitizer treatments but not inoculated with any pathogen, were homogenized with 225 mL of 0.1 % peptone water in a stomacher (Marconi®, Piracicaba, São Paulo, Brazil) for 2 min at normal speed. Appropriate decimal dilutions were prepared, and aliquots were transferred to specific culture media for the determination of the counts of each microbial group. The plating of the aliquots was performed in duplicates, and the results were expressed in colony-forming units per gram (CFU/g). Mesophilic aerobic bacteria were plated on plate count agar (Himedia®, Mumbai, India) and incubated at 35 °C for 48 h. Molds and yeasts were plated on potato dextrose agar (Himedia®, Mumbai, India) and incubated at 25 °C for 7 days. Coliform counts were carried out at 35 °C using Petrifilm plates (3M®, Sant Paul, MN, USA), incubated at 35 °C for 48 h, enumerating *E. coli* as blue colonies with gas and total coliforms at 35 °C as red colonies with gas.

2.4. Inoculation of Salmonella enterica Typhimurium ATCC 14028 on Kale Surface

This experimental step was carried out separately from the natural contamination evaluation and was performed according to previous studies [11,22], with some modifications. Suspensions of vegetative Salmonella enterica Typhimurium ATCC 14028 cells were activated two consecutive times in 10 mL of brain heart infusion broth (BHI; Himedia®, Mumbai, India) and incubated at 37 °C for 24 h until populations of 10^8 and 10^9 CFU/mL were reached. Kale leaves (25 g) were cut into 3 cm × 3 cm pieces and placed in previously sterilized plastic bags. Then, the inoculum (BHI inoculated with vegetative cells) was added along with 225 mL of 0.1 % peptone water, and the plastic bags containing the inoculum and the vegetable material were stirred gently for 5 min. The content was kept in static contact with the cell suspension for 60 min at 24 ± 1 °C to promote higher cell adhesion to the vegetable. After, the cell suspension was drained, and the kale contaminated with the pathogen was subjected to the same sanitization treatments described before. After each treatment, the samples were transferred to sterile plastic bags containing 0.1% peptone water and then homogenized in a stomacher (Marconi®, Piracicaba, São Paulo, Brazil) for 2 min at normal speed. Finally, 1 mL of homogenized sample was used to prepare decimal dilutions, and aliquots were plated on Hektoen agar (Himedia®, Mumbai, India). After plating and incubation at 37 °C for 24 h, colony counting was performed.

2.5. Scanning Electronic Microscopy after Inoculation of Salmonella enterica Typhimurium ATCC 14028 and Decontamination Procedures

Scanning electron microscopy was performed to confirm the adhesion of Salmonella cells and the possible effect of their sanitization on the kale surface. First, the preparation of samples for microscopic observation was conducted. Briefly, kale inoculated with Salmonella Typhimurium ATCC 14028 strain and sanitized with different procedures were cut into 1.0 cm × 1.0 cm pieces using a previously sterilized scalpel. Kale slices were rinsed in 0.1 mol/L phosphate-buffered saline (PBS) at pH 6.8 to 7.2 to remove residues from sanitizers and unbound cells. The fixation step consisted of treatment with 5% glutaraldehyde (Vetec®, São Paulo, Brazil) and 0.1 mol/L PBS buffer in a 1:1 ratio for 1 h at room temperature. Thereafter, the samples were washed 6 times in 0.1 mol/L PBS buffer at pH 6.8 to 7.2, and each wash lasted for 10 min. The dehydration stage consisted of serial treatments in 30, 50, 70, 80, and 95 °C Gay-Lussac (GL) ethanol for 10 min each and three 15 min treatments in 100 °C GL ethanol. Samples were transferred to a critical point dryer model AutoSamdri-815 A (Tousimins®, Rockville, MD, USA) to complete the dehydration process. Finally, they were submitted to a Desk V sputter coater (Denton Vacuum®, Cherry Hill, NJ, USA) for deposition of a thin layer of gold and then analyzed by scanning electron microscopy (model JSM-6610, Scanning Electron Microscope® LV, JEOL, Tokyo, Japan).

2.6. Physicochemical and Bioactive Compound Analysis

The total titratable acidity, pH, total soluble solids, and vitamin C content were determined according to the AOAC [23]. The total phenolic compounds were measured using the Folin–Ciocalteu reagent (Sigma Aldrich®, St. Louis, MO, USA) [24]. The effect of sanitizing treatments on antioxidant capacity was evaluated using 1,1-diphenyl-2-picrylhydrazyl (DPPH) (Sigma Aldrich®, St. Louis, MI, USA), according to Blois [25]. The percentage of mass lost by each sample was determined according to Equation (1):

$$\% \text{ Mass loss }(t) = \frac{[M_0 - M_t] \times 100}{M_0} \quad (1)$$

where % Mass loss (t) is the mass lost after time t; M_0 is the initial mass of the sample; and M_t is the weight of the sample at time t.

These analyses identified possible changes in the quality of the vegetables after sanitization treatments compared to non-sanitized samples (control). The samples used in this series of tests were not inoculated with the pathogen.

2.7. Statistical Analysis

The efficiency of the sanitization treatments in changing the microbiological and physicochemical characteristics was evaluated soon after sanitization and at the end of the storage period using the paired t-test. For the evaluation of the treatments at the time point, the data were analyzed by analysis of variance (ANOVA), and the means were compared by Tukey's test at 5% probability. All results were analyzed with the InfoStat Statistical Software for students (version 2012, Cordoba National University, Cordoba, Argentina).

3. Results and Discussion

3.1. Effect of Sanitization Treatments on the Natural Contaminating Microbiota

The results of the entire discussion in this section are shown in Table 1. None of the samples was naturally contaminated with *E. coli*. All applied treatments, except US 50 kHz, significantly reduced the count of aerobic mesophiles ($p \leq 0.05$). The sanitization treatments that promoted the highest reductions in aerobic mesophile counts at day 1 compared to sodium dichloroisocyanurate were US combined with 2000 mg/L AA (2.5 log CFU/g reduction) and US combined with 20 mg/L PAA (2.6 log CFU/g reduction) ($p \leq 0.05$).

Table 1. Average and standard deviation of aerobic mesophilic bacteria, molds, yeasts, and coliforms at 35 °C counts (log CFU/g) in kale samples after sanitization and over storage at 7 °C.

Treatment	Aerobic Mesophilic Bacteria		Molds and Yeasts		Coliforms at 35 °C	
	Day 1	Day 6	Day 1	Day 6	Day 1	Day 6
Non-sanitizer	6.8 ± 0.3 aA	8.4 ± 0.1 bA	5.4 ± 0.3 aA	5.5 ± 0.5 aA	3.8 ± 0.3 aA	2.00 ± 0.19 b
SDC 200 mg/L	5.4 ± 0.4 aBC	8.1 ± 0.3 bA	4.2 ± 0.3 aBC	4.5 ± 0.5 aAB	2.6 ± 0.2 B	Nd
AA 1000 mg/L	5.0 ± 0.5 aBCD	7.7 ± 0.3 bA	3.6 ± 0.3 aBCD	4.0 ± 0.2 aAB	2.2 ± 0.3 BC	Nd
AA 2000 mg/L	4.5 ± 0.5 aCD	5.8 ± 0.8 bAB	3.5 ± 0.2 aCD	3.9 ± 0.3 aAB	2.3 ± 0.3 B	Nd
PAA 20 mg/L	5.0 ± 0.3 aBCD	6.7 ± 0.9 bAB	3.6 ± 0.3 aBCD	3.7 ± 0.3 aB	2.0 ± 0.5 BC	Nd
US 50 kHz	5.7 ± 0.4 aAB	7.4 ± 0.6 bAB	4.4 ± 0.3 aB	4.5 ± 0.6 aAB	2.9 ± 0.6 AB	Nd
US 50 kHz + AA 1000 mg/L	5.0 ± 0.2 aBCD	5.0 ± 0.8 aB	3.1 ± 0.3 aD	3.9 ± 0.5 aB	2.1 ± 0.2 BC	Nd
US 50 kHz + AA 2000 mg/L	4.3 ± 0.4 aD	5.0 ± 0.6 aB	3.3 ± 0.2 aD	3.7 ± 0.3 aB	2.2 ± 0.2 BC	Nd
US 50 kHz + PA 20 mg/L	4.2 ± 0.1 aD	4.6 ± 0.3 aB	3.0 ± 0.2 aD	3.4 ± 0.2 aB	1.2 ± 0.3 C	Nd

Means on the same line followed for the same lowercase letter do not differentiate between each other ($p > 0.05$), for each microorganism, in the Tukey test after three replications. Means in the same column followed for the same capital letter do not differentiate between each other ($p > 0.05$) in the Tukey test after three replications. SDC: sodium dichloroisocyanurate; AA: acetic acid; PAA: peracetic acid; US: ultrasound; Nd: not detected in the smallest plated dilution (10^{-1}).

US alone did not promote a statistically significant reduction in aerobic mesophiles compared to non-sanitized kale samples ($p > 0.05$). Other studies have also shown that US alone is less effective in microbial inactivation, but when applying US combined with chemicals, the bactericidal effect increases [2,10,12,26]. Duarte et al. [2] verified that US alone and US combined with benzalkonium chloride promoted reduction equal to 0.6 and 2.5 log CFU/g of mesophilic aerobic bacteria counts on purple cabbage, respectively. Alvarenga et al. [26] verified that the application of ultrasound alone reduced the aerobic mesophile count by 1.09 log CFU/g in strawberries. The limited efficiency of US treatment isolated in the sanitization step of fruits and vegetables could be attributed to the capacity of microorganisms to penetrate these foods, becoming inaccessible to sound waves and the cavitation process produced by the equipment [7]. Furthermore, treatment conditions, such as time, ultrasound equipment (frequency and potency), and the type and amount of food treated, could impact sanitization efficiency [2].

Increasing the concentration of AA had a small impact on the reduction in aerobic mesophilic bacteria, molds, yeasts, and coliforms at 35 °C. The results demonstrate that it is still possible to produce microbiologically safe vegetables minimizing the concentration of chemical sanitizers. Furthermore, the use of a lower concentration may be more viable from an economic point of view. Similar results were described previously [11] after the sanitization of arugula leaves with AA for aerobic mesophile counts. However, the authors described an improvement in the reduction in molds and yeasts after the increase in organic acid concentration.

In the present study, the treatments in which the organic acids were applied in isolation or combined with US did not differ statistically ($p > 0.05$), indicating that there was no incremental reduction in the counts of microorganisms. In this way, the combined treatments had similar effects to the chemical treatments alone. Therefore, no synergistic effect was observed on the first day when ultrasound was combined with AA and PAA for all microorganism groups evaluated. However, treatments combining US with organic acids produced samples with stable aerobic mesophile counts over the storage period, different from what was observed for other treatments. These results indicate that treatments with US and organic acids are able to prevent microbial multiplication during the storage period.

Concerning molds and yeasts, all the treatments applied provided significant reductions ($p \leq 0.05$), and further reduction beyond what is achieved by sodium dichloroisocyanurate ($p \leq 0.05$) was observed when US was applied in combination with both concentrations of AA and PAA. On the sixth day of storage, kale samples that had been submitted to the different sanitization treatments had a statistically significant number of molds and yeasts.

Some postharvest practices such as peeling and slicing can damage the vegetal tissue, favors the release of cellular fluids, and provides water and nutrients to microorganisms' growth [7]. Despite this fact, the counts on the sixth day of storage were statistically the same as the counts determined one day after sanitization ($p > 0.05$). This result indicates that there was no resumption of growth of this group of microorganisms during storage.

Cao et al. [27] achieved a reduction in molds and yeasts after the application of US at 40 kHz for 10 min in strawberries. In the present study, five minutes of sanitization was sufficient to reduce the counts of molds and yeasts of kale, and this result may be associated with the use of US at a higher frequency (50 kHz). However, it is essential to note that the effects of ultrasound depend, in addition to the treatment conditions applied (e.g., temperature, the amplitude of the wave, time, and frequency), on the type, volume, and composition of food being sanitized; its properties (e.g., surface roughness); and the interactions between the surface and microorganisms [9,22].

For coliforms at 35 °C, US with PAA afforded a statistically superior reduction ($p \leq 0.05$) compared to treatment with sodium dichloroisocyanurate, 2000 mg/L AA, and US alone (Table 1) on the first day of analysis. The reduction after the application of US combined with PAA was 2.6 log CFU/g. On the other hand, the US and SDC 200 mg/L promoted the lowest reductions, equal to 0.8 and 1.2 log CFU/g, respectively. On day 6, no coliform at 35 °C was identified in the sanitized samples, even in the smallest plated dilution (10^{-1}). This result may suggest that the sanitization treatments caused irreparable damage to the cells, and they could not remain throughout the storage time; that is, the treatments were all effective in the inactivation of coliforms at 35 °C.

Leafy vegetable composition favors the presence and growth of many pathogenic and non-pathogenic microorganisms, including bacteria, yeasts, and molds. Considering the long distance between farms and commercialization, the procedures applied during processing must guarantee the safety of the product [7].

In the present work, the reductions obtained for the natural microbial groups after sanitization with organic acids were similar to or higher than those observed with conventional chlorinated compound treatment.

Chlorinated compounds are oxidative disinfectants, and in contact with water, these compounds mainly produce hypochlorous acid, but other reactive chlorine species as well. These subproducts damage multiple cellular components of the cell, and this is the mechanism of action [7]. However, chlorinated compounds generally have limited antimicrobial action, and under typical conditions, their reduction values reach approximately 2 log CFU/g. Chlorinated compounds are widely used by the food industry, mainly because they are inexpensive, but there are concerns related to the interaction with the organic components of food and the formation of toxic and potentially carcinogenic trihalomethanes, which have the potential to cause skin irritation [9,28].

In contrast, PAA is an oxidative disinfectant, environmentally friendly, and effective in inactivating aerobic and pathogenic bacteria and molds as well as yeasts at concentrations even lower than those required for chlorinated solutions, without a report on cytotoxic effects [7]. Moreover, PAA cannot be deactivated by enzymes such as catalase and peroxidase [29], and it decomposes into nontoxic residues (AA and oxygen) [9]. Furthermore, its concentration is comparatively stable in the presence of high organic loads because of its slow reaction degree with organic matter in the wash water [16]. The antimicrobial effect of PAA is related to the release of reactive oxygen species and DNA and lipid damage [9].

Organic acids are naturally occurring and generally recognized as safe (GRAS) compounds. The mechanism of microbial inactivation is attributed to their pKa values since these compounds are considered weak organic acids, and their bactericidal effect is higher than strong organic acids. It occurs because they are more lipophilic, and it facilitates the penetration in the bacterial cells, causing faster acidification of the interior, reducing the pH, and inhibiting essential metabolic reactions [9].

3.2. Inactivation of Salmonella enterica Typhimurium and Microscopic Visualization of Cells Adhered to the Surface of Kale after Sanitization

The count of S. enterica Typhimurium cells on the non-sanitized kale samples was 6.3 ± 0.1 log CFU/g, indicating that this pathogenic bacterium can adhere to the surface of kale. A point that needs to be addressed is that biofilm formation starts with microbial adhesion to the plant. Bacteria that have flagella, for example, may show higher swimming motility for surface colonization after adhesion. Salmonella spp. is a group of bacteria capable of forming biofilms, and the presence of fimbriae in this group is crucial to intensifying the degree of virulence of the pathogen after biofilm formation [7,30]. Furthermore, pathogens produce some compounds such as polysaccharides (e.g., cellulose), lipopolysaccharides, and colonic acid to colonize plant surfaces [7].

After the sanitization of kale, all treatments reduced the counts by significantly more than 1.0 log CFU/g, and the average reduction in the pathogen in proposed treatments varied between 1.6 and 3.8 log CFU/g (Figure 2). Furthermore, all alternative treatments applied in contaminated samples reduced the pathogen equal to or higher than sodium dichloroisocyanurate. In the present study, the application of 50 kHz US without any chemical treatment caused a reduction that was statistically equal to treatments with sodium dichloroisocyanurate, PAA, and AA without the US ($p > 0.05$). Different results were observed by Lepaus et al. [11], who evaluated the efficacy of 200 mg/L sodium dichloroisocyanurate and 200 mg/L sodium hypochlorite in the sanitization of strawberries, cucumbers, and arugula leaves. The results demonstrated that chlorinated compounds were less effective than organic acids and hydrogen peroxide in reducing natural microbiota and Salmonella Enteritidis. São José and Vanetti [31] also observed that 200 mg/L sodium dichloroisocyanurate did not inactivate Salmonella enterica Typhimurium inoculated in cherry tomatoes. Alenyorege et al. [1] evaluated different conditions for fresh-cut Chinese cabbage decontamination and observed that US (40 kHz, 125.45 W/L for 15 min) promoted reduction equal to 5.6 and 4.7 log CFU/g for E. coli and L. innocua, respectively.

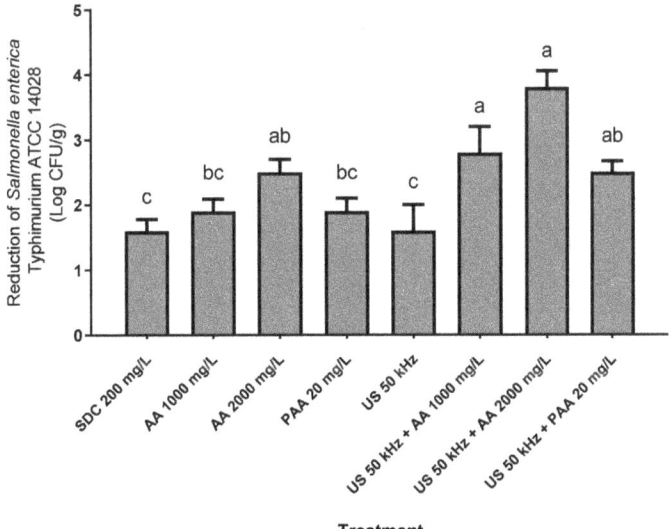

Figure 2. Decimal reductions in the count of *Salmonella enterica* Typhimurium ATCC 14028 (log CFU/g) adhered to the surface of kale compared with the non-sanitized sample. Treatments (columns) with the same letter do not differ significantly by Tukey's test after three replications ($p > 0.05$). SDC: sodium dichloroisocyanurate; AA: acetic acid; PAA: peracetic acid; US: ultrasound.

In the present study, the treatments using US combined with AA at 2000 mg/L, US combined with AA at 1000 mg/L, and US associated with PAA 20 mg/L afforded reductions of 3.8, 2.8 log, and 2.5 CFU/g, respectively. Similar results are presented by Rosário et al. [32], who observed that the US + peracetic acid occasioned a reduction of 2.1 log of CFU/g of Salmonella enterica subsp. enterica on strawberries.

However, in the present study, no synergic effect was observed in the combination of US and organic acids on the Salmonella enterica Typhimurium inactivation. In contrast, Turhan et al. [33] observed that organic acid treatment combined with US produced mainly synergistic effects, and the highest value on the surface of lettuce was achieved when US and citric acid for 30 min were applied. According to Mendoza et al. [7], using a single technique is not sufficient for biofilm removal. Turhan et al. [33] mentioned that the combination of ultrasound with other processes produces synergistic and antagonistic impacts against pathogens. Ultraviolet light treatment (30 mW/cm^2) combined with 80 mg/L of PAA, a recommended commercial practice for cleaning fresh produce, achieved significantly higher Salmonella spp. reduction in lettuce [16]. Rosário et al. [32] also observed that adhered Salmonella enterica could efficiently be removed from the surface of strawberries through the application of 40 mg/L PAA combined with 40 kHz US for 5 min. The intense pressure generated during US treatment can contribute to the penetration of chemical oxidants through the cell membrane, and the cavitation process can aid in the breakdown of microorganisms, which culminates in higher efficiency of the sanitization treatment [34].

Microscopy was applied to confirm microbial adhesion and identify the morphological appearance of the pathogen after the sanitization procedures. The adhesion of the Salmonella cells and the effect of proposed treatments on the surface of kale were confirmed by scanning electron microscopy (Figures 3 and 4).

In Figure 3, some characteristics were observed in the bacterial adhesion after sanitization procedures and support the result described before for the Salmonella inactivation study. Visually, treatments with sodium dichloroisocyanurate, AA 1000 mg/L, and 50 kHz US appeared to have little impact on pathogen inactivation (Figure 3). Therefore, these conditions could not efficiently remove the Salmonella cells adhered to the kale surface. The survival of pathogenic microorganisms after sanitization treatments becomes a risk to food safety. However, when the combination of US with AA 2000 mg/L and PAA was applied, it was observed that these treatments reduced adherent Salmonella cells at the same time that the cellular morphology of the pathogen was altered (Figure 4). These results demonstrate the importance of properly evaluating and choosing the method that will be used.

Li et al. [35], when applying US (55 °C; 3–15 min) in Staphylococcus aureus cells, observed a great impact on the cell morphology, as well as leakage of cellular content, cell wall disintegration, and plasma membranes. US promotes microbial inactivation, generates membrane damage, disrupts the cell wall, and damages DNA [36]. Furthermore, the collapse of cavitation bubbles can cause pore formation and partial disruption of cell walls and cytoplasmic membranes [35].

Sanitizers that do not promote adequate inactivation of microorganisms on the surface of vegetables allow the possibility of microbial growth during storage. Despite the reduction in the number of natural microbiota cells, in the present study, the inability of some sanitizers to effectively remove the pathogen adhered to the vegetable surface promotes the possibility of cell survival and proliferation during storage. For future research, it is recommended that an analysis is also carried out during storage.

The ability of pathogens to adhere to leaf epidermises and form biofilms are challenges for many sanitization procedures and is thus detrimental to product safety. In addition, intrinsic food parameters (e.g., roughness and hydrophobicity), microorganisms (e.g., presence of fimbriae, flagella, and pili) [37], and the interactions between them (e.g., the free energy of hydrophobic interaction and free energy of adhesion of surfaces) can influence the decontamination efficiency [22]. According to Pimentel-Filho et al. [38], surfaces are susceptible to biofilm formation, and the surface physiochemical aspects present an important function in bacterial adhesion. Hydrophobic interactions between cell surface and substrate facilitate cells to surmount electrostatic repulsive forces and favor adhesion [39]. For this reason, it is essential to study processes that are capable of eliminating microorganisms and do not bring risks to consumer health and the environment.

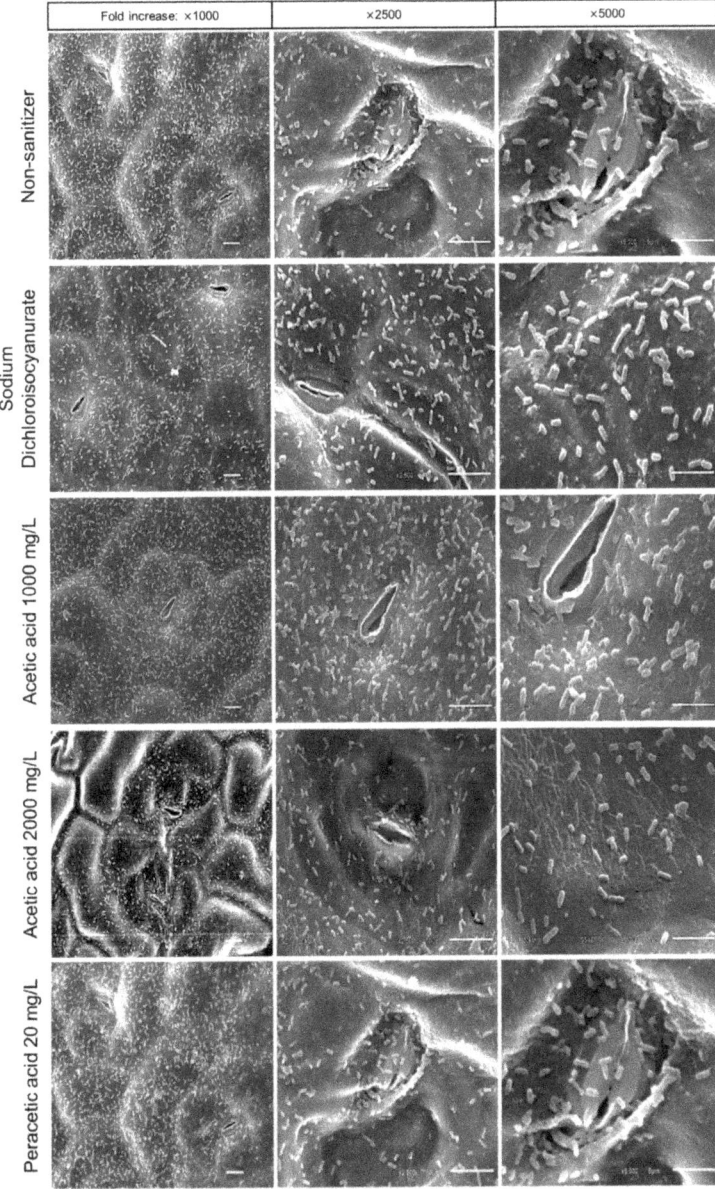

Figure 3. Scanning electronic microscopy micrographs of kale samples intentionally contaminated with *S*. Typhimurium cells ATCC 14028 and submitted to different sanitization treatments (Non sanitizer, sodium dichloroisocyanurate; Acetic Acid 1000 mg/L; Acetic Acid 2000 mg/L, Peracetic Acid 20 mg/L) for 5 min.

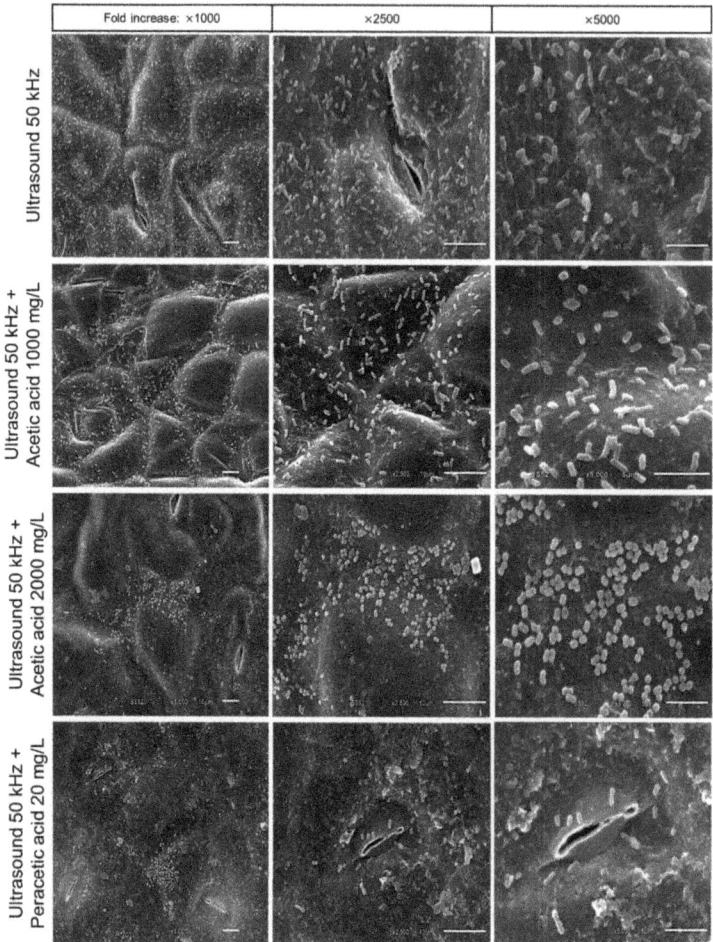

Figure 4. Scanning electronic microscopy micrographs of kale samples intentionally contaminated with *S.* Typhimurium cells ATCC 14028 and submitted to different sanitization treatments (Ultrasound 50 kHz, Ultrasound 50 kHz + Acetic Acid 1000 mg/L; Ultrasound 50 kHz + Acetic Acid 2000 mg/L; Ultrasound 50 kHz+ Peracetic Acid 20 mg/L) for 5 min.

3.3. Impact of Sanitization Treatments on the Physicochemical and Nutritional Parameters

Decontamination treatments in fresh produce could affect food quality, and methods should preserve characteristics at appropriate levels [33]. Regarding the physicochemical parameters (Table 2), both concentrations of treatments with AA, alone or associated with ultrasound, significantly reduced ($p \leq 0.05$) the pH of the samples concerning non-sanitized kale. However, the other treatments presented results statistically similar to the chlorinated compound for this variable ($p > 0.05$). No statistical difference in pH values between samples occurred between the first and sixth day of storage, except for the kale treated with 2000 mg/L of AA. Furthermore, this treatment, isolated and combined with ultrasound, significantly increased ($p \leq 0.05$) the titratable total acidity (TTA) values compared to the non-sanitized kale sample on the first day of storage. Through to the sixth day, no variations in TTA occurred.

Table 2. Average and standard deviation of pH, total titratable acidity, and total soluble solids in kale samples after sanitization and over storage at 7 °C.

Treatment	pH		TTA (mg Citric Acid/100 mg)		TSS (°Brix)	
	Day 1	Day 6	Day 1	Day 6	Day 1	Day 6
Non-sanitizer	6.7 ± 0.3 aAB	6.8 ± 0.1 aA	0.1 ± 0.01 aC	0.08 ± 0.01 aB	7.5 ± 0.5 aA	6.1 ± 0.3 aA
SDC 200 mg/L	6.8 ± 0.2 aA	6.8 ± 0.2 aA	0.1 ± 0.01 aC	0.10 ± 0.01 aB	6.2 ± 0.4 aABC	6.2 ± 0.3 aA
AA 1000 mg/L	5.3 ± 0.1 aBC	5.6 ± 0.4 aAB	0.4 ± 0.01 aABC	0.26 ± 0.02 aAB	4.9 ± 0.3 aBC	5.9 ± 0.3 aA
AA 2000 mg/L	5.0 ± 0.1 aC	4.7 ± 0.1 bB	0.5 ± 0.08 aA	0.49 ± 0.11 aA	5.7 ± 0.3 aBC	5.8 ± 0.1 aA
PAA 20 mg/L	6.8 ± 0.2 aAB	6.9 ± 0.2 aA	0.2 ± 0.02 aBC	0.10 ± 0.02 aB	6.1 ± 0.1 aABC	6.1 ± 0.5 aA
US 50 kHz	6.8 ± 0.2 aAB	7.0 ± 0.2 aA	0.1 ± 0.01 aC	0.14 ± 0.03 aB	6.9 ± 0.4 aAB	5.8 ± 0.4 aA
US 50 kHz + AA 1000 mg/L	5.8 ± 0.7 aABC	6.1 ± 0.8 aAB	0.3 ± 0.12 aABC	0.28 ± 0.10 aAB	5.4 ± 0.1 aBC	5.8 ± 0.2 aA
US 50 kHz + AA 2000 mg/L	5.1 ± 0.4 aC	4.9 ± 0.1 aB	0.4 ± 0.06 aAB	0.53 ± 0.07 aA	5.9 ± 0.2 aABC	6.4 ± 0.4 aA
US 50 kHz + PAA 20 mg/L	6.8 ± 0.2 aAB	6.9 ± 0.1 aA	0.1 ± 0.01 aC	0.13 ± 0.03 aB	5.6 ± 0.4 aBC	5.7 ± 0.6 aA

Means on the same line followed for the same lowercase letter do not differentiate between each other ($p > 0.05$) in the Tukey test after three replications. Means in the same column followed for the same capital letter do not differentiate between each other ($p > 0.05$) in the Tukey test after three replications. TTA: total titratable acidity; SDC: sodium dichloroisocyanurate; AA: acetic acid; PAA: peracetic acid; US: ultrasound.

São José et al. [40] observed a decrease in pH values of watercress and parsley treated with sodium dichloroisocyanurate (200 mg/L), hydrogen peroxide (5%), and PAA (40 mg/L) after 10 min of treatment. However, the pH of strawberries decreased only after sanitization with PAA. Similarly, another study demonstrated that strawberry and cucumber samples were more resistant to the 5 min immersion sanitization method with AA (1 and 2%), lactic acid (1 and 2%), and hydrogen peroxide (3%) than arugula leaves [18].

The changes in these parameters may indicate that the time of the procedure or the immersion sanitization method for leafy vegetables may have contributed to the incorporation of the sanitizing solution, reducing the pH value of the sample. Therefore, monitoring the pH values after sanitization, as well as the conditions applied during the procedure (e.g., time, temperature, concentration), is very relevant because it avoids the fast deterioration of the product during storage due to changes in the pH values after processing.

A statistically significant decrease ($p \leq 0.05$) in the total soluble solids content occurred after the application of both concentrations of AA, as well as the treatments combining US with the organic acids. There was no variation in this parameter over the six days of refrigerated storage.

In the present study, the mass loss over refrigerated storage was not significantly affected by the applied treatments ($p > 0.05$), and the mean of samples was 1.56 ± 0.42 (data not shown). This result indicates that these values were not enough to cause a significant decline in product quality. Excessive water loss due to the transpiration of the vegetal tissue can contribute to the loss of mass and lead to nutritional losses, wilting, and changes in texture and aroma. According to Turhan et al. [33], US treatment can affect the firmness of fruits and vegetables, and different results could be associated with the processing conditions, food matrix, variety, maturity stage, intensity, and time of US.

No reduction in vitamin C content was observed in sanitized kale samples compared to non-sanitized vegetables ($p > 0.05$) on the first (mean 51.8 ± 5.0 mg of ascorbic acid/100 g) and last day (mean 47.9 ± 7.4 mg of ascorbic acid/100 g) of storage (Table 3).

Different from this result, Wu et al. [41] observed that the contents of ascorbic acid significantly decreased during 12 days of storage of bok choy treated with aqueous chlorine dioxide in combination with US treatment. Vitamin C is considered the least stable of all vitamins and can easily be destroyed during processing and storage, so it is a natural indicator of the quality of food-processing techniques. Despite this fact, in the present study, the vitamin C content was maintained in all the evaluated sanitization treatments. This maintenance in ready-to-eat products is necessary since consumers are aware that fresh and vitamin-rich products have health benefits.

Table 3. Average and standard deviation of vitamin C (mg of ascorbic acid/100 g), phenolic compounds (mg gallic acid equivalent/100 g), and antioxidant activity (%) in kale samples after sanitization and storage at 7 °C.

Treatment	Vitamin C		Total Phenolic Compounds		Antioxidant Activity	
	Day 1	Day 6	Day 1	Day 6	Day 1	Day 6
No sanitizer	56.0 ± 4.0 aA	50.3 ± 6.9 aA	21.9 ± 2.8 aA	25.1 ± 3.2 aA	92.6 ± 6.4 aA	56.5 ± 11.9 bA
SD 200 mg/L	52.9 ± 2.9 aA	45.2 ± 2.0 aA	21.4 ± 1.2 aA	23.9 ± 4.6 aA	87.9 ± 5.2 aA	41.9 ± 7.7 bA
AA 1000 mg/L	51.5 ± 1.9 aA	41.4 ± 8.2 aA	19.4 ± 2.5 aA	24.1 ± 5.1 aA	84.6 ± 13.6 aA	29.3 ± 6.1 bA
AA 2000 mg/L	52.1 ± 4.3 aA	47.3 ± 4.7 aA	24.3 ± 0.7 aA	23.0 ± 5.5 aA	76.0 ± 7.5 aA	33.1 ± 3.3 bA
PAA 20 mg/L	48.5 ± 9.1 aA	48.2 ± 10.8 aA	22.5 ± 3.5 aA	20.1 ± 2.4 aA	89.9 ± 11.1 aA	40.1 ± 7.0 bA
US 50 kHz	48.3 ± 2.9 aA	46.5 ± 8.6 aA	20.2 ± 5.3 aA	23.0 ± 5.3 aA	96.9 ± 2.4 aA	40.9 ± 8.0 bA
US 50 kHz + AA 1000 mg/L	51.9 ± 3.5 aA	52.9 ± 6.6 aA	23.7 ± 1.9 aA	23.8 ± 5.1 aA	96.3 ± 3.3 aA	35.0 ± 6.5 bA
US 50 kHz + AA 2000 mg/L	56.3 ± 6.3 aA	53.6 ± 7.3 aA	21.1 ± 1.1 aA	25.7 ± 2.5 aA	96.0 ± 3.9 aA	33.8 ± 5.5 bA
US 50 kHz + PAA 20 mg/L	48.4 ± 10.5 aA	46.0 ± 11.1 aA	19.2 ± 1.4 aA	26.2 ± 6.9 aA	96.6 ± 2.1 aA	41.8 ± 5.6 bA

Means on the same line followed for the same lowercase letter do not differentiate between each other ($p > 0.05$) in the Tukey test after three replications. Means in the same column followed for the same capital letter do not differentiate between each other ($p > 0.05$) in the Tukey test after three replications. SDC: sodium dichloroisocyanurate; AA: acetic acid; PAA: peracetic acid; US: ultrasound.

The sanitization treatments and storage time did not cause significant changes in the level of total phenolic compounds in kale samples ($p > 0.05$) (Table 3). After sanitization procedures, the mean value for this nutritional parameter was 21.5 ± 2.3 mg gallic acid equivalent/100 g fresh sample. At the end of storage, the value was 23.9 ± 4.5 mg gallic acid equivalent/100 g fresh sample. This result is considered positive because even the application of chemical sanitizers in combination with US or on their own guaranteed that the content of these bioactive compounds in kale was maintained.

The antioxidant activity of kale was preserved after sanitization without differences between treatments (mean 90.8 ± 6.2%; $p > 0.05$), but during storage, it fell significantly ($p \leq 0.05$) in all treatments (mean 39.2 ± 6.8%) (Table 3). The most important goal of sanitization is to reduce the number of pathogenic and spoilage microorganisms, but it is crucial to maintain the physicochemical and nutritional properties of the food.

If the operating conditions are inadequate, the nutritional quality of vegetables may be affected. Thus, it is important to adjust the time and sanitizer concentrations to obtain a balance that results in decontamination. Therefore, the procedures applied in this study and the packaging used were adequate to maintain the mass of the samples throughout the storage period, as well as the vitamin C and bioactive compounds. Storage in a temperature and humidity-controlled environment is recommended.

4. Conclusions

Considering the hypotheses tested in this study, we concluded that AA and PAA are good alternatives to chlorine-based compounds for kale sanitization, as they obtained a sanitizing effect similar or superior to these compounds in microbial reduction. However, no synergistic effect occurred after the combination of US and AA and PAA for all microorganisms. Moreover, despite the reduction in pH by treatments with alternative strategies, the treatments contributed to the preservation of the physicochemical and nutritional properties of kale. Finally, we concluded that the proposed treatments have the potential to be applied in the sanitization step of kale, but other methods, such as spraying techniques, are recommended over immersion for this crop. Studies that evaluate different conditions of treatment with ultrasound, concentrations of organic acids, and sanitization time are suggested. Furthermore, further studies should be conducted to evaluate the sensory quality of sanitized vegetables to verify if characteristics are retained at satisfactory levels after the application of decontamination strategies.

Author Contributions: Conceptualization, M.C.d.M.M.M. and J.F.B.d.S.J.; methodology, M.C.d.M.M.M.; formal analysis, B.M.L. and M.C.d.M.M.M.; writing—original draft preparation, M.C.d.M.M.M., P.C.B. and J.F.B.d.S.J.; writing—review and editing, M.C.d.M.M.M., B.M.L. and J.F.B.d.S.J.; supervision, P.C.B. and J.F.B.d.S.J.; project administration, J.F.B.d.S.J.; funding acquisition, J.F.B.d.S.J. All authors have read and agreed to the published version of the manuscript.

Funding: This research was funded by Fundação de Amparo à Pesquisa e Inovação do Espírito Santo (FAPES), project number 554/2015, approved on FAPES notice number 006/2014). Furthermore, FAPES granted a scholarship for the first author.

Institutional Review Board Statement: Not applicable.

Informed Consent Statement: Not applicable.

Data Availability Statement: The data that support the findings of this study are available on request from the corresponding author.

Acknowledgments: We are grateful to Fundação de Amparo à Pesquisa e Inovação do Espírito Santo for the scholarship of the first author. We thank Fundo de Apoio à Pesquisa of Universidade Federal do Espírito Santo (FAP/UFES) for the support. We thank Ultrastructure Cell Laboratory Carlos Alberto Redins (LUCCAR) of the Federal University of Espírito Santo for support in the scanning electron microscopy analyses. The authors gratefully acknowledge the Coordination of Superior Level Staff Improvement (Coordenação de Aperfeiçoamento de Pessoal de Nível Superior-CAPES) for supporting the Post-Graduate Program in Nutrition and Health. We thank the Dean of Research and Graduate Studies of the Federal University of Espírito Santo (PRPPG-UFES) for their support in the development of research and publication of scientific papers.

Conflicts of Interest: The authors declare no conflict of interest.

Sample Availability: Samples of the compounds are not available from the authors.

References

1. Alenyorege, E.A.; Ma, H.; Aheto, J.H.; Ayim, I.; Chikari, F.; Osae, R.; Zhou, C. Response Surface Methodology Centred Optimization of Mono-Frequency Ultrasound Reduction of Bacteria in Fresh-Cut Chinese Cabbage and Its Effect on Quality. *LWT* **2020**, *122*, 108991. [CrossRef]
2. Duarte, A.L.A.; do Rosário, D.K.A.; Oliveira, S.B.S.; de Souza, H.L.S.; de Carvalho, R.V.; Carneiro, J.C.S.; Silva, P.I.; Bernardes, P.C. Ultrasound Improves Antimicrobial Effect of Sodium Dichloroisocyanurate to Reduce *Salmonella* Typhimurium on Purple Cabbage. *Int. J. Food Microbiol.* **2018**, *269*, 12–18. [CrossRef] [PubMed]
3. Marques, C.S.; Grillo, R.P.; Bravim, D.G.; Pereira, P.V.; Oliveira Villanova, J.C.; Pinheiro, P.F.; Souza Carneiro, J.C.; Bernardes, P.C. Preservation of Ready-to-Eat Salad: A Study with Combination of Sanitizers, Ultrasound, and Essential Oil-Containing β-Cyclodextrin Inclusion Complex. *LWT* **2019**, *115*, 108433. [CrossRef]
4. Alenyorege, E.A.; Ma, H.; Ayim, I. Inactivation Kinetics of Inoculated *Escherichia coli* and *Listeria innocua* in Fresh-Cut Chinese Cabbage Using Sweeping Frequency Ultrasound. *J. Food Saf.* **2019**, *39*, e12696. [CrossRef]
5. Vargas, L.; Kapoor, R.; Nemzer, B.; Feng, H. Application of Different Drying Methods for Evaluation of Phytochemical Content and Physical Properties of Broccoli, Kale, and Spinach. *LWT* **2022**, *155*, 112892. [CrossRef]
6. Mir, S.A.; Shah, M.A.; Mir, M.M.; Dar, B.N.; Greiner, R.; Roohinejad, S. Microbiological Contamination of Ready-to-Eat Vegetable Salads in Developing Countries and Potential Solutions in the Supply Chain to Control Microbial Pathogens. *Food Control* **2018**, *85*, 235–244. [CrossRef]
7. Mendoza, I.C.; Luna, E.O.; Pozo, M.D.; Vásquez, M.V.; Montoya, D.C.; Moran, G.C.; Romero, L.G.; Yépez, X.; Salazar, R.; Romero-Peña, M.; et al. Conventional and Non-Conventional Disinfection Methods to Prevent Microbial Contamination in Minimally Processed Fruits and Vegetables. *LWT* **2022**, *165*, 113714. [CrossRef]
8. CDC Lettuce, Other Leafy Greens, and Food Safety. Available online: https://www.cdc.gov/foodsafety/communication/leafy-greens.html (accessed on 9 June 2020).
9. Bhilwadikar, T.; Pounraj, S.; Manivannan, S.; Rastogi, N.K.; Negi, P.S. Decontamination of Microorganisms and Pesticides from Fresh Fruits and Vegetables: A Comprehensive Review from Common Household Processes to Modern Techniques. *Compr. Rev. Food Sci. Food Saf.* **2019**, *18*, 1003–1038. [CrossRef]
10. Pelissari, E.M.R.; Covre, K.V.; do Rosario, D.K.A.; de São José, J.F.B. Application of Chemometrics to Assess the Influence of Ultrasound and Chemical Sanitizers on Vegetables: Impact on Natural Microbiota, Salmonella Enteritidis and Physicochemical Nutritional Quality. *LWT* **2021**, *148*, 111711. [CrossRef]
11. Lepaus, B.M.; Rocha, J.S.; de São José, J.F.B. Organic Acids and Hydrogen Peroxide Can Replace Chlorinated Compounds as Sanitizers on Strawberries, Cucumbers and Rocket Leaves. *Food Sci. Technol.* **2020**, *40*, 242–249. [CrossRef]

12. De São José, J.F.B.; Medeiros, H.S.; de Andrade, N.J.; Ramos, A.M.; Vanetti, M.C.D. Effect of Ultrasound and Chemical Compounds on Microbial Contamination, Physicochemical Parameters and Bioactive Compounds of Cherry Tomatoes. *Italian J. Food Sci.* **2018**, *30*, 467–486. [CrossRef]
13. Pahariya, P.; Fisher, D.J.; Choudhary, R. Comparative Analyses of Sanitizing Solutions on Microbial Reduction and Quality of Leafy Greens. *LWT* **2022**, *154*, 112696. [CrossRef]
14. Nicolau-Lapeña, I.; Abadias, M.; Bobo, G.; Aguiló-Aguayo, I.; Lafarga, T.; Viñas, I. Strawberry Sanitization by Peracetic Acid Washing and Its Effect on Fruit Quality. *Food Microbiol.* **2019**, *83*, 159–166. [CrossRef]
15. Zhou, B.; Luo, Y.; Nou, X.; Mwangi, E.; Poverenov, E.; Rodov, V.; Demokritou, P.; Fonseca, J.M. Effects of a Novel Combination of Gallic Acid, Hydrogen Peroxide and Lactic Acid on Pathogen Inactivation and Shelf-Life of Baby Spinach. *Food Control* **2023**, *143*, 109284. [CrossRef]
16. Lippman, B.; Yao, S.; Huang, R.; Chen, H. Evaluation of the Combined Treatment of Ultraviolet Light and Peracetic Acid as an Alternative to Chlorine Washing for Lettuce Decontamination. *Int. J. Food Microbiol.* **2020**, *323*, 108590. [CrossRef] [PubMed]
17. Alenyorege, E.A.; Ma, H.; Ayim, I.; Lu, F.; Zhou, C. Efficacy of Sweep Ultrasound on Natural Microbiota Reduction and Quality Preservation of Chinese Cabbage during Storage. *Ultrason. Sonochem.* **2019**, *59*, 104712. [CrossRef] [PubMed]
18. Coswosck, K.H.C.; Giorgette, M.A.; Lepaus, B.M.; da Silva, E.M.M.; Sena, G.G.S.; de Almeida Azevedo, M.C.; de São José, J.F.B. Impact of Alternative Sanitizers on the Physicochemical Quality, Chlorophyll Content and Bioactive Compounds of Fresh Vegetables. *Food Sci. Technol.* **2021**, *41*, 328–334. [CrossRef]
19. Ranjha, M.M.A.N.; Irfan, S.; Lorenzo, J.M.; Shafique, B.; Kanwal, R.; Pateiro, M.; Arshad, R.N.; Wang, L.; Nayik, G.A.; Roobab, U.; et al. Sonication, a Potential Technique for Extraction of Phytoconstituents: A Systematic Review. *Processes* **2021**, *9*, 1406. [CrossRef]
20. Zia, S.; Khan, M.R.; Shabbir, M.A.; Aslam Maan, A.; Khan, M.K.I.; Nadeem, M.; Khalil, A.A.; Din, A.; Aadil, R.M. An Inclusive Overview of Advanced Thermal and Nonthermal Extraction Techniques for Bioactive Compounds in Food and Food-Related Matrices. *Food Rev. Int.* **2022**, *38*, 1166–1196. [CrossRef]
21. Downes, F.P.; Ito, K. *Compendium of Methods for Microbiological Examination of Foods*; American Public Health Association: Washington, DC, USA, 2001.
22. De São José, J.F.B.; de Medeiros, H.S.; Bernardes, P.C.; de Andrade, N.J. Removal of *Salmonella enterica* Enteritidis and *Escherichia coli* from Green Peppers and Melons by Ultrasound and Organic Acids. *Int. J. Food Microbiol.* **2014**, *190*, 9–13. [CrossRef]
23. AOAC International. *Official Methods of Analysis of AOAC International*; AOAC International: Gaithersburg, MD, USA, 2005.
24. Singleton, V.L.; Orthofer, R.; Lamuela-Raventós, R.M. Analysis of Total Phenols and Other Oxidation Substrates and Antioxidants by Means of Folin-Ciocalteu Reagent. *Methods Enzymol.* **1999**, *299*, 152–178. [CrossRef]
25. Blois, M.S. Antioxidant Determinations by the Use of a Stable Free Radical. *Nature* **1958**, *181*, 1199–1200. [CrossRef]
26. Alvarenga, P.D.L.; Mileib Vasconcelos, C.; de São José, J.F.B. Application of Ultrasound Combined with Acetic Acid and Peracetic Acid: Microbiological and Physicochemical Quality of Strawberries. *Molecules* **2020**, *26*, 16. [CrossRef] [PubMed]
27. Cao, S.; Hu, Z.; Pang, B.; Wang, H.; Xie, H.; Wu, F. Effect of Ultrasound Treatment on Fruit Decay and Quality Maintenance in Strawberry after Harvest. *Food Control* **2010**, *21*, 529–532. [CrossRef]
28. Ju, S.Y.; Ko, J.J.; Yoon, H.S.; Seon, S.J.; Yoon, Y.R.; Lee, D.I.; Kim, S.Y.; Chang, H.J. Does Electrolyzed Water Have Different Sanitizing Effects than Sodium Hypochlorite on Different Vegetable Types? *Br. Food J.* **2017**, *119*, 342–356. [CrossRef]
29. Bang, H.J.; Park, S.Y.; Kim, S.E.; Md Furkanur Rahaman, M.; Ha, S.D. Synergistic Effects of Combined Ultrasound and Peroxyacetic Acid Treatments against *Cronobacter sakazakii* Biofilms on Fresh Cucumber. *LWT* **2017**, *84*, 91–98. [CrossRef]
30. Yaron, S.; Römling, U. Biofilm Formation by Enteric Pathogens and Its Role in Plant Colonization and Persistence. *Microb. Biotechnol.* **2014**, *7*, 496–516. [CrossRef]
31. De São José, J.F.B.; Vanetti, M.C.D. Effect of Ultrasound and Commercial Sanitizers in Removing Natural Contaminants and *Salmonella enterica* Typhimurium on Cherry Tomatoes. *Food Control* **2012**, *24*, 95–99. [CrossRef]
32. Rosário, D.K.A.; da Silva Mutz, Y.; Peixoto, J.M.C.; Oliveira, S.B.S.; de Carvalho, R.V.; Carneiro, J.C.S.; de São José, J.F.B.; Bernardes, P.C. Ultrasound Improves Chemical Reduction of Natural Contaminant Microbiota and *Salmonella enterica* subsp. Enterica on Strawberries. *Int. J. Food Microbiol.* **2017**, *241*, 23–29. [CrossRef]
33. Turhan, E.U.; Polat, S.; Erginkaya, Z.; Konuray, G. Investigation of Synergistic Antibacterial Effect of Organic Acids and Ultrasound against Pathogen Biofilms on Lettuce. *Food Biosci.* **2022**, *47*, 101643. [CrossRef]
34. De São José, J.F.B.; de Andrade, N.J.; Ramos, A.M.; Vanetti, M.C.D.; Stringheta, P.C.; Chaves, J.B.P. Decontamination by Ultrasound Application in Fresh Fruits and Vegetables. *Food Control* **2014**, *45*, 36–55. [CrossRef]
35. Li, J.; Suo, Y.; Liao, X.; Ahn, J.; Liu, D.; Chen, S.; Ye, X.; Ding, T. Analysis of *Staphylococcus aureus* Cell Viability, Sublethal Injury and Death Induced by Synergistic Combination of Ultrasound and Mild Heat. *Ultrason. Sonochem.* **2017**, *39*, 101–110. [CrossRef]
36. Tan, M.S.F.; Rahman, S.; Dykes, G.A. Sonication Reduces the Attachment of *Salmonella* Typhimurium ATCC 14028 Cells to Bacterial Cellulose-Based Plant Cell Wall Models and Cut Plant Material. *Food Microbiol.* **2017**, *62*, 62–67. [CrossRef] [PubMed]
37. Giaouris, E.E.; Simões, M.V. Pathogenic Biofilm Formation in the Food Industry and Alternative Control Strategies. *Foodborne Dis.* **2018**, *15*, 309–377. [CrossRef]

38. De Jesus Pimentel-Filho, N.; de Freitas Martins, M.C.; Nogueira, G.B.; Mantovani, H.C.; Vanetti, M.C.D. Bovicin HC5 and Nisin Reduce *Staphylococcus aureus* Adhesion to Polystyrene and Change the Hydrophobicity Profile and Gibbs Free Energy of Adhesion. *Int. J. Food Microbiol.* **2014**, *190*, 1–8. [CrossRef] [PubMed]
39. Bayoudh, S.; Othmane, A.; Bettaieb, F.; Bakhrouf, A.; Ouada, H.B.; Ponsonnet, L. Quantification of the Adhesion Free Energy between Bacteria and Hydrophobic and Hydrophilic Substrata. *Mater. Sci. Eng. C* **2006**, *26*, 300–305. [CrossRef]
40. De São José, J.F.B.; Vanetti, M.C.D. Application of Ultrasound and Chemical Sanitizers to Watercress, Parsley and Strawberry: Microbiological and Physicochemical Quality. *LWT* **2015**, *63*, 946–952. [CrossRef]
41. Wu, W.; Gao, H.; Chen, H.; Fang, X.; Han, Q.; Zhong, Q. Combined Effects of Aqueous Chlorine Dioxide and Ultrasonic Treatments on Shelf-Life and Nutritional Quality of Bok Choy (*Brassica chinensis*). *LWT* **2019**, *101*, 757–763. [CrossRef]

Article

Chemical Composition Assessment of Structural Parts (Seeds, Peel, Pulp) of *Physalis alkekengi* L. Fruits

Venelina Popova [1], Zhana Petkova [2], Nadezhda Mazova [3], Tanya Ivanova [1], Nadezhda Petkova [4], Magdalena Stoyanova [5], Albena Stoyanova [1], Sezai Ercisli [6], Zuhal Okcu [7], Sona Skrovankova [8,*] and Jiri Mlcek [8]

[1] Department of Tobacco, Sugar, Vegetable and Essential Oils, University of Food Technologies, 4002 Plovdiv, Bulgaria
[2] Department of Chemical Technology, Faculty of Chemistry, University of Plovdiv "Paisii Hilendarski", 4000 Plovdiv, Bulgaria
[3] Department of Engineering Ecology, University of Food Technologies, 4002 Plovdiv, Bulgaria
[4] Department of Organic Chemistry and Inorganic Chemistry, University of Food Technologies, 4002 Plovdiv, Bulgaria
[5] Department of Analytical Chemistry and Physical Chemistry, University of Food Technologies, 4002 Plovdiv, Bulgaria
[6] Department of Horticulture, Atatürk University, 25240 Erzurum, Turkey
[7] Department of Gastronomy, Faculty of Tourism, Ataturk University, 25240 Erzurum, Turkey
[8] Department of Food Analysis and Chemistry, Tomas Bata University in Zlin, 76001 Zlin, Czech Republic
* Correspondence: skrovankova@utb.cz; Tel.: +420-57603-1524

Citation: Popova, V.; Petkova, Z.; Mazova, N.; Ivanova, T.; Petkova, N.; Stoyanova, M.; Stoyanova, A.; Ercisli, S.; Okcu, Z.; Skrovankova, S.; et al. Chemical Composition Assessment of Structural Parts (Seeds, Peel, Pulp) of *Physalis alkekengi* L. Fruits. *Molecules* 2022, 27, 5787. https://doi.org/10.3390/molecules27185787

Academic Editor: Sabina Lachowicz-Wiśniewska

Received: 9 August 2022
Accepted: 2 September 2022
Published: 7 September 2022

Publisher's Note: MDPI stays neutral with regard to jurisdictional claims in published maps and institutional affiliations.

Copyright: © 2022 by the authors. Licensee MDPI, Basel, Switzerland. This article is an open access article distributed under the terms and conditions of the Creative Commons Attribution (CC BY) license (https://creativecommons.org/licenses/by/4.0/).

Abstract: In recent years there has been an extensive search for nature-based products with functional potential. All structural parts of *Physalis alkekengi* (bladder cherry), including fruits, pulp, and less-explored parts, such as seeds and peel, can be considered sources of functional macro- and micronutrients, bioactive compounds, such as vitamins, minerals, polyphenols, and polyunsaturated fatty acids, and dietetic fiber. The chemical composition of all fruit structural parts (seeds, peel, and pulp) of two phenotypes of *P. alkekengi* were studied. The seeds were found to be a rich source of oil, yielding 14–17%, with abundant amounts of unsaturated fatty acids (over 88%) and tocopherols, or vitamin E (up to 5378 mg/kg dw; dry weight). The predominant fatty acid in the seed oils was linoleic acid, followed by oleic acid. The seeds contained most of the fruit's protein (16–19% dw) and fiber (6–8% dw). The peel oil differed significantly from the seed oil in fatty acid and tocopherol composition. Seed cakes, the waste after oil extraction, contained arginine and aspartic acid as the main amino acids; valine, phenylalanine, threonine, and isoleucine were present in slightly higher amounts than the other essential amino acids. They were also rich in key minerals, such as K, Mg, Fe, and Zn. From the peel and pulp fractions were extracted fruit concretes, aromatic products with specific fragrance profiles, of which volatile compositions (GC-MS) were identified. The major volatiles in peel and pulp concretes were β-linalool, α-pinene, and γ-terpinene. The results from the investigation substantiated the potential of all the studied fruit structures as new sources of bioactive compounds that could be used as prospective sources in human and animal nutrition, while the aroma-active compounds in the concretes supported the plant's potential in perfumery and cosmetics.

Keywords: *Physalis alkekengi*; bladder cherry fruit; seeds; peel; pulp; oil; composition; bioactive compounds; concretes

1. Introduction

Physalis alkekengi L. (family Solanaceae), also known as the Chinese lantern, Japanese lantern, bladder cherry, winter cherry, and by many other common names, is a species indigenous to Asia and Southern Europe, further naturalized worldwide [1,2]. Nowadays, the Chinese lantern is encountered as cultivated and ornamental varieties or as a wild-growing plant in various climatic zones, from Central and Southern Europe to South and Northeast Asia.

The species is the only one in the genus *Physalis*, which is found as wild populations in Bulgaria, growing in different regions at altitudes up to 1200–1500 m [3,4]. The local name of the species is 'mekhunka', and its preservation and use are regulated by the Medicinal Plants Act [5]. Survey data [6] have documented a steady export of about 760 kg fresh *P. alkekengi* fruit per year in the period 2001–2005. In Bulgaria, the Chinese lantern grows as a perennial plant with widely spreading roots and a slightly branched or unbranched stem with a height between 40 and 60 cm. The fruit ripen in August–September, presenting as small, oval, brightly colored berries containing numerous tiny seeds, and completely covered by the characteristic wide orange-red papery calyx (husk).

Despite its identification as a medicinal plant, there are practically no data from investigations of *P. alkekengi* phytochemical composition in Bulgaria, nor from studies of its biological activities or range of application. Ivanov et al. [7] reported that the fruits contain red pigments, physalin, citric, malic and tartaric acids, vitamin C, and bitter substances. The seeds alone can contain up to 25% of the oil [7]. According to numerous studies on the species in its natural areas of distribution, over 100 bioactive metabolites have been identified in the fruit and other aerial parts of the plant, including alkaloids, nucleosides, peptides, terpenoids, megastigmanes, aliphatic derivatives, organic acids, coumarins, sucrose esters, polysaccharides, and carotenoid derivatives [8–17].

The Chinese lantern has been recognized for centuries as a medicinal plant in the traditional medical practices of many countries, due to its anti-inflammatory, antibacterial, antiseptic, sedative, laxative, diuretic, hypoglycemic, spasmolytic, and other effects, as well as for the relief of malaria and syphilis symptoms [1,18,19]. In Chinese medicine, *P. alkekengi* (Physalis calyx seu Fructus) is a remedy for a number of diseases—from sore throat, eczemas, and rheumatism to hepatitis, urinary disorders, and tumors [1,2,10,14,18]. In turn, Bulgarian folk medicine recommends the use of fresh or dried fruit for the treatment of liver diseases, combining hepatitis and ascites [3]. Dried fruits are also used as a painkiller for kidney and bladder stones, inflammations of the urinary tract, and hemorrhoids. In topical application, fresh juice or whole fruits relieve skin irritations, wounds, and inflammation. A daily consumption of at least 10–15 fresh berries (or the equivalent 20 mL freshly squeezed juice) is highly recommended [3].

P. alkekengi is also recognized as a functional food, being a rich source of valuable nutrients—vitamins A and C, minerals, unsaturated fatty acids, phenolics, phytosterols, and pectic substances [18,20–22]. It should be noted that only fully ripened berries are suitable for consumption (unripe fruit and all aerial parts of the plant are toxic if swallowed), having a juicy texture, fresh flavor, and a slightly bitter taste, which normally disappears after fruit freezing.

There is relatively limited information about the cosmetic uses of *P. alkekengi* in the form of aqueous, ethanolic, and other extracts from the fruit and calyces, which are incorporated in different cosmetic formulations, taking advantage of its beneficial effects on the skin (protective, soothing, anti-ageing, anti-pigmentation, and other effects) [1,23]. Those and related investigations have supported the inclusion of *P. alkekengi* fruit and calyx extracts (CAS No 90082-67-0) in the Cosmetic Ingredient Database (CosIng) of the European Commission, in the category of cosmetic ingredients with skin conditioning functions [24].

To the best of our knowledge, there are not enough data on the distribution of phytonutrients and other chemical compounds among the structural parts of the fruit (peel, pulp, and seeds), nor on the characteristics of *P. alkekengi* growing in different regions of Bulgaria. The main objective of this study was therefore to provide a comparative assessment of the chemical composition of the structural parts of the fruit in two Bulgarian phenotypes of *Physalis alkekengi* L., thus supplementing the already existing knowledge about the species and expanding its possible prospective use in nutrition and cosmetics.

2. Materials and Methods

2.1. Plant Material

Two phenotypes of *Physalis alkekengi* found in Bulgaria were analyzed in this study. Fully ripe fruits (about 150 pieces for each phenotype) were collected in August—September 2020 from wild plant populations in Central Southern Bulgaria (PA-SB phenotype; the city of Plovdiv; 42°14′26″ N 24°70′24″ E), and North-Eastern Bulgaria (PA-NB phenotype; near Ivanski village, Shumen region; 43°07′24″ N 27°04′35″ E). Species identification was confirmed by the botanist at the Department of Botany, Plovdiv University, Bulgaria.

According to the objectives of the study, fresh fruits (Figure 1) were de-husked and carefully divided into structural parts (seeds, peel, and pulp), that were analyzed individually or in mixture due to determination requirements.

Figure 1. Fresh fruits of *P. alkekengi*: (**a**) PA-SB phenotype from Central Southern Bulgaria; (**b**) PA-NB phenotype from North-Eastern Bulgaria. Photos by authors.

2.2. Basic Evaluation of Fresh Berries and Their Structural Parts

The proportion (%) of the structural fruit parts (peel, pulp, and seeds) in fresh berries was obtained by gravimetrical determination of each element's fresh weight (fw) (Mettler-Toledo, Switzerland; precision ± 0.0001 g) for 100 randomly selected fruits. Seed absolute weight (g) was obtained as mean by weighing of 1000 seeds on a precision balance (Mettler-Toledo, Switzerland; precision ± 0.0001 g).

The moisture content of each fruit sample in the study (seeds, peel, and pulp) was determined by drying in a laboratory drying oven (Robotika, Velingrad, Bulgaria) at 103 ± 2 °C to the constant weight [25]. The results from the chemical analyses in the study were further re-calculated and presented on a dry weight (dw) basis.

In the first step of chemical analysis, the samples were analyzed in two structural forms, as seed samples and peel/pulp samples.

Cellulose content (crude fiber) in the seed and peel/pulp samples was determined by a slight modification of a method described earlier [26]. The hydrolysis of the plant material (1.0 g) was carried out with 16.5 mL 80% CH_3COOH and 1.5 mL concentrated HNO_3 for 1.5 h at 100 °C. The filtrated residue was dried at 103 ± 2 °C for 24 h, cooled in a dessicator, and weighed for the quantitative determination of cellulose. The results are presented on a dry weight basis.

Protein content in the seed and peel/pulp fractions was analyzed by the Kjeldahl method [25] using an UDK 152 unit (Velp Scientifica Srl, Usmate Velate, Italy). The conversion to protein content of the determined nitrogen content, present as ammonia in the digested sample, was by the multiplication factor of 6.25.

Oil content in the seed and in the peel/pulp parts was determined after extraction with *n*-hexane (Soxhlet, for 8 h), followed by evaporation of the solvent on a rotary vacuum evaporator (at 40 °C) and under a stream of nitrogen [27].

2.3. Determination of Fatty Acids in Seed and Peel Oils

In the second step of the analysis, the oils from the fruit seeds and peel were isolated separately and then analyzed for the composition and content of the fatty acids and tocopherols. The oil fractions from seeds and peel were obtained by extraction with *n*-hexane, as described above [27].

Fatty acids in the extracted seed and peel oils were determined after transmethylation with 2% H_2SO_4 in CH_3OH at 50 °C [28,29]. The GC analysis was performed on a Hewlett Packard 5890A unit, with a flame ionization detector (FID) (Santa Clara, CA, USA) and a capillary Supelco 2560 column, 75 m × 0.25 mm × 18 μm (i.d.). The column temperature increase was from 130 °C (4 min) to 240 °C (5 min) at 15 °C/min; injector/detector temperatures were 250 °C; the flow rate of the carrier gas (hydrogen) was 0.8 mL/min; the split was 50:1. Fatty acid identification was completed by referring to the retention times of fatty acid methyl esters (FAME) in a standard mixture of 37 components (Supelco, Bellefonte, PA, USA).

2.4. Determination of Tocopherols in Seed and Peel Oils

Tocopherols were analyzed directly in the seed and peel oils, without saponification, by HPLC method. A 2% solution of the respective oil in *n*-hexane was prepared, and 20 μL was injected in the Merck-Hitachi unit (Merck, Darmstadt, Germany); column 250 mm × 4 mm Nucleosil Si 50-5; fluorescent detector Merck-Hitachi F-1000 (Merck, Darmstadt, Germany). The flow rate of the mobile phase (*n*-hexane: dioxan 96:4, v/v) was 1.0 mL/min; detector excitation was at 295 nm, emission at 330 nm. Tocopherol identification was based on comparison with reference standards (DL-α-, DL-β-, DL-γ- and DL-δ-tocopherols, 98% purity) [30].

2.5. Determination of Amino Acids in Seed Cakes

In the third step, the seed cakes remaining after seed oil extraction were analyzed for the composition and content of amino acids and mineral elements.

Amino acids were detemined after hydrolysis of the seed cake material and completed with 6 N HCl at 105 °C for 24 h in sealed glass ampules. Ampule content was then evaporated at 40–50 °C under vacuum, and the residue was dissolved in 20 mM HCl and filtered. The free amino acids resulting from protein hydrolysis were derivatized using the AccQ-Fluor kit, WATO52880 (Waters Corporation, Milford, MA, USA). The HPLC separation of the resulting AccQ-Fluor amino acid derivatives was performed on an Elite LaChrome (Hitachi, Tokyo, Japan) unit, with a reverse phase C18 AccQ-Tag (3.9 mm × 150 mm) column (at 37 °C). WATO52890 buffer (Waters Corporation, Milford, MA, USA) and 60% acetonitrile were the eluting phases; the injected sample volume was 20 μL, heated to 55 °C. The unit was equipped with a diode array detector (DAD) (Hitachi, Tokyo, Japan), and the detection was performed at the wavelength 254 nm.

2.6. Determination of Mineral Elements in Seed Cakes

Mineral elements in the seed cakes were determined after mineralization at 450 °C; the resultant ash was first dissolved in concentrated HCl and then in 0.1 mol/L HNO_3 [31]. The atomic absorption spectrometry (AAS) was performed on a Perkin Elmer/HGA 500 instrument (Norwalk, CT, USA). The detection wavelengths were: Na, 589.6 nm; K, 766.5 nm; Mg, 285.2 nm; Ca, 317.0 nm; Zn, 213.9 nm; Cu, 324.7 nm; Fe, 238.3 nm; Mn, 257.6 nm; Pb, 283.3 nm; Cd, 228.8 nm; Cr, 357.9 nm. The elemental identification was completed by comparison with standard metal salt solutions, and the estimation of metal ion concentration by using calibration curves (built for 1 μg/mL standard salt solutions).

2.7. Determination of Volatiles in Fruit Concretes

Finally, the fruit concretes were analyzed. These were obtained from *P. alkekengi* peel and pulp as concentrated aromatic products, which are commonly used in perfumery and cosmetics.

In the obtainment of fruit concretes, the samples were subjected to a double extraction with n-hexane, at a temperature of 40 °C and a solid-to-solvent ratio of 1:10 (w/v). The duration of the first and second extraction was 60 min and 30 min, respectively. The extracts were then combined and concentrated on a rotary vacuum evaporator until complete solvent removal at a temperature 40 °C. The yield of peel and pulp concretes was determined gravimetrically (%, w/w) and calculated on a dry weight basis. The initial moisture content of the extracted plant materials was as follows: peel—53.85 ± 0.45% (PA-SB) and 49.73 ± 0.41% (PA-NB); pulp—79.43 ± 0.62% (PA-SB) and 82.99 ± 0.71% (PA-NB). The color and appearance of the concretes were determined by visual assessment.

The GC analysis for the determination of the volatile composition of the obtained fruit concretes was performed on an Agilent 7890A instrument (Agilent Technologies Inc., Santa Clara, CA, USA), with the following parameters: HP-5ms column, 30 m × 250 mm × 0.25 μm (i.d.); oven temperature increased at a rate of 5 °C/min from 35 °C (3 min) to 250 °C (3 min), total run time 49 min; carrier gas (helium) at 1 mL/min constant rate; 30:1 split mode. The GC-MS analysis employed an Agilent 5975C inert XL EI/CI mass selective detector (MSD) (Agilent Technologies Inc., Santa Clara, CA, USA), under the same operational conditions as in the GC analysis. Mass spectra acquisition was at 70 eV in electron impact (EI) mode; MS scan was from 50 to 550 m/z. The ionization source temperature was 230 °C, and the MS quad and the injector temperatures were 150 °C and 250 °C, respectively. Mass spectra were read using the built-in toolkit of 2.64 AMDIS (Automated Mass Spectral Deconvolution and Identification System; NIST, Gaithersburg, MD, USA) software. The identification of volatiles was based on comparison of their retention (Kovats) indices (RI) and MS fragmentation patterns with spectral library data [32,33]. Components were listed in ascending order of the RI, calculated using a standard calibration mixture of n-alkanes (C_8–C_{40}) in hexane, under the same operational conditions. Compound concentrations were calculated as percentage of the total ion current (TIC), after normalization of the recorded peak areas.

2.8. Statistics

All measurements in the study were performed in triplicate (n = 3), except for the fruit parts proportion (n = 100). The results are presented as the mean value with the corresponding standard deviation (SD). ANOVA and Tukey multiple comparison test were used as statistical tools in the assessment of significant differences at $p < 0.05$.

3. Results and Discussion

3.1. Basic Evaluation of Fruit Structural Parts

The proportions of fruit structural parts (peel, seeds, and pulp, respectively) in the analyzed fresh fruits of two *P. alkekengi* phenotypes (denoted as PA-SB and PA-NB) are presented in Table 1. Fruit pulp accounted for about 70% of fresh fruit weight, while seeds constituted about a quarter of the fruit weight, with no significant deviations between the phenotypes. Seed dimensions varied, however, with seed absolute weight being 1.68 ± 0.01 g (mean of 1000 seeds results) for PA-SB phenotype, and 1.52 ± 0.01 g (per 1000 seeds) for PA-NB. The average number of seeds in a single berry was 195 ± 1.80 for PA-SB, and 107 ± 0.90 for PA-NB phenotype.

Table 1. Proportion of fruit structural parts of two *P. alkekengi* phenotypes (PA-SB and PA-NB).

Fruit Part	PA-SB (% fw)	PA-NB (% fw)
Peel	4.99 ± 0.05 [a]	5.43 ± 0.05 [a]
Seeds	23.58 ± 0.18 [b]	26.15 ± 0.23 [b]
Pulp	71.43 ± 0.73 [c]	68.42 ± 0.64 [c]

Results: mean value ± standard deviation (n = 100). Different letters in the same row indicate significant differences ($p < 0.05$).

The basic macro component characteristics (cellulose, protein, oil) of the studied fruit fractions (the isolated seeds and the combined peel/pulp fraction) are presented in Table 2. The data indicate some differences between the phenotypes, with slightly higher oil amounts in PA-NB, and higher cellulose and protein amounts in the PA-SB phenotype. As seen from the data, the seeds were the primary site of oil accumulation in the fruit, although the combined peel/pulp samples also had detectable amounts of the oil fraction. The seeds of *P. alkekengi* were a sufficiently rich source of oil, yielding 14–17% oil, thus approximating the data for a related *Physalis* species, *P. peruviana* pomace oil (19.3%) detected by Ramadan [34], as well as those for soybean oil (18%) [35], grapeseed oil (8–20%) [36], or *P. alkekengi* seed oil [37].

Table 2. Macro component characteristics of fruit structural parts of two *P. alkekengi* phenotypes (PA-SB and PA-NB).

Component	PA-SB		PA-NB	
	Seeds (% dw)	Peel/Pulp (% dw)	Seeds (% dw)	Peel/Pulp (% dw)
Cellulose	8.06 ± 0.07 [d]	1.44 ± 0.01 [a]	6.12 ± 0.05 [c]	2.15 ± 0.01 [b]
Protein	19.14 ± 0.14 [d]	2.51 ± 0.01 [b]	16.22 ± 0.12 [c]	1.94 ± 0.01 [a]
Oil	14.13 ± 0.12 [c]	1.27 ± 0.01 [a]	17.57 ± 0.15 [d]	1.81 ± 0.01 [b]

Results: mean value ± standard deviation (n = 3). Different letters in the same row indicate significant differences ($p < 0.05$).

Similarly, the seeds contained most of the fruit's protein and fiber. The cellulose content was several times higher in the seeds (6–8% dw) than in the peel/pulp residues (1.5–2% dw). The reported data were close to the cellulose contents detected in the whole fruits of different cape gooseberry (*P. peruviana*) phenotypes measured by Petkova et al. [38], as well as in different fruit and vegetable pomaces, such as apples and tomatoes [39].

3.2. Determination of Fatty Acids in Seed and Peel Oils

The significant oil yield from the seeds of *P. alkekengi* fruit was the reason for subjecting the extracted seed oil to a more detailed analysis in order to reveal its micro component characteristics.

As seen in Table 2, the combined peel/pulp fraction was also associated with the presence of oil fraction, although in a minor amount. The individual analysis of the two fruit parts constituting the combined sample revealed that oil content in the pulp was minimal (below 1% dw) in both phenotypes, while the peel contained 2.54 ± 0.02% and 2.05 ± 0.02% oil (dw) in the PA-NB and PA-SB phenotypes, respectively. Therefore, it was considered interesting to identify the composition of the peel fraction oil, as well, in view of a more complete insight into the composition of *P. alkekengi* fruits.

The results regarding the fatty acid composition of the seed and peel oils of the two phenotypes are presented in Table 3, and an example of the obtained FAME chromatograms is shown in Figure 2. The data proved significant variations in the number and distribution of the identified fatty acids in the seed and peel oils, while the differences between the phenotypes were less pronounced.

The fatty acid composition of *P. alkekengi* seed oil was dominated by unsaturated fatty acids in a ratio of about 7:1 to saturated ones; this was the same for both phenotypes. The ratio between polyunsaturated and monounsaturated fatty acids was also favorable and comparable for the phenotypes, being about 5:1. The proportions of unsaturated and saturated fatty acids, however, were reversed and much more unfavorable in the peel oils, being 32:68 in PA-SB and 44:56 in PA-NB, respectively.

The predominant fatty acid in the seed oils was linoleic acid (73.67% in PA-SB and 74.43% in PA-NB), followed by oleic and palmitic acids, while the major fatty acids in peel oils were palmitic (57.88% in PA-SB and 36.21% in PA-NB) and oleic (24.02% in PA-SB and 30.08% in PA-NB) acids.

Table 3. Fatty acid composition of seed and peel oils of two *P. alkekengi* phenotypes (PA-SB and PA-NB).

Fatty Acid		PA-SB		PA-NB	
		Seeds (% dw)	Peel (% dw)	Seeds (% dw)	Peel (% dw)
$C_{10:0}$	Capric	nd [1]	nd	0.10 ± 0.0 [a]	0.09 ± 0.0 [a]
$C_{11:0}$	Undecylic	nd	0.33 ± 0.0 [b]	nd	0.17 ± 0.0 [a]
$C_{12:0}$	Lauric	nd	1.08 ± 0.0 [a]	nd	1.11 ± 0.01 [a]
$C_{13:0}$	Tridecylic	nd	0.09 ± 0.0 [a]	nd	0.42 ± 0.0 [b]
$C_{14:0}$	Myristic	nd	3.89 ± 0.01 [b]	0.11 ± 0.0 [a]	5.01 ± 0.02 [c]
$C_{14:1}$	Myristoleic	nd	nd	nd	0.82 ± 0.0 [a]
$C_{15:0}$	Pentadecylic	0.21 ± 0.0 [a]	nd	nd	4.44 ± 0.01 [b]
$C_{16:0}$	Palmitic	11.28 ± 0.10 [b]	57.88 ± 0.42 [d]	10.49 ± 0.10 [a]	36.21 ± 0.29 [c]
$C_{16:1}$	Palmitoleic	0.12 ± 0.0 [a]	1.79 ± 0.01 [b]	0.18 ± 0.0 [a]	8.22 ± 0.02 [c]
$C_{17:0}$	Margaric	nd	0.41 ± 0.0 [a]	nd	0.57 ± 0.0 [a]
$C_{17:1}$	Heptadecenoic	nd	0.69 ± 0.0 [b]	0.11 ± 0.0 [a]	0.49 ± 0.0 [b]
$C_{18:0}$	Stearic	0.51 ± 0.0 [a]	4.32 ± 0.02 [b]	0.42 ± 0.0 [a]	8.37 ± 0.02 [c]
$C_{18:1}$	Oleic	13.88 ± 0.11 [a]	24.02 ± 0.31 [b]	13.39 ± 0.12 [a]	30.08 ± 0.21 [c]
$C_{18:2}$ (n-6)	Linoleic	73.67 ± 0.72 [b]	1.00 ± 0.01 [a]	74.43 ± 0.71 [b]	0.83 ±0.0 [a]
$C_{18:3}$ (n-3)	Linolenic	0.21 ± 0.0 [a]	2.38 ± 0.01 [c]	nd	2.09 ± 0.02 [b]
$C_{20:2}$ (n-6)	Eicosadienoic	nd	2.12 ± 0.01 [c]	0.28 ± 0.0 [a]	1.08 ± 0.01 [b]
$C_{20:3}$ (n-6)	Eicosatrienoic	0.12 ± 0.0 [a]	nd	0.29 ± 0.0 [b]	nd
$C_{20:4}$ (n-6)	Eicosatetraenoic	nd	nd	0.10 ± 0.0	nd
$C_{20:5}$ (n-3)	Eicosapentaenoic	nd	nd	0.10 ± 0.0	nd
Saturated fatty acids		12.00	68.00	11.12	56.40
Unsaturated fatty acids, of which		88.00	32.00	88.88	43.60
Monounsaturated fatty acids		14.00	26.50	13.68	39.60
Polyunsaturated fatty acids		74.00	5.50	75.20	4.00

nd [1] = not detected; Results: mean value ± standard deviation ($n = 3$). Different letters in the same row indicate significant differences ($p < 0.05$).

Unlike the similar distribution of fatty acids in the seed oils, there were differences between the two phenotypes in the individual fatty acid composition of the oil derived from fruit peel alone, especially with regard to palmitic, palmitoleic, stearic, and some other acids. Although the results revealed a more favorable composition in the isolated seed oils, they supported the feasibility of oil extraction from *P. alkekengi* fruit, using both whole fruit and seeds alone. Regarding the individual fatty acid composition of the seed oils, our results differed only numerically from the data in the previous study [37], which identified linoleic (86.9%) and palmitic (6.6%) acids as the major fatty acids in the seed oil, at a ratio of unsaturated to saturated fatty acids of about 14:1. Our results were fully compliant with the data available for *P. peruviana* oils [34], which pointed out that pulp/peel oil was characterized by high amounts of saturated fatty acids, while ω-3 acids (α-linolenic) were found in lower levels.

3.3. Determination of Tocopherols in Seed and Peel Oils

The tocopherol composition of the extracted oils is presented in Table 4. As seen in the table, there was impressive differentiation in the content of the bioactive tocopherols on the bases of fruit structural parts and phenotype. Seed oils, despite the phenotype-related differences observed, contained considerably more tocopherols than peel oils. The data showed about a 2.5 times higher concentration of the biologically active tocopherols in the oil isolated from the seeds of PA-SB berries than for the PA-NB phenotype.

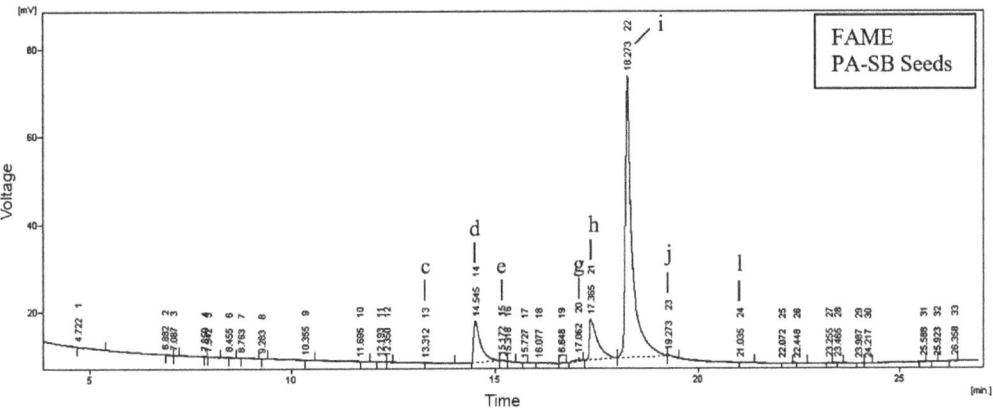

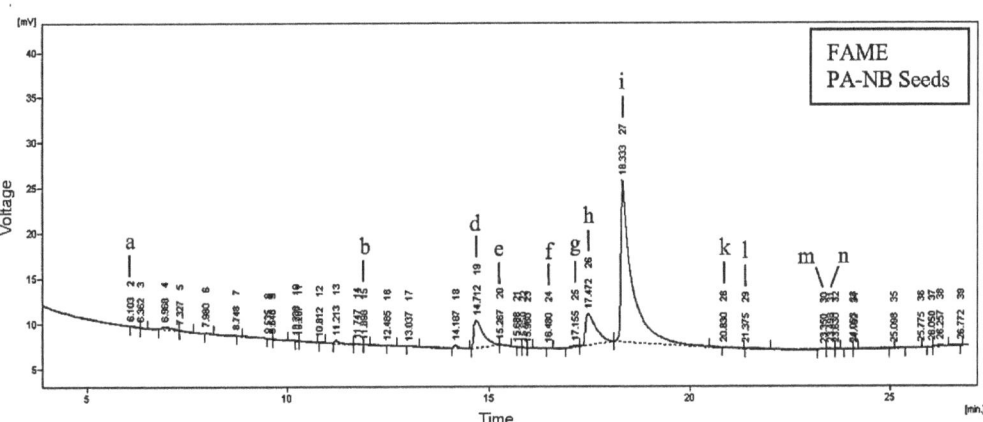

Figure 2. Chromatograms of the fatty acid composition of seed oils of two *P. alkekengi* phenotypes (PA-SB and PA-NB): a—capric acid; b—myristic acid; c—pentadecylic acid; d—palmitic acid; e—palmitoleic acid; f—heptadecenoic acid; g—stearic acid; h—oleic acid; i—linoleic acid; j—linolenic acid; k—eicosadienoic acid; l—eicosatrienoic acid; m—eicosapentaenoic acid; n—eicosatetraenoic acid.

Table 4. Tocopherol composition of seed and peel oils of two *P. alkekengi* phenotypes (PA-SB and PA-NB).

Tocopherols	PA-SB		PS-NB	
	Seeds	Peel	Seeds	Peel
α-Tocopherol (% of the total tocopherols)	1.01 ± 0.01 [a]	nd [1]	nd	100 ± 0.01 [b]
β-Tocopherol (% of the total tocopherols)	70.63 ± 0.68 [b]	30.32 ± 0.20 [a]	76.61 ± 0.71 [c]	nd
γ-Tocopherol (% of the total tocopherols)	28.44 ± 0.22 [b]	69.74 ± 0.31 [c]	23.42 ± 0.21 [a]	nd
Total tocopherols (mg/kg dw)	5378 ± 51.00 [d]	340.00 ± 17.00 [b]	2009 ± 20.00 [c]	216.00 ± 11.00 [a]

[1] nd = not detected; Results: mean value ± standard deviation (*n* = 3); Different letters in the same row indicate significant differences ($p < 0.05$).

The tocopherol fraction of both seed oils was predominated by β-tocopherol (70.63%, and 76.61%, of the total tocopherols, PA-SB and PA-NB, respectively), and γ-tocopherol (28.44% and 23.42% of the total content, PA-SB and PA-NB, respectively). The peel oils showed completely different tocopherol profiles—γ-tocopherol was dominating in PA-SB

(69.74%), followed by β-tocopherol (30.32%), while α-tocopherol was detected as a single representative (100%) in PA-NB peel oil.

In the study of Ramadan [34], β- and γ-tocopherols were also the major tocopherols in the seed oil of *P. peruviana*, while γ- and α-tocopherols were the main components in the pulp/peel oil. No further parallel to other data could be made concerning the peel oil composition or the tocopherols, as to the best of our knowledge, there have been no other previous investigations in that direction.

3.4. Determination of Amino Acids in Seed Cakes

Accounting for the significant ratio of seeds in fresh fruit weight (Table 1) and their macronutrient indices (Table 2), the seed cakes resulting from oil extraction, otherwise considered a waste product, were considered a potentially valuable plant material worthy of recovery. Therefore, an attempt to provide new data in favor of their potential use and nutritional value was made in this study, by identifying some individual micro components in the seed cakes—amino acids and minerals. The results of the amino acid profile of seed cake proteins are presented in Table 5.

Table 5. Amino acid composition of seed cakes from *P. alkekengi* fruit phenotypes (PA-SB and PA-NB).

Amino Acid	PA-SB (mg/g dw)	PA-NB (mg/g dw)
Aspartic acid	12.23 ± 0.11 [a]	12.16 ± 0.11 [a]
Serine	8.41 ± 0.07 [b]	6.08 ± 0.06 [a]
Glutamic acid	7.12 ± 0.07 [a]	11.97 ± 0.10 [b]
Glycine	10.94 ± 0.08 [a]	14.83 ± 0.11 [b]
Histidine	7.18 ± 0.07 [b]	5.90 ± 0.05 [a]
Arginine	15.75 ± 0.11 [b]	13.17 ± 0.11 [a]
Threonine [1]	4.17 ± 0.03 [b]	2.89 ± 0.03 [a]
Alanine	9.31 ± 0.08 [a]	14.34 ± 0.12 [b]
Proline	2.54 ± 0.01 [a]	4.27 ± 0.02 [b]
Cysteine	0.24 ± 0.0 [b]	0.14 ± 0.0 [a]
Tyrosine	3.20 ± 0.02 [a]	6.19 ± 0.04 [b]
Valine [1]	4.53 ± 0.02 [b]	3.56 ± 0.02 [a]
Methionine [1]	0.77 ± 0.0 [a]	0.70 ± 0.0 [a]
Lysine [1]	2.94 ± 0.01 [a]	3.49 ± 0.02 [b]
Isoleucine [1]	3.37 ± 0.01 [a]	4.41 ± 0.01 [b]
Leucine [1]	0.56 ± 0.0 [a]	0.86 ± 0.0 [b]
Phenylalanine [1]	3.19 ± 0.02 [a]	4.63 ± 0.02 [b]

Results: mean value ± standard deviation ($n = 3$). Different letters in the same row indicate significant differences ($p < 0.05$). [1] essential amino acid.

As seen from the data, the dominant amino acids in the seed cakes of both the *P. alkekengi* phenotypes were arginine, aspartic acid, glycine, and alanine, with numerical deviations between the samples. Respectively, there were differences in the distributions of the rest of the amino acids, explicable by the varying conditions of the plant vegetation (phenotype). The ratio of essential amino acids (EAA) was relatively low and comparable in the two phenotypes, being about 0.4:1. Valine, phenylalanine, threonine, and isoleucine were present in slightly higher amounts than the other EAAs.

3.5. Determination of Mineral Elements in Seed Cakes

The results from the determination of 11 key macro and micro mineral elements in the seed cakes (Table 6) confirmed significant differences between the phenotypes only with regard to two of the macro minerals, sodium and calcium; the content of sodium was substantially higher in the PA-SB phenotype, and that of calcium was higher in the seed cakes of PA-NB fruit. The macrominerals potassium and magnesium were found in identical amounts in both phenotypes. The identified microminerals (Fe, Mn, Cu, Zn, Pb) showed comparable distributions between the phenotypes, with the single exception of Cr in PA-NB.

Table 6. Mineral composition of seed cakes from *P. alkekengi* fruit of two phenotypes (PA-SB and PA-NB).

Mineral Element	PA-SB (mg/kg dw)	PA-NB (mg/kg dw)
Potassium (K)	4122.28 ± 19.43 [a]	4668.32 ± 21.23 [a]
Sodium (Na)	182.33 ± 1.33 [b]	80.79 ± 0.33 [a]
Calcium (Ca)	529.15 ± 1.87 [a]	1586.23 ± 4.78 [b]
Magnesium (Mg)	2418.97 ± 11.24 [a]	2318.26 ± 11.09 [a]
Iron (Fe)	61.71 ± 0.21 [b]	50.38 ± 0.19 [a]
Manganese (Mn)	25.40 ± 0.09 [a]	23.42 ± 0.08 [a]
Copper (Cu)	9.98 ± 0.03 [a]	10.30 ± 0.04 [a]
Zinc (Zn)	29.64 ± 0.09 [b]	25.24 ± 0.08 [a]
Lead (Pb)	2.67 ± 0.0 [a]	2.93 ± 0.0 [a]
Cadmium (Cd)	nd [1]	nd
Chromium (Cr)	nd	2.32 ± 0.0

Results: mean value ± standard deviation (n = 3). Different letters in the same row indicate significant differences ($p < 0.05$). [1] nd = not detected.

Evaluating the potential of the seed cakes for nutritional purposes, they could be considered a good source of potassium (supplying about 15% of the reference dietary intake (RDI) for men and about 12% for women; the RDIs being 2700 and 3500 mg, respectively), magnesium (about 60% of RDI for men and about 70% of RDI for women; 350 and 300 mg), as well as of iron (about 60% of RDI for men and about 30% of RDI for women; 8 and 18 mg), zinc (about 20% of RDI for men and about 30% of RDI for women; 11 and 8 mg), and other key microelements (copper and manganese) [40].

3.6. Determination of Volatile Components in Fruit Concretes

Another important aspect of the study was the objective of investigating the potential of *P. alkekengi* fruit for obtaining fruit concrete, a type of established aromatic product, widely used in perfumery and cosmetics [41]. Concretes are obtained by extracting fresh or dry plant materials with a non-polar solvent (e.g., hexane, petroleum ether, or benzene), followed by the complete removal of the solvent; the resulting concentrated aromatic product carries a specific fragrance profile, adding to the diversification of the fragrance nuances provided by the other types of aromatic products from the respective plant material, such as essential oils, absolute, resinoids, tinctures, and other extracts [41].

Accounting for the fact that the fruit pulp and peel are generally the vegetal material strongly related to fruit flavor, those parts of the *P. alkekengi* fruit were processed individually by *n*-hexane double extraction to obtain the respective fruit concretes. All concretes obtained in the study represented thick waxy masses with dark orange–yellow color.

The results revealed significant differences in the yield of fruit concretes, both between the phenotypes and between the fruit structural parts compared in the study. As seen from the data in Table 7, the yield of peel concrete was significantly higher than that of pulp concrete, thus clearly differentiating the potential of the two fruit structures with regard to their processing efficiency. In turn, there were significant differences in peel concrete yield on a phenotype basis; it was about three time higher in PA-NB phenotype, which was obviously related to the impact of environmental characteristics on plant metabolism.

The results from the identification by GC-MS analysis of individual aromatic volatiles in the obtained concretes and the total ion current (TIC) chromatograms with the major compounds (over 3%) are presented in Table 8 and Figure 3, respectively.

Table 7. Primary characteristics of fruit concretes obtained from two phenotypes of *P. alkekengi* (PA-SB and PA-NB).

Index	PA-SB (% DW; w/w)	PS-NB (% DW; w/w)
Yield of pulp concrete	0.02 ± 0.00 [a]	0.03 ± 0.00 [a]
Yield of peel concrete	0.72 ± 0.01 [a]	2.16 ± 0.01 [b]

Results: mean value ± standard deviation ($n = 3$). Different letters in the same row indicate significant differences ($p < 0.05$).

Table 8. Volatile components obtained by GC-MS in the concretes from two phenotypes of *P. alkekengi* (PA-SB and PA-NB) fruit.

Volatiles	RT [1]	RI [2]	PA-SB (% or TIC [2])		PA-NB (% or TIC [3])	
			Peel	Pulp	Peel	Pulp
α-Pinene	9.45	933	4.63 ± 0.03 [a]	5.30 ± 0.04 [b]	7.57 ± 0.06 [d]	6.85 ± 0.05 [c]
Camphene	9.98	945	0.31 ± 0.0 [a]	0.53 ± 0.0 [b]	0.69 ± 0.0 [c]	0.28 ± 0.0 [a]
Sabinene	10.76	969	0.05 ± 0.0 [a]	0.22 ± 0.0 [b]	0.08 ± 0.0 [a]	0.15 ± 0.0 [b]
β-Pinene	10.90	975	0.09 ± 0.0 [a]	0.41 ± 0.0 [c]	0.06 ± 0.0 [a]	0.12 ± 0.0 [b]
Myrcene	11.33	987	0.44 ± 0.0 [b]	0.59 ± 0.0 [c]	0.23 ± 0.0 [a]	0.25 ± 0.0 [a]
p-Cymene	12.45	1020	0.54 ± 0.0 [a]	1.04 ± 0.09 [c]	0.74 ± 0.0 [b]	0.56 ± 0.0 [a]
Limonene	12.60	1023	0.43 ± 0.0 [a]	1.21 ± 0.03 [d]	1.07 ± 0.09 [c]	0.89 ± 0.0 [b]
γ-Terpinene	13.54	1055	2.34 ± 0.01 [a]	7.77 ± 0.06 [c]	5.52 ± 0.05 [b]	5.46 ± 0.05 [b]
Camphenilone	14.40	1077	0.35 ± 0.0 [b]	0.41 ± 0.0 [b]	0.12 ± 0.0 [a]	0.09 ± 0.0 [a]
β-Linalool	14.92	1095	49.57 ± 0.47 [a]	61.91 ± 0.60 [c]	57.76 ± 0.50 [b]	78.29 ± 0.70 [d]
Nonanal	15.02	1101	0.11 ± 0.0 [b]	0.58 ± 0.0 [d]	0.04 ± 0.0 [a]	0.38 ± 0.0 [c]
Camphor	16.29	1140	7.53 ± 0.06 [d]	3.26 ± 0.03 [b]	5.53 ± 0.05 [c]	2.31 ± 0.02 [a]
1-Terpinen-4-ol	17.31	1174	0.04 ± 0.0 [a]	0.18 ± 0.0 [b]	nd [4]	nd
α-Terpineol	17.75	1185	0.09 ± 0.0 [a]	0.32 ± 0.0 [b]	nd	nd
Nerol	19.33	1126	1.48 ± 0.01 [b]	1.73 ± 0.01 [c]	2.80 ± 0.02 [d]	0.06 ± 0.0 [a]
n-Tridecane	21.34	1300	nd	nd	6.18 ± 0.05 [b]	1.72 ± 0.01 [a]
Isoamyl benzyl ether	21.40	1310	nd	nd	nd	1.15 ± 0.01
Neryl acetate	22.83	1349	11.42 ± 0.10 [c]	3.71 ± 0.03 [b]	1.06 ± 0.01 [a]	nd
Sibirene	23.55	1398	nd	0.72 ± 0.0 [a]	5.85 ± 0.05 [b]	nd
β-Caryophyllene	23.98	1419	nd	0.33 ± 0.0	nd	nd
Germacrene D	25.24	1483	6.85 ± 0.06 [c]	0.88 ± 0.0 [b]	0.23 ± 0.03 [a]	nd
β-Selinene	25.40	1491	0.05 ± 0.0 [a]	0.43 ± 0.0 [b]	nd	nd
α-Zingiberene	25.61	1493	nd	0.27 ± 0.0	nd	nd
Bicyclogermacrene	25.91	1501	nd	0.55 ± 0.0	nd	nd
δ-Cadinene	26.55	1522	nd	1.07 ± 0.01	nd	nd
α-Cadinene	26.80	1536	nd	0.32 ± 0.05	nd	nd
1-epi-Cubenol	29.11	1627	nd	0.41 ± 0.05	nd	nd
tau-Cadinol	29.30	1640	3.69 ± 0.03 [b]	0.54 ± 0.0 [a]	nd	nd
(2E,6E)-Methyl farnesoate	35.12	1785	1.11 ± 0.01	nd	nd	nd
(2Z,6E)-Farnesyl acetate	36.19	1820	2.58 ± 0.02	nd	nd	nd
(5Z,9E)-Farnesyl acetone	38.30	1889	2.09 ± 0.02 [b]	3.87 ± 0.03 [c]	2.88 ± 0.02 [a]	nd
Phytol	40.74	1940	2.93 ± 0.02	nd	nd	nd

[1] RT—retention time, min; [2] RI—retention index (Kovat's index); [3] TIC—total ion current; Results: mean value ± standard deviation ($n = 3$). Different letters in the same row indicate significant differences ($p < 0.05$). [4] nd = not detected.

Fifteen individual volatiles were identified by the applied GC-MS analysis in the pulp concrete of the PA-NB phenotype, accounting for 98.62% of the total content. The major components (in amounts over 3%) were: β-linalool (78.3%), α-pinene (6.9%), and γ-terpinene (5.5%). In the pulp concrete from the second phenotype, PA-SB, the number of identified compounds was 27 (98.56% of the total content), among which the major components were: β-linalool (61.9%), γ-terpinene (7.8%), α-pinene (5.3%), (5Z,9E)-farnesyl acetone (3.9%), neryl acetate (3.7%), and camphor (3.3%). The concrete obtained from the peel of the PA-NB phenotype contained 18 identified components (representing 98.41% of the total content). The major constituents in the product (over 3%) were: β-linalool (57.8%), α-pinene (7.6%), *n*-tridecane (6.2%), sibirene (5.9%), camphor (5.5%), and γ-terpinene (5.5%). In turn, a total of 23 individual components were identified in the peel concrete from the PA-SB phenotype (98.72% of the content), with the major constituents being: β-linalool (49.6%), neryl acetate (11.4%), camphor (7.5%), germacrene D (6.9%), α-pinene (4.6%), and *tau*-cadinol (3.7%).

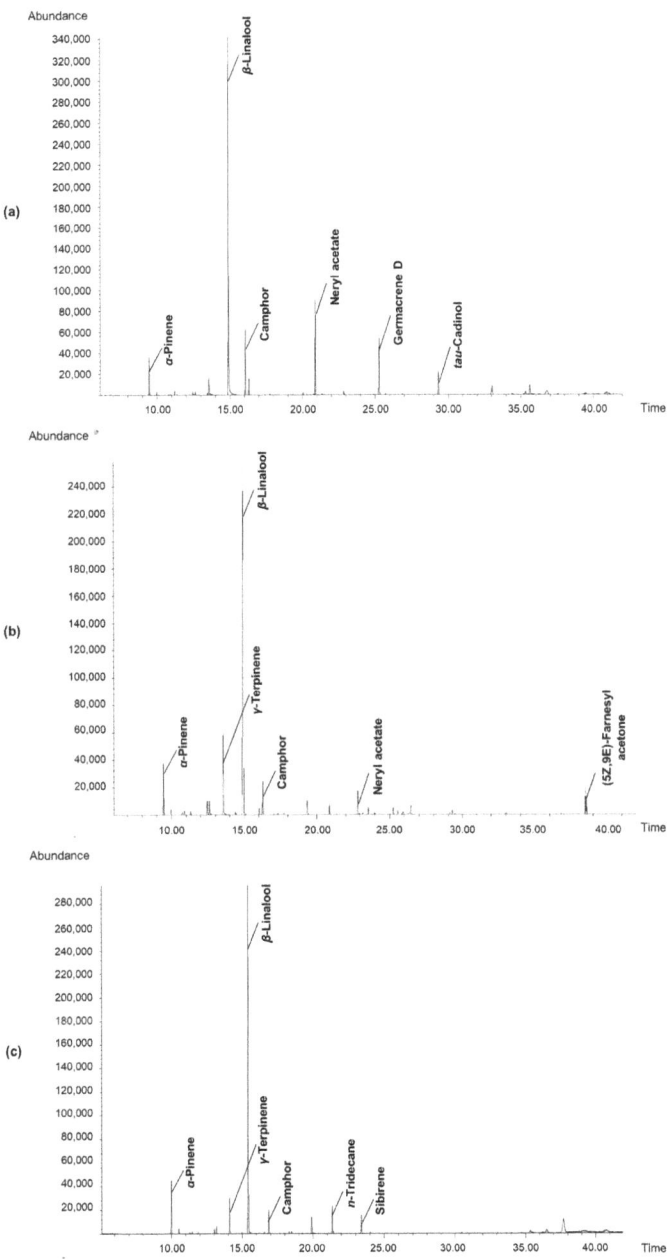

Figure 3. *Cont.*

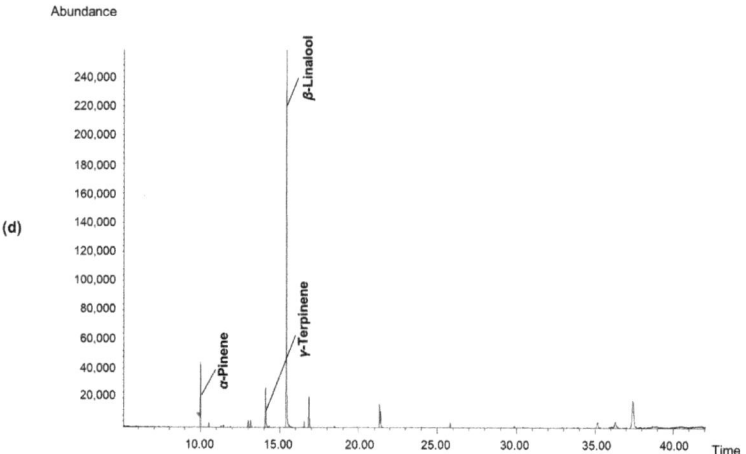

Figure 3. Chromatograms of volatiles in concretes obtained from two phenotypes of *P. alkekengi* fruit (PA-SB and PA-NB): (**a**) concrete from the peel of PA-SB phenotype; (**b**) concrete from the pulp of PA-SB phenotype; (**c**) concrete from the peel of PA-NB phenotype; (**d**) concrete from the pulp of PA-NB phenotype.

The comparison of the data obtained for the respective aromatic products from two phenotypes revealed significant differences in the contents of a number of minor and major components with known fragrance properties; for instance, components in the pulp concretes included nerol (1.7% in PA-SB and 0.1% in PA-NB), neryl acetate (3.7% in PA-SB, not detected in PA-NB), and (5Z,9E)-farnesyl acetone (3.9% in PA-SB, not detected in PA-NB); components in the peel concretes included neryl acetate (11.4% in PA-SB and only 1.1% in PA-NB), sibirene (5.9% in PA-NB, not detected in PA-SB), germacrene D (6.9% in PA-SB and only 0.2% in PA-NB), and *tau*-cadinol (3.7% in PA-SB, not detected in PA-NB).

The classification of the identified fruit concrete constituents revealed the presence of compounds belonging to different chemical classes (Figure 4). The distribution of the identified compounds (equaled to 100%) did not suggest considerable differences between the two structural elements of the fruit (pulp, peel), or between the phenotypes, although some specifics also existed. In all extracts, the volatile composition was dominated by oxygenated monoterpene derivatives (72.6% and 81.9% in the pulp and 71.4% and 68.4% in the peel for the PA-SB and PA-NB phenotypes, respectively), followed by monoterpene hydrocarbons (16.3% and 14.2% in fruit pulp; 8.4% and 15.5% in the peel). Sesquiterpene hydrocarbons and oxygenated sesquiterpene derivatives were not detected in the concrete from the pulp of PA-NB phenotype, while diterpene representatives were found only in the concrete from the peel of PA-SB phenotype (3.0% of the identified content). The results revealed that the differences between the two phenotypes were more pronounced with regard to the extractible aromatic compounds in the peel compared with the fruit pulp. Thus, it could be presumed that fruit origin (phenotype) would not be a decisive factor for the organoleptic properties of the final product in *P. alkekengi* juice/pulp production, but would probably affect the usability of the resultant waste (peel or seed/peel residue); of course, additional and more targeted investigations are needed to support such assumptions, utilizing a wider range of sampling, parameter selection, and statistical tools.

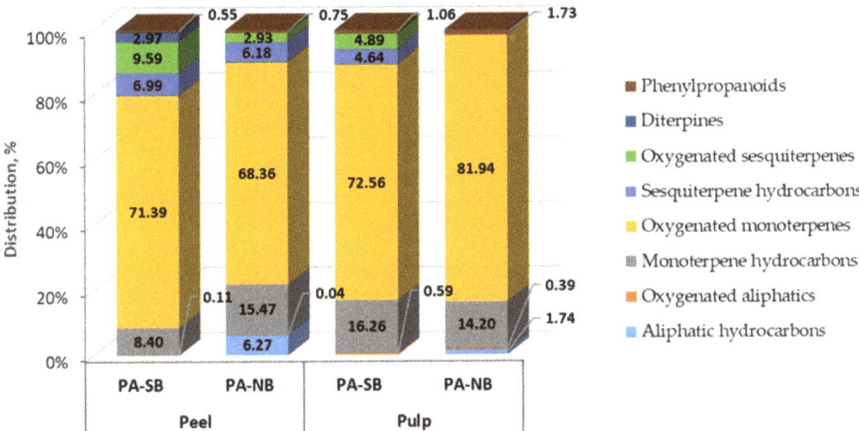

Figure 4. Distribution of volatiles by chemical groups in concretes obtained from two phenotypes of *P. alkekengi* fruit (PA-SB and PA-NB).

4. Conclusions

The results achieved by this study conducted on *P. alkekengi* fruit from Bulgaria add new details to the existing knowledge about the species. The study revealed that each of the fruit structural parts (peel, pulp, and seeds) or the resultant by-products (the seed cakes), rarely analyzed individually, had its specific features in terms of the assessed chemical characteristics, thus suggesting different options for their prospective use as sources of functional macro- and micronutrients—fiber, protein and some essential amino acids, oils and unsaturated fatty acids, tocopherols, and minerals. The study also provides new data on the obtaining and identification of the volatile profiles of *P. alkekengi* fruit concretes, thus contributing to the expansion of the range of available aromatic products. The presence of a number of aroma-active volatile compounds in the obtained peel and pulp concretes spoke in favor of a tangible potential for their future consideration as ingredients, e.g., in perfumery and cosmetic formulations. The results on the phenotype-related differences presented in the study also suggested that fruit origin could, to a lesser or greater extent, affect the chemical composition of the assessed individual parts of *P. alkekengi* fruit and the aromatic products derived from them, especially if considering fruit properties on a wider geographical basis, in which genotype would most probably be another decisive factor. Based on these considerations, the outcomes from the study could be the grounds for future investigations in the indicated directions.

Author Contributions: Conceptualization, V.P., A.S. and S.E.; Formal analysis, V.P., Z.P., N.M., T.I., N.P. and M.S.; Funding acquisition, J.M.; Supervision, S.E., Z.O., S.S. and J.M.; Writing—original draft, V.P., A.S. and S.E.; Writing—review & editing, Z.O., S.S. and J.M. All authors have read and agreed to the published version of the manuscript.

Funding: Financial support was provided by Tomas Bata University in Zlin, Faculty of Technology (IGA FT 2022/004).

Institutional Review Board Statement: Not applicable.

Informed Consent Statement: Not applicable.

Data Availability Statement: The data presented in this study are available on request from the authors.

Acknowledgments: The authors express their gratitude to Ivanka Dimitrova-Dyulgerova, Department of Botany, Plovdiv University "Paisii Hilendarski", Bulgaria, for the support in the collection and identification of the plant material.

Conflicts of Interest: The authors declare no conflict of interest.

References

1. Namjoyan, F.; Jahangiri, A.; Azemi, M.E.; Arkian, E.; Mousavi, H. Inhibitory effects of *Physalis alkekengi* L., *Alcea rosea* L., *Bunium persicum* B. Fedtsch. and *Marrubium vulgare* L. on mushroom tyrosinase. *Jundishapur J. Nat. Pharm. Prod.* **2015**, *10*, e23356. [CrossRef]
2. Li, A.; Chen, B.; Li, G.; Zhou, M.; Li, Y.; Ren, D.; Lou, H.; Wang, X.; Shen, T. *Physalis alkekengi* L. var. *Franchetii* (Mast.) Makino: An ethnomedical, phytochemical and pharmacological review. *J. Ethnopharmacol.* **2018**, *210*, 260–274. [CrossRef] [PubMed]
3. Petkov, V. *Contemporary Phytotherapy*; Medicina i Fizkultura: Sofia, Bulgaria, 1982. (In Bulgarian)
4. Delipavlov, D.; Cheshmedzhiev, I.; Popova, M.; Terzijski, D.; Kovatchev, I. *Key to the Plants in Bulgaria*; Agricultural University Academic Press: Plovdiv, Bulgaria, 2003. (In Bulgarian)
5. Parliament of the Republic of Bulgaria. Medicinal Plants Act of the 38th Parliament of the Republic of Bulgaria of 23 March 2000. *State Gaz.* **2000**, *29*, 8. Available online: https://dv.parliament.bg/DVWeb/index.faces (accessed on 13 April 2022). (In Bulgarian).
6. Evstatieva, L.; Hardalov, R.; Stoyanova, K. Medicinal plants in Bulgaria: Diversity, legislation, conservation and trade. *Phytol. Balc.* **2007**, *13*, 415–427. Available online: http://www.bio.bas.bg/~{}phytolbalcan/PDF/13_3/13_3_19_Evstatieva_&_al.pdf (accessed on 13 April 2022).
7. Ivanov, I.; Landzhev, I.; Neshev, G. *Herbs in Bulgaria and Their Use*; Zemizdat: Sofia, Bulgaria, 1977. (In Bulgarian)
8. Pintea, A.; Varga, A.; Stepnowski, P.; Socaciu, C.; Culea, M.; Diehl, H. Chromatographic analysis of carotenol fatty acid esters in *Physalis alkekengi* and *Hippophae rhamnoides*. *Phytochem. Anal.* **2005**, *16*, 188–195. [CrossRef] [PubMed]
9. Chen, C.-Y.; Peng, W.-H.; Tsai, K.-D.; Hsu, S.-L. Luteolin suppresses inflammation-associated gene expression by blocking NF-κB and AP-1 activation pathway in mouse alveolar macrophages. *Life Sci.* **2007**, *81*, 1602–1614. [CrossRef] [PubMed]
10. Yang, J.; Sun, Y.; Cao, F.; Yang, B.; Kuang, H. Natural products from *Physalis alkekengi* L. var. *franchetii* (Mast.) Makino: A review on their structural analysis, quality control, pharmacology, and pharmacokinetics. *Molecules* **2022**, *27*, 695. [CrossRef]
11. Tong, H.; Liang, Z.; Wang, G. Structural characterization and hypoglycemic activity of a polysaccharide isolated from the fruit of *Physalis alkekengi* L. *Carbohydr. Polym.* **2008**, *71*, 316–323. [CrossRef]
12. Zarei, A.; Ashtiyani, S.; Mohamadi, A.; Gabari, A. The effects of *Physalis alkekengi* extract on lipids concentrations in rats. *J. Arak Uni. Med. Sci.* **2011**, *14*, 36–42.
13. Chen, L.-X.; Xia, G.-Y.; Liu, Q.-Y.; Xie, Y.-Y.; Qiu, F. Chemical constituents from the calyces of *Physalis alkekengi* var. *franchetii*. *Biochem. Syst. Ecol.* **2014**, *54*, 31–35. [CrossRef]
14. Shu, Z.; Xing, N.; Wang, Q.; Li, X.; Xu, B.; Li, Z.; Kuang, H. Antibacterial and anti-inflammatory activities of *Physalis alkekengi* var. *franchetii* and its main constituents. *Evid. Based Complementary Altern. Med.* **2016**, 4359394. [CrossRef]
15. Zhang, C.; Khan, W.; Bakht, J.; Nair, M. New antiinflammatory sucrose esters in the natural sticky coating of tomatillo (*Physalis philadelphica*), an important culinary fruit. *Food Chem.* **2016**, *196*, 726–732. [CrossRef] [PubMed]
16. Liu, X.; Bian, J.; Li, D.; Liu, C.; Xu, S.; Zhang, G.; Zhang, L.; Gao, P. Structural features, antioxidant and acetylcholinesterase inhibitory activities of polysaccharides from stem of *Physalis alkekengi* L. *Ind. Crops Prod.* **2019**, *129*, 654–661. [CrossRef]
17. Vicas, L.G.; Jurca, T.; Baldea, I.; Filip, G.A.; Olteanu, D.; Clichici, S.V.; Pallag, A.; Marian, E.; Micle, O.; Crivii, C.B.; et al. *Physalis alkekengi* L. extract reduces the oxidative stress, inflammation and apoptosis in endothelial vascular cells exposed to hyperglycemia. *Molecules* **2020**, *25*, 3747. [CrossRef]
18. Sharma, N.; Bano, A.; Dhaliwal, H.; Sharma, V. Perspectives and possibilities of Indian species of genus *Physalis* (L.)—A comprehensive review. *Eur. J. Pharm. Med. Res.* **2015**, *2*, 326–353.
19. Bahmani, M.; Rafieian-Kopaei, M.; Naghdi, N.; Nejad, A.; Afsordeh, O. *Physalis alkekengi*: A review of its therapeutic effects. *J. Chem. Pharm. Sci.* **2016**, *9*, 1472–1475. Available online: http://eprints.umsha.ac.ir/id/eprint/2360 (accessed on 13 April 2022).
20. Lu, Z.; Li, W.; Wang, P.J.; Liu, X.Y. Extraction of the oil from ground cherry seeds by Soxhlet method with different solvents and composition analysis. *Chem. Adhesion* **2011**, *1*, 18.
21. Hu, X.-F.; Zhang, Q.; Zhang, P.-P.; Sun, L.-J.; Liang, J.-C.; Morris-Natschke, S.; Chen, Y.; Lee, K.-H. Evaluation of in vitro/in vivo anti-diabetic effects and identification, of compounds from *Physalis alkekengi*. *Fitoterapia* **2018**, *127*, 129–137. [CrossRef]
22. Wen, X.; Erşan, S.; Li, M.; Wang, K.; Steingass, C.; Schweiggert, R.; Ni, Y.; Carle, R. Physicochemical characteristics and phytochemical profiles of yellow and red physalis (*Physalis alkekengi* L. and *P. pubescens* L.) fruits cultivated in China. *Food Res. Int.* **2019**, *120*, 389–398. [CrossRef]
23. Liu, X.-G.; Jiang, F.-Y.; Gao, P.-Y.; Jin, M.; Yang, D.; Nian, Z.-F.; Zhang, Z.-X. Optimization of extraction conditions for flavonoids of *Physalis alkekengi* var. *franchetii* stems by response surface methodology and inhibition of acetylcholinesterase activity. *J. Mex. Chem. Soc.* **2015**, *59*, 59–66. [CrossRef]
24. Cosmetic Ingredient Database (CosIng) of the European Commission. Available online: https://ec.europa.eu/growth/sectors/cosmetics_en (accessed on 18 March 2022).
25. Association of Official Analytical Chemists (AOAC). *Official Methods of Analysis*, 20th ed.; AOAC International: Geithersburg, MD, USA, 2016.
26. Brendel, O.; Iannetta, P.; Stewart, D. A rapid and simple method to isolate pure alpha-cellulose. *Phytochem. Anal.* **2000**, *11*, 7–10. [CrossRef]
27. Standard ISO 659:2014; Oilseeds. Determination of Oil Content (Reference Method). International Organization for Standardization: Geneva, Switzerland, 2014.

28. *Standard ISO 12966-1:2014*; Animal and Vegetable Fats and Oils. Gas Chromatography of Fatty Acid Methyl Esters—Part 1: Guidelines on Modern Gas Chromatography of Fatty Acid Methyl Esters. International Organization for Standardization: Geneva, Switzerland, 2014.
29. *Standard ISO 12966-2:2017*; Animal and Vegetable Fats and Oils. Gas chromatography of Fatty Acid Methyl Esters—Part 2: Preparation of Methyl Esters of Fatty Acids. International Organization for Standardization: Geneva, Switzerland, 2017.
30. *Standard ISO 9936:2016*; Animal and Vegetable Fats and Oils. Determination of Tocopherol and Tocotrienol Contents by High-performance Liquid Chromatography. International Organization for Standardization: Geneva, Switzerland, 2016.
31. *Standard ISO 14084:2003*; Foodstuffs—Determination of Trace Elements—Determination of Lead, Cadmium, Zinc, Copper and Iron by Atomic Absorption Spectrometry (AAS) after Microwave Digestion. International Organization for Standardization: Geneva, Switzerland, 2003.
32. Adams, R. *Identification of Essential Oil Components by Gas Chromatography/Mass Spectrometry*, 4th ed.; Allured Publishing Co.: Carol Stream, IL, USA, 2007.
33. Shen, V.K.; Siderius, D.W.; Krekelberg, W.P.; Hatch, H.W. (Eds.) *NIST Standard Reference Simulation Website, NIST Standard Reference Database 173*; National Institute of Standards and Technology: Gaithersburg, MD, USA, 2017. [CrossRef]
34. Ramadan, M.F. Bioactive phytochemicals, nutritional value, and functional properties of Cape gooseberry (*Physalis peruviana*): An overview. *Food Res. Int.* **2011**, *44*, 1830–1836. [CrossRef]
35. Popov, A.; Ilinov, P. *Chemistry of Lipids*; Nauka i Iskustvo: Sofia, Bulgaria, 1986. (In Bulgarian)
36. Heuzé, V.; Tran, G. Grape Seeds and Grape Seed Oil Meal. Feedipedia, a Programme by INRA, CIRAD, AFZ and FAO. 2017. Available online: https://feedipedia.org/node/692 (accessed on 18 March 2022).
37. Asilbekova, D.; Ulchenko, N.; Glushenkova, A. Lipids from *Physalis alkekengi*. *Chem. Nat. Compd* **2016**, *52*, 96–97. [CrossRef]
38. Petkova, N.; Popova, V.; Ivanova, T.; Mazova, N.; Panayotov, N.; Stoyanova, A. Nutritional composition of different Cape gooseberry genotypes (*Physalis peruviana* L.)—A comparative study. *Food Res.* **2021**, *5*, 191–202. [CrossRef]
39. Szymańska-Chargot, M.; Chylińska, M.; Gdula, K.; Kozioł, A.; Zdunek, A. Isolation and characterization of cellulose from different fruit and vegetable pomaces. *Polymers* **2017**, *9*, 495. [CrossRef] [PubMed]
40. Ministry of Health of the Republic of Bulgaria. Ordinance No 1 of 22 January 2018 on the Physiological Norms of Nutrition for the Population. *State Gaz.* **2018**, *11*, 2. Available online: https://dv.parliament.bg/DVWeb/broeveList.faces (accessed on 13 April 2022). (In Bulgarian)
41. Baser, K.H.C.; Buchbauer, G. *Handbook of Essential Oils: Science, Technology, and Applications*; CRC Press: Boca Raton, FL, USA, 2010.

Article

Effectiveness of E-Beam Radiation against *Saccharomyces cerevisiae*, *Brettanomyces bruxellensis*, and Wild Yeast and Their Influence on Wine Quality

Magdalena Błaszak [1], Barbara Jakubowska [1], Sabina Lachowicz-Wiśniewska [2,*], Wojciech Migdał [3], Urszula Gryczka [3] and Ireneusz Ochmian [4,*]

1. Department of Chemistry, Microbiology and Environmental Biotechnology, West Pomeranian University of Technology in Szczecin, Słowackiego 17 Street, 71-434 Szczecin, Poland; blaszak.magdalena@zut.edu.pl (M.B.); bjakubowska@zut.edu.pl (B.J.)
2. Department of Health Sciences, Calisia University, 4 Nowy Świat Street, 62-800 Kalisz, Poland
3. Institute of Nuclear Chemistry and Technology, 16 Dorodna Street, 03-195 Warsaw, Poland; w.migdal@ichtj.waw.pl (W.M.); u.gryczka@ichtj.waw.pl (U.G.)
4. Department of Horticulture, West Pomeranian University of Technology Szczecin, Słowackiego 17 Street, 71-434 Szczecin, Poland
* Correspondence: s.lachowicz-wisniewska@akademiakaliska.edu.pl (S.L.-W.); ireneusz.ochmian@zut.edu.pl (I.O.)

Citation: Błaszak, M.; Jakubowska, B.; Lachowicz-Wiśniewska, S.; Migdał, W.; Gryczka, U.; Ochmian, I. Effectiveness of E-Beam Radiation against *Saccharomyces cerevisiae*, *Brettanomyces bruxellensis*, and Wild Yeast and Their Influence on Wine Quality. *Molecules* 2023, 28, 4867. https://doi.org/10.3390/molecules28124867

Academic Editor: Carmen González-Barreiro

Received: 1 April 2023
Revised: 12 June 2023
Accepted: 16 June 2023
Published: 20 June 2023

Copyright: © 2023 by the authors. Licensee MDPI, Basel, Switzerland. This article is an open access article distributed under the terms and conditions of the Creative Commons Attribution (CC BY) license (https://creativecommons.org/licenses/by/4.0/).

Abstract: The simplest way to eliminate microorganisms in the must/wine is through sulfuration, as it allows the introduction of pure yeast varieties into the must, which guarantees a high-quality wine. However, sulfur is an allergen, and an increasing number of people are developing allergies to it. Therefore, alternative methods for microbiological stabilization of must and wine are being sought. Consequently, the aim of the experiment was to evaluate the effectiveness of ionizing radiation in eliminating microorganisms in must. The sensitivity of wine yeasts, *Saccharomyces cerevisiae*, *S. cerevisiae* var. *bayanus*, *Brettanomyces bruxellensis*, and wild yeasts to ionizing radiation was com-pared. The effects of these yeasts on wine chemistry and quality were also determined. Ionizing radiation eliminates yeast in wine. A dose of 2.5 kGy reduced the amount of yeast by more than 90% without reducing the quality of the wine. However, higher doses of radiation worsened the organoleptic properties of the wine. The breed of yeast used has a very strong influence on the quality of the wine. It is justifiable to use commercial yeast breeds to obtain standard-quality wine. The use of special strains, e.g., *B. bruxellensis*, is also justified when aiming to obtain a unique product during vinification. This wine was reminiscent of wine produced with wild yeast.. The wine fermented with wild yeast had a very poor chemical composition, which negatively affected its taste and aroma. The high content of 2-methylbutanol and 3-methylbutanol caused the wine to have a nail polish remover smell.

Keywords: polyphenols; color; yeast; quality of wine; wine preservation; environmental protection

1. Introduction

Sulfurization of must eliminates the population of microorganisms during the early stages of vinification. The introduction of pure yeast varieties into the musts assures product quality, due to the well-known metabolic profile of these varieties. The most commonly used wine yeast strain is *Saccharomyces cerevisiae* [1] which ensures a cultivar-specific product due to the repeatability of the chemical composition and the bouquet of the wine [2].

The unique properties of natural and regional wines can be attributed to the compounds produced by unusual microorganisms or hybrids of various wine and beer yeasts. A notable example is Château Musar, which owes its uniqueness to the presence of *Brettanomyces bruxellensis* yeast, contributing to its aromatic profile primarily through

4-ethylphenol and 4-ethylguaiacol. These wines are appreciated by a small group of connoisseurs due to their characteristics of table aroma [3,4].

Natural wines are the result of spontaneous fermentation of grape must, facilitated by a diverse consortium of microorganisms naturally present on the grapes. The predominant wine yeast, *S. cerevisiae*, is involved in the formation of natural wines [5]. The wine yeast *S. cerevisiae* is usually very little in natural musts, just less than 1% of the total population of active microorganisms. Grape must contain a huge biodiversity of microorganisms, including wild yeasts of the genus *Hanseniaspora, Candida, Metschnikowia, Pichia, Rhodotorula,* and *Torulaspora*, as well as various bacteria and molds [6]. In recent years, natural wines have become an increasingly important range of wines on the market. They hold great appeal to both connoisseurs and ordinary consumers due to their unique and distinctive qualities [7].

However, spontaneous fermentation does not always produce the expected results. The predominance of wild yeasts that intensely produce higher alcohols can lead to wine spoilage, intense chemical aromas, and vinegary taste sensations. Yeast is responsible for the production of several hundred chemical compounds, and any imbalances can pose risks of unfavorable flavors and aromas [3,5,8].

The addition of sulfur compounds in food products exposes allergies in sensitive consumers [9]. Therefore, alternative methods that do not leave behind preservatives in the final product are being actively pursued. While food irradiation has been commercially employed since the 1950s, the technology is still under development. This applies to the irradiation of new product categories such as wine [9], raw milk [10], as well as lyophilized fruit, vegetables, meat [11] and honey [12].

Furthermore, new areas of application such as irradiation to prevent the spread of pests, especially in the case of tropical fruits and vegetables [13], extending the shelf life of fresh products packed in modified atmospheres [14], and preparing sterilized, shelf-stable food for patients with compromised immune systems or NASA astronauts [15]. One of the significant advantages of food irradiation technology is that it is a nonthermal process, capable of replacing chemical methods or steaming, as demonstrated in the case of dried herbs [16]. Food irradiation is a process of exposing food to ionizing energy to eliminate insects, fungi, or bacteria that can cause human diseases or spoilage, as well as delay the ripening of fresh products. The approved sources of ionizing radiation for food irradiation, as listed in General Standard for Irradiated Foods [17], include gamma rays from radionuclides such as Co-60, X-rays generated from machine sources operating at or below 5 MeV energy level, and electrons generated from machine sources operating at or below 10 MeV energy level.

During the radiation process, microorganisms are effectively killed, and the ripening, germination, and spoilage of vegetables are inhibited [18,19]. The effectiveness of ionizing radiation depends on the species of microorganisms, the dose, and the intensity of the radiation [20]. Mold fungi such as *Fusarium oxysporum, Phytophthoracitricola, Pythium ultimum,* and *Botrytis cinerea* have shown sensitivity to irradiation within the range of 1.5–6 kGy [21]. Three species of *Escherichia coli* O157:H7 suspended in apple juice were sensitive even to a dose of 1 kGy, while complete elimination was achieved at a dose of 2 kGy [20]. Spore-forming bacteria from the genus *Clostridium* and *Enterobacteriaceae* proved resistant to irradiation, with a dose of 4–5 kGy reducing their populations by 90%. Complete elimination of these microorganisms required a minimum dose of 10 kGy [22]. The worldwide development of food irradiation can be attributed to the growing utilization of machine sources of ionizing radiation instead of radioisotopes. Recently, the use of low-energy electrons has gained significant attention. This technology, due to the limited penetration of electrons with energy below 300 keV, is used for surface microbial decontamination [23].

The aim of the experiment was to determine the optimal irradiation dose for the elimination of *S. cerevisiae, S. cerevisiae* var. *bayanus, B. bruxellensis,* and wild yeasts while ensuring the preservation of wine quality. The analysis encompassed an evaluation of the wine's chemical composition, color, and organoleptic characteristics.

2. Results

Four doses of ionizing radiation (1, 2.5, 5.0, and 7.5 kGy) were used to reduce the yeast content in the wine. Upon applying the lowest dose of 1 kGy, a decrease in yeast content was observed (Figure 1). The level of reduction varied among the strains, with *S. cerevisiae* ES181 and *B. bruxellensis* exhibiting greater resistance to the first dose, as their numbers decreased by 40 and 50%, respectively.

In contrast, the strains *S. cerevisiae* ES123 and *S. cerevisiae* var. *bayanus* proved higher sensitivity to irradiation. In these cases, the yeast abundance was nearly 16 and 20%, respectively, relative to the initial value before the application of the physical factor. Notably, a dose of 2.5 kGy significantly reduced the yeast abundance, with only the *S. cerevisiae* ES181 strain remaining at a similar level, representing approximately 60% of the initial value.

The other strains were significantly reduced to levels of a few percent compared to the initial value. Once again, *S. cerevisiae* var. *bayanus* proved the highest sensitivity. A subsequent dose of 5 kGy resulted in nearly complete yeast mortality, with only a 3% reduction observed in the case of *S. cerevisiae* ES181. Finally, a dose of 7.5 kGy left minimal traces of alive yeast cells in each wine sample.

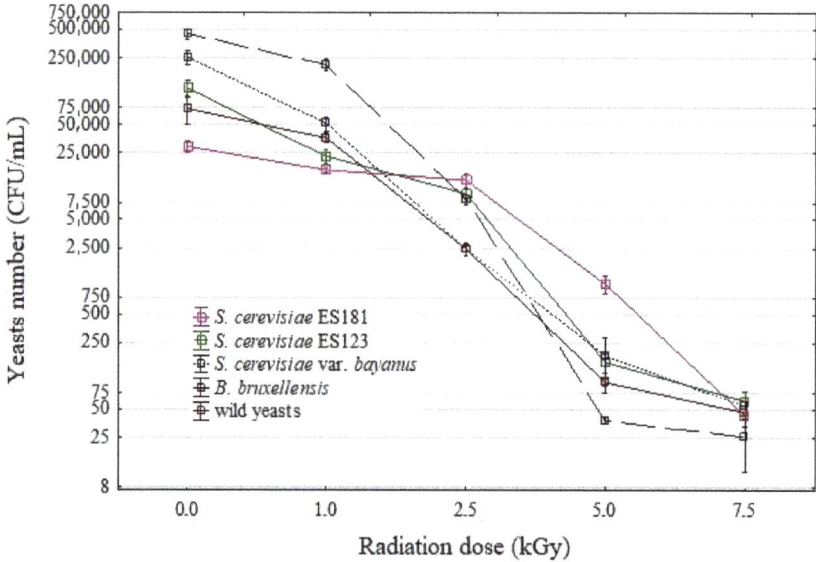

Figure 1. Effect of radiation on the yeast content of wine. The mean and standard deviation are marked.

2.1. Quality of Irradiated Wine—Selected Parameters

Regardless of the yeast used, ionizing radiation from a dose of 2.5 kGy already had a negative effect on wine quality. This is indicated by a decreasing polyphenol content and changes in the color of wine (Figures 2–4. The most significant changes were observed in anthocyanins and phenolic acids (Figure 3). On the other hand, the lowest dose of 1 kGy had no significant effect on the changes in wine color (Figures 3 and 5), which could be considered advantageous. However, this dose did not achieve a satisfactory reduction in yeast content (Figure 1), rendering its practical use impractical. The higher doses used resulted in a substantial reduction in yeast abundance (Figure 1).

The loss of polyphenols and changes in the color of the wine, characterized by a darkening effect, was observed gradually. With each radiation dose, there was a diminishing presence of polyphenols and red coloring pigments in the wine. A dose of 5 kGy reduced the content of polyphenols by about 20% except for the wine with wild yeast, which exhibited no change in polyphenol content. Anthocyanins and phenolic acids were particularly sensitive to radiation, while other tested polyphenols were moderately sensitive (Figures 3 and 5). Following a dose of 7.5 kGy, the content of polyphenols decreased by approximately 40%. Flavan-3-ols and stilbenes also experienced a reduction of about 20%, while flavonols demonstrated a slight increase, though statistically insignificant (Figure 3).

The changes in polyphenol content and wine color were similar in wines where commercial yeast was used (Figure 2). The polyphenol content experienced a decrease, resulting in lighter and less intensely colored wines. Among the wines subjected to a radiation dose of 7.5 kGy, those inoculated with *S. cerevisiae* ES181 yeast displayed the least reduction in polyphenol content (27.8%), while *S. cerevisiae* 2 wine exhibited the greatest reduction (41.3%). The greatest changes were found in the category of compounds classified as anthocyanins, which are mainly responsible for the color of the wine (Supplementary Table S1).

This observation is further supported by the change in color, as indicated by the L* parameter (Figure 4a). The wines exhibited a lighter shade, and the values of the chromatic color parameters a* and b* decreased (Figure 4b). The smallest changes in the a* and b* color parameter values were observed when the lowest radiation dose of 1 kGy was applied, with values similar to those of nonirradiated wine. In contrast, the most significant changes in color parameters occurred after the subsequent dose of 2.5 kGy.

The greatest color change was observed between these radiation doses, with subsequent doses yielding less pronounced color changes. In the wine produced with natural yeast, even after a radiation dose of 5.0 kGy, the total polyphenol content remained at a similar level to that of untreated wine. However, analysis of individual polyphenolic compounds revealed a decrease in anthocyanin content and an increase in phenolic acids (Supplementary Table S1).

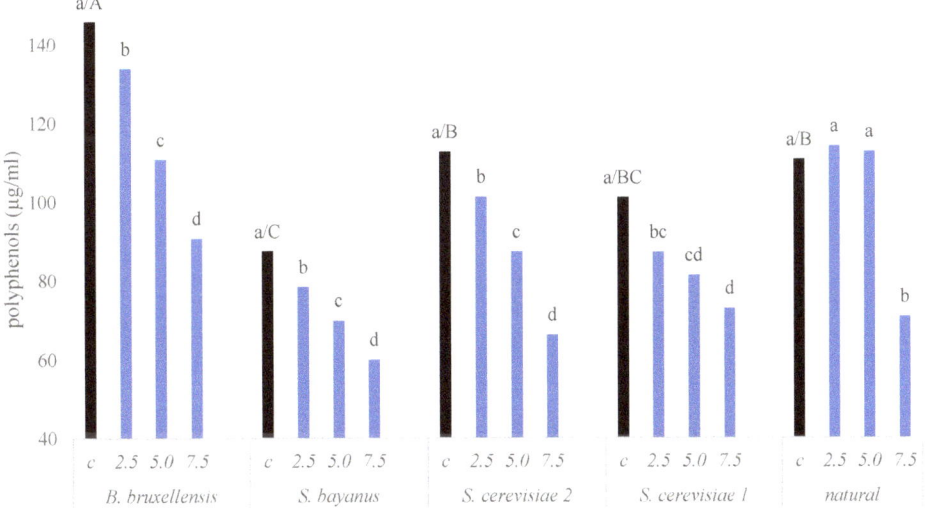

Figure 2. Changes in the content of polyphenolic compounds, depending on the yeast used and the dose of ionizing radiation given. Lowercase letters indicate homogeneous groups within the yeasts and uppercase letters indicate groups between the yeasts.

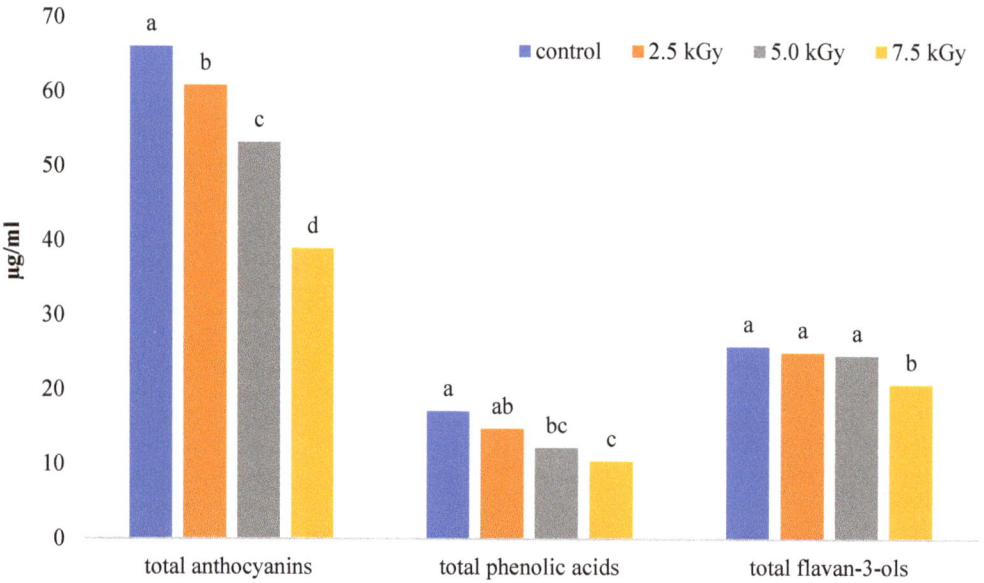

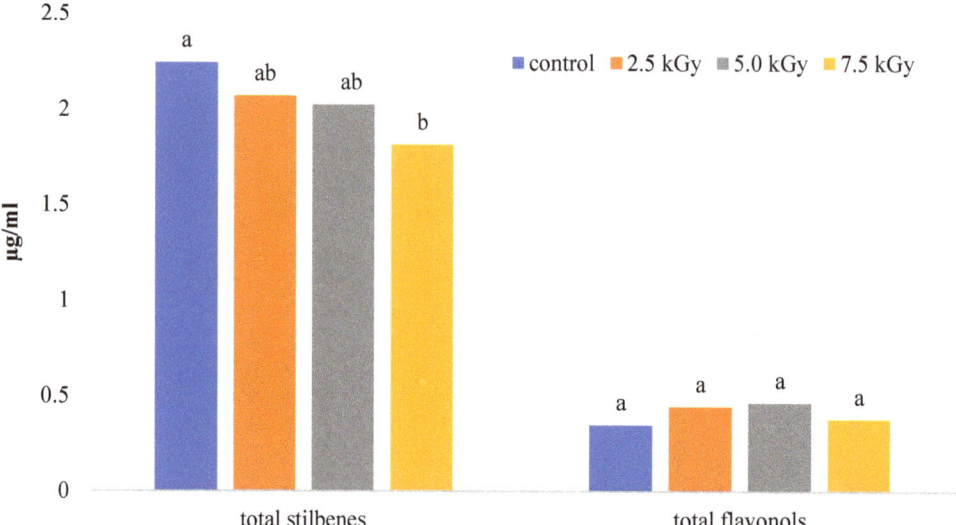

Figure 3. Content of polyphenolic compounds in relation to the ionizing radiation dose. Lowercase letters indicate homogeneous groups within a group of polyphenolic compounds.

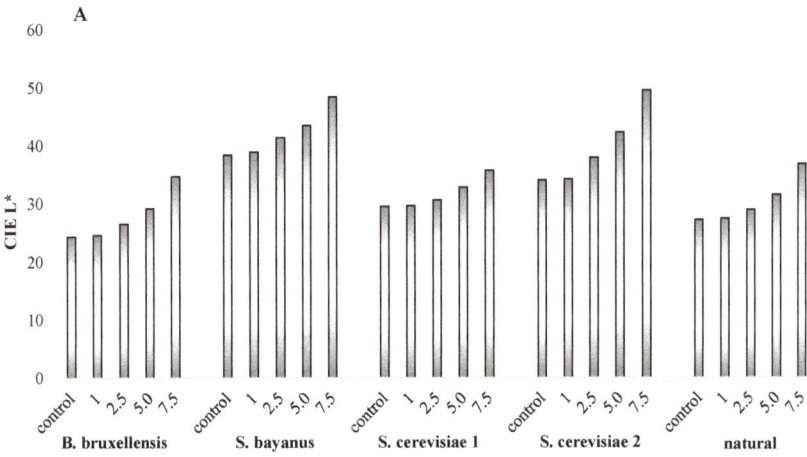

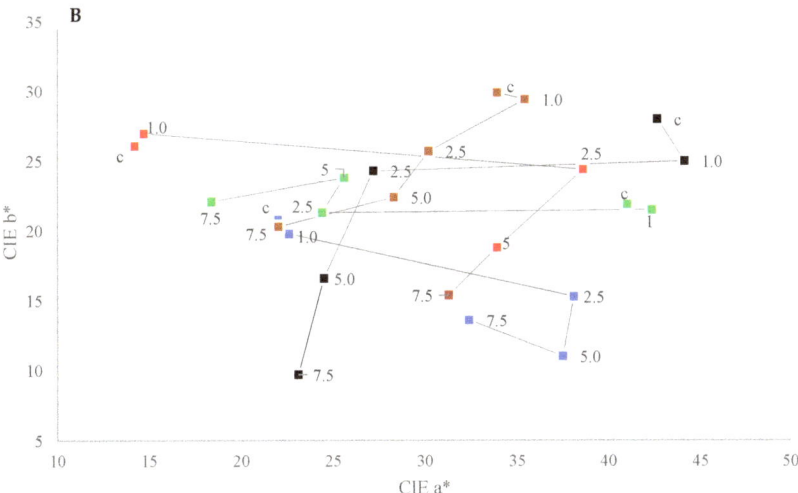

Figure 4. Changes in wine color parameters after the application of ionizing radiation: a monochromatic parameter CIE L* (**A**), chromatic parameter CIE a* and b* (**B**). Types of yeast: *B. bruxellensis*; *S. bayanus*; *S. cerevisiae* ES181; *S. cerevisiae* ES123; wild yeast.

2.2. Influence of Yeast Bread on Color, Polyphenol Content, and Selected Chemical Compounds Included in the Wine Aroma

Four yeast strains, including three strains of *S. cerevisiae* and *B. bruxellensis*, were utilized for fermentation. After fermentation, the wine was analyzed for various parameters, including polyphenols, glycerol, alcohols (isoamyl, iso-butanol, n-propanol, ethyl acetate, isoamyl acetate, isobutyl acetate, acetoin, acetaldehyde, 2 and 3-methylbutanol), and color related factors. The wine treated with *B. bruxellensis* had the highest amount of polyphenols (approximately 140 μg/mL), this wine was also characterized by its dark red color (Figures 2, 5 and 6). The natural wine was also characterized by a high polyphenol content (about 30 μg/mL less than in the wine after using *B. bruxellensis*). In contrast, the wine after using *S. bayanus* had the least polyphenol content and appeared the brightest (Figures 3 and 5a). *S. bayanus* and *S. cerevisiae* ES181 influenced that the wine had the least intense red color.

In the wines examined, a total of 32 polyphenolic compounds were identified and classified into five groups: seven anthocyanins, eight phenolic acids, six flavonols, seven flavan-3-ols, and five stilbenes. Among these compounds, anthocyanins are primarily responsible for the color of pink and red wines. They represented the largest group of polyphenolic compounds in the wines studied (Figure 3). The wines fermented *with B. bruxellensis* exhibited the highest number of anthocyanins (91.78 µg/mL), while the wines with *S. bayanus* had the lowest concentration of anthocyanins (40.31 µg/mL) (Supplementary Table S1).

The contents of the individual compounds in this group varied and depended on the yeast used. Malvidin 3,5-O-diglucoside, malvidin 3-O-glucoside, and cyanidin 3,5-O-diglucoside (Supplementary Table S1) were found to be the highest in all wines. In red grape cultivars, malvidin derivatives can contribute up to 85% of all anthocyanins [24]. The content of flavan-3-ols in the wines was relatively unaffected by the yeast used, ranging from 23 to 29 µg/mL. Despite the high polyphenol content, wines fermented with wild yeast exhibited the lowest levels of phenolic acids (6.33 µg/mL). In the other wines, the content of these compounds ranged from 16.37 to 22.43 µg/mL. Flavonols and stilbenes were the smallest groups of compounds present in the wines.

The highest amount of glycerol was present in natural wine (5.3 g/L) and the wine made after using *B. bruxellensis* and significantly less in wines with *S. cerevisiae* strains (Table 1). The least amount of higher alcohols were found in wines after the use of *S. cerevisiae* yeast (about 130 mg/L on average), while more than twice as much was found in natural wine and the wine after the use of the *B. bruxellensis* strain (Table 1). Isoamyl alcohol was highest in all wines, ranging from 88 to 252 mg/L (*S. cerevisiae* 2 and wild yeast, respectively). Propanol and isobutanol were also present in the highest concentrations in natural wines and wines containing *B. bruxellensis*.

According to Table 1, aryl alcohols (the sum of 2-methylbutanol and 3-methylbutanol) were highest in natural wine, with a concentration of about 100 mg/L, while wines made with classical *S. cerevisiae* strains had lower levels, averaging around 69 mg/L. Acetaldehyde content varied depending on the yeast used, ranging from approximately 29 (*S. cerevisiae* 2) to 63 mg/L (wild yeast), with the highest concentration observed in *B. bruxellensis* wine at 67 mg/L. Ethyl acetate was most abundant in wine produced with wild yeast (53.2 mg/L), but high levels were also found in wine fermented with *B. bruxellensis* (47.1 mg/L). *S. cerevisiae* wines also contained ethyl acetate and acetaldehyde, but at lower levels compared to the other yeast strains.

The blind organoleptic evaluation of the wines showed that all wines received high scores in terms of clarity (Figure 5). However, the other parameters, such as taste and aroma, were strongly influenced by the breed of yeast used. The wine made with wild yeast received the lowest rating, particularly in terms of taste and aroma. This is consistent with its chemical composition and high content of undesirable substances, as shown in Table 1. The wines fermented with *S. cerevisiae* yeast were rated the highest in overall quality.

Table 1. Fermentation byproducts.

Fermentation Byproducts	B. bruxellensis	S. bayanus	S. cerevisiae 1 (ES 181)	S. cerevisiae 2 (ES 123)	Wild Yeasts
Isoamyl (mg/L)	177 d	126 c	88 a	97 b	252 e
Iso-butanol (mg/L)	40.1 e	26.3 c	23.0 b	19.6 a	35.5 d
N-propanol (mg/L)	52.3 c	28.9 b	18.7 a	16.2 a	78.6 d
Total	269.4 C	181.2 B	129.7 A	132.8 A	366.1 D
Glycerol (g/L)	4.53 d	4.28 c	3.90 b	2.81 a	5.34 e
Ethyl acetate (mg/L)	47.1 c	37.8 b	26.5 a	22.9 a	53.2 d
Isoamyl acetate (mg/L)	0.18 c	0.14 b	0.13 ab	0.11 a	0.22 d
Isobutyl acetate (mg/L)	0.55 cd	0.49 bc	0.68 de	0.72 e	0.24 a
Acetoine (mg/L)	0.31 cd	0.35 d	0.12 a	0.17 b	0.27 c
Acetaldehyde (mg/L)	67.4 d	49.7 c	33.5 b	29.6 a	63.0 d
2 and 3-Methyl-butanol (mg/L)	98.4 c	78.4 b	66.8 a	72.5 ab	112.2 d

Explanation: Mean values denoted by the same letter do not differ statistically significantly at 0.05, according to the t-Tukey test.

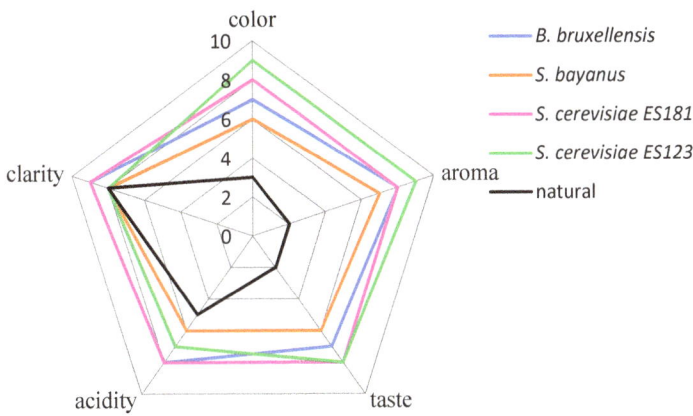

Figure 5. The organoleptic test results.

Figure 6. Colors of must/wine after the maceration period.

3. Discussion

3.1. Effects of Ionizing Radiation on Yeast and Polyphenols

In a study by Błaszak et al. [9], ionizing radiation at a dose of 1 kGy reduced the number of yeast by approximately 95%, and complete elimination of yeast was observed after the application of 2.5 kGy. Similarly, in the experiment presented here, 1 kGy significantly reduced yeast viability, with between 40 and 85% remaining (depending on the yeast strains used). The radiation dose of 2.5 kGy had a significant effect on reducing yeast abundance, with only a few percent of yeast remaining compared to the control wine. The exception was with the strain *S. cerevisiae* ES181, as radiation at a dose of 2.5 kGy only reduced the number of these yeasts by 40%, with minimal impact on the degradation of polyphenols. Therefore, based on the results of Błaszak et al. [9] and those presented in this work, a radiation dose of 2.5 kGy is the best option for wine preservation. This dose effectively kills yeast while preserving phenols.

Electron beam irradiation affected the degradation of tannins in the species, simultaneously increasing the content of phenolic acids. Additionally, the changes in anthocyanin content within the wine corresponded to a change in its color. Morata et al. [25] also observed a decrease in polyphenol content and color changes in wine exposed to a higher dose of irradiation.

The electron beam irradiation of spices affected the degradation of tannins while increasing the content of phenolic acids. Phenolic acids underwent degradation only after the application of a high dose of radiation—10 kGy [26]. The change in anthocyanin content in this wine also manifested as a change in wine color. The wine became brighter, which may indicate its oxidation. The degradation products of these compounds caused by radiation or the formation of free radicals may be formed [27]. Morata et al. [25] also observed a decrease in polyphenol content and changes in wine color after subjecting it to higher radiation doses.

3.2. Influence of Yeast on the Bouquet of Wine

In line with the Georgian/natural/homegrown wine tradition, some winemakers have started to explore alternative yeast strains beyond the standard *Saccharomyces* for wine production. These wild yeasts contribute to spontaneous fermentation, imparting unique flavors and aromas to the wines [28]. However, non *Saccharomyces* yeasts produce various metabolites that can worsen the quality of the wine, including undesirable flavors such as acetic acid, acetoin, ethyl acetate, and acetaldehyde, as well as off-putting aromas associated with vinyl phenols and ethyl phenols, which are linked to the development of *Brettanomyces/Dekkera* yeast strains. The use of such yeast raises concerns regarding both the safety and validity of their application [29,30].

During the process of alcoholic fermentation, glycerol emerges as the third most quantitatively produced compound, following ethanol and carbon dioxide. Its content ranges from 1 to 15 g/dm^3 [31]. Glycerol is odorless and therefore has no effect on the aroma of the wine. However, owing to its thick and oily texture, it influences the flavor of the wine, imparting a smooth, velvety aftertaste, and enhancing the overall richness and fullness of flavor. Glycerol concentrations tend to be considerably higher in red wines compared to white wines. The sensory threshold for glycerolize, the perception of glycerol, has been established at 5.2 g/dm^3 [32,33].

The level of glycerol in wine products depends on many factors, including the type of yeast strain employed, sugar levels, must pH, SO_2 content, and fermentation temperature [34]. In our study, all wines were produced from the same must under identical fermentation conditions, maintaining a temperature of 12 ± 0.5 °C. The only variable was the breed of yeast strain. Remarkably, the wine fermented with wild yeast exhibited the highest glycerol content, as indicated in Table 1). The high glycerol content in the wines gave pronounced teardrops, which are characteristic forms formed on the walls of the glass when droplets flow back into the glass. Wines produced with *S. cerevisiae* had the least amount of glycerol.

One of the main groups of compounds synthesized by yeast is higher alcohols, also known as fusel alcohols. These alcohols exhibit an intense aroma that plays an important role in shaping the bouquet of wine. At low concentrations (below 300 mg/dm^3), they have a positive effect on the aroma, while higher concentrations can mask the inherent aroma of the beverage [29].

Only the wine produced with wild yeast exceeded levels of 300 mg/L for these alcohols. This confirms the opinion that high concentrations can be unfavorable, as this wine received the worst judgment from testers in terms of both aroma and taste (Figures 1 and 2). Once again, wines made with *S. cerevisiae* yeast had the least amount of these alcohols (Table 1). Among these alcohols, isoamyl alcohol was highest in all wines, ranging from 88 to 252 mg/L (*S. cerevisiae* 2 and wild yeast, respectively).

Only the wine produced with wild yeast exceeded 300 mg/L of these alcohols. This confirms the opinion that high concentrations can be unfavorable, as this wine received the worst judgment from testers in terms of both aroma and taste (Figures 1 and 2). Once again, wines made with *S. cerevisiae* yeast had the least amount of these alcohols (Table 1). Among these alcohols, isoamyl alcohol was highest in all wines, ranging from 88 to 252 mg/L (*S. cerevisiae* 2 and wild yeast, respectively).

According to other authors [35], isoamyl alcohol usually is above 70% of the total of these compounds, with a content range of 12–310 mg/L. In the wines studied, it ranged from 67 to 73%. The presence of propanol in wines is typically between 10 and 125 mg/dm^3, while isobutanol ranges from 15 to 175 mg/L [1,35].

In the wines studied, these alcohols were found at the lower end of these ranges. The amount of fusel alcohols produced during fermentation is significantly influenced by the breed of yeast while reducing their concentration can be achieved by using nitrogenous fermentation media [36].

Apart from isobutanol, another compound that has a particularly negative effect on the sensory properties of fermented beverages is aryl alcohols (the sum of 2 and 3-methylbutanol).

They contribute to an unpleasant solvent-like aroma and taste [37]. Once again, wines fermented with wild yeast were characterized by the highest content of these compounds.

Aldehydes present in grapes play a crucial role in the formation of cultivar aromas, and their levels in young wines typically do not exceed 75 mg/L. Among this group of compounds, acetaldehyde stands out as the most significant. In our study, its concentration ranged from 29.6 (*S. cerevisiae* 2) to 63 mg/L (wild yeast) and 67.4 mg/L (*B. bruxellensis*). Acetaldehyde plays a vital role in the stabilization of wines, especially red wines, during aging. It accelerates the polymerization reaction of anthocyanins and phenols, contributing to the overall quality and structure of the wine [38].

Ethyl acetate, when present in low concentrations, enhances the fruity aromas in beverages, such as pear, peach, pineapple, and raspberry. However, at higher concentrations, it can develop an odor reminiscent of varnish or nail polish remover. The acetic acid esters that contribute to aroma are isoamyl acetate—banana and isobutyl acetate—fruit [39]. Once again, the wine fermented with wild yeast displayed notable distinctions in terms of these compounds. Isoamyl acetate exhibited the highest concentration in this wine, while isobutyl acetate demonstrated the lowest concentration.

Acetoin affects the buttery aroma of wine and is produced by lactic acid bacteria (*Oenococcus oeni*) from citric acid, thereby enhancing the buttery aroma of the wine [40]. The presence of yeast cultures other than *Saccharomyces* can influence the production of various metabolites that affect wine quality, including acetoin, ethyl acetate, and acetaldehyde, as well as compounds associated with unpleasant odors such as ethyl phenols, which are associated with the development of Brettanomyces/Dekkera [41,42]. Our study aligns with these findings, as wines fermented with *S. cerevisiae* exhibited lower levels of acetoin, as well as ethyl acetate and acetaldehyde (Table 1).

The fermentation process with *B. bruxellensis* yeast was observed to be the fastest and most rapid compared to other inoculated yeasts. The rapid fermentation may have increased the ethanol content, facilitating compounds' extraction from the fruit. The fastest change in extract content was observed. Notably, this particular wine displayed the highest polyphenol content (Figure 3) and had the darkest color (Figure 4). In contrast, wines fermented with *S. bayanus* yeast had the least polyphenols and appeared brightest. These wines also characterized the lowest values for the a* and b* color parameters (Figure 4). In some fermentation processes, such as beer production, the presence of *B. bruxellensis* can be considered beneficial, as this strain contributes to the specific characteristics and aromas of these specialty beverages. Likewise, unique wines that owe their specific aroma to compounds like 4-ethylphenol, 4-ethylguaiacol, and tetrahydropyridine produced by *B. bruxellensis* also find connoisseurs [43,44].

The fermentation process of wine with wild yeast can also be influenced by the cultivation method. In our experiment, the fruit was harvested from an organic plantation, where there was/may have been greater microbial biodiversity. The must on wild yeast fermented rapidly and turbulently. The polyphenol content is comparable to wines made by *S. cerevisiae* and higher than in wines made by *S. bayanus*.

An organoleptic evaluation of the five parameters performed blind showed that all wines scored highly only in terms of clarity (Figure 5). However, the other parameters were strongly influenced by the breed of yeast used. By far the lowest rating was given to the wine made with wild yeast. In particular, taste and aroma were rated very low. This was confirmed by its chemical composition and high content of undesirable substances (Table 1). The wines in which *S. cerevisiae* yeast was used were rated best. Spontaneous fermentation has a beneficial effect when carried out in vineyards that have relied on indigenous wild yeast strains for generations.

Over the years or even decades, a distinctive and unique microbiome is produced in the vineyard, and so fermentation also has a predictable course to a certain extent. Such natural wines are of great interest to connoisseurs and are very expensive [45].

Apart from possessing strong antioxidant properties, polyphenols also influence the sensory characteristics of food products [46]. The polyphenol content and the colour of the wine were decisively influenced by the yeast breeds used (Figures 3 and 4). The yeasts used during alcoholic fermentation have a strong influence on the rate of alcohol production. During fermentation from the skins, extraction of the constituents in the skins also occurs [29]. It was observed that the process of turbulent fermentation was the fastest and most rapid in most inoculating with *B. bruxellensis* yeast. The rapid fermentation may have increased the ethanol content, facilitating compounds' extraction from the fruit.

The fastest change in extract content was observed in this setting/ must/wine. This wine had the highest polyphenol content (Figure 3) and was the darkest (Figure 4). The wine fermented with *S. bayanus* yeast was the least polyphenols and the lightest. This wine also had the lowest a* and b* color parameters (Figure 4).

In the wines studied, 32 polyphenolic compounds were identified and classified into 5 groups: 7 anthocyanins, 8 phenolic acids, 6 flavonols, 7 flavan-3-ols and 5 stilbenes. In rosé and red wines, anthocyanins are mainly responsible for the colour. They constituted the largest group of polyphenolic compounds in the wines studied (Figure 3). Their content reflects the content of total polyphenols. They were highest in wines fermented with *B. bruxellensis* (91.78 µg/mL) and lowest in wines with *S. bayanus* (40.31 µg/mL) (Supplementary Table S1). The content of the individual compounds in this group varied and depended on the yeast used.

Malvidin 3,5-*O*-diglucoside, malvidin 3-*O*-glucoside and cyanidin 3,5-*O*-diglucoside (Supplementary Table S1) were the most abundant in all wines. In red grape cultivars, malvidin derivatives may account for as much as 85% of all anthocyanins [24]. Flavan-3-ols were another group of compounds, but their content in wine was little affected by yeast (23–29 µg/mL). Despite the high polyphenol content, wines with wild yeast had the least phenolic acids (6.33 µg/mL). In the other wines, the content of these compounds ranged from 16.37 to 22.43 µg/mL. Flavonols and stilbenes were the smallest groups of compounds.

The fermentation process of wine with wild yeast can also be influenced by the cultivation method. In our experiment, the fruit was harvested from an organic plantation, where there was/may have been greater microbial biodiversity. The must on wild yeast fermented rapidly and turbulently. The polyphenol content is comparable to wines made on *S. cerevisiae* and higher than those made on *S. bayanus*. In the experiment of Sterczyńska et al. [47], the fruit was harvested from a conventional plantation. The fermentation process on wild yeast was slower, the wine had less alcohol and polyphenols compared to wine made on *S. cerevisiae*.

4. Materials and Methods

4.1. Characteristics of the Area of Research and Plant Material

The grape was harvested at the research station West Pomeranian University of Technology in Szczecin located in the north-western part of Poland. The majority of the West Pomeranian Province belongs to the 7A zone on Heinz and Schreiber's "Map of zones of plant resistance to frost". However, in the area of Szczecin and in the nearby northern region, minimal temperatures range from $-12\,°C$ to $-15\,°C$, which corresponds to values typical of zone 7B. The average temperature during the growing season (April-October) between 1951 and 2012 was 13.7 °C and rainfall was 391 mm [48]. The soil in the vineyard was agricultural soil with a natural profile, developed from silt-loam, pH 6.9 higher water capacity and optimal mineral content [49].

The vines were grafted on SO_4 rootstock and planted in 2016 with a North-South row orientation at 1.01 m × 2.18 m. The vines were pruned with a Guyot (one arm) training system and vertically positioned with eight shoots, each had two clusters. Standard vineyard management methods for organic plantations were used in both growing seasons.

4.2. Description of the Variety and Production of Wine

The study involved the dark-skinned vine cultivar Souvignier Gris, which is a German cultivar with increasing interest in its cultivation in cool climate areas. The vine is valued especially due to its high fungus- and cold-resistant. 'Souvignier Gris' is a cross between the grapes Cabernet Sauvignon and Bronner. Berries of 'Souvignier Gris' were harvested in October (25.4 Brix) and immediately crushed in order to prepare grape must. Grape must be inoculated with commercial yeast or left to wild yeast. The dry active wine yeast (30 g/hL) was prepared with 150 mL of water at 35 °C and added to the grape must.

The wine was prepared in steel containers with a volume of 50 L. The fruits were separated from the stalks, crushed and then macerated for 3 days at 14 °C. Then they must be pressed on the wine press. The experiment with e-beam irradiation of the wine was performed four weeks after the initiation of alcoholic fermentation.

4.3. Yeast: Assessment of Their Numbers in the Wine

The test yeast, *S. cerevisiae* (ES181; ES 123—ES Viniarske Potrebys.r.o.), and *S. cerevisiae* var. *bayanus* (Fermivin LS2—Brovin) are strains that are widely used for the vinification of grape musts on the industrial scale. These yeast strains are characterized by high tolerance of the alcohol content (16.5%) and sugar content (300 g/L) in the culture medium. *B. bruxellensis* (WLP650—White Labs) was also used in the experiments. Inoculation was carried out 12 h after the wine decontamination process was performed. Radiation was applied at doses of 1.0, 2.5, 5.0, and 7.5 kGy. The yeast counts in the wine were recorded in accordance with ISO 21527–1:2008 [50]. A specialized yeast culture medium was used—YPG Agar (Sigma-Aldrich, Melbourne, Australia). Wine samples were taken after serial decimal dilutions and were then added to the microbial medium (deep inoculation) and then incubated for 3 days at 25 °C; the colony-forming units (CFU) were then counted using the eCount Colony Counter (AllChem, Beirut, Lebanon). The experiments were performed in triplicate.

4.4. Irradiation

The Institute of Nuclear Chemistry and Technology (Warsaw, Poland) has unique devices and elaborate procedures for the process of irradiation, ensuring high efficiency of sterilization and microbiological decontamination. The Accelerator ELEKTRONIKA 10/10 is a high-power radiation device that allows electron beams with 9 MeV energy and average power of up to 10 kW to be obtained. These parameters allow the irradiation process to be performed at a commercial scale. The main parameters of the ELEKTRONIKA 10/10 accelerator: are pulse electron beam mode; electron energy of 8–10 MeV; average beam power of 10 kW; dose rate of 700 Gy/s.

The linear electron accelerator Elektronika 10-10 was used for wine irradiation with doses in the range of 1.0–7.5 kGy (in triplicate). The defined doses were controlled using graphite calorimeters from RISO High Dose Reference Laboratory, in Denmark. For the determination of electron energy and dose uniformity RISO B3 dosimetric foil was measured with a flatbed scanner and RisoScan software was used. The uncertainty of dose measurements was about 10% due to the dosimetric system and the instability of irradiation conditions. The electron energy used for the irradiation of yeasts in wine was 9 MeV. Wine samples were packed in 5 mL plastic tubes. Each tube was placed horizontally to the beam to ensure as high as possible dose uniformity, determined as dose maximum to dose minimum ratio. In the conditions of the experiments, the dose uniformity was 1.1. The doses presented are the average doses calculated based on the depth dose profile of the beam in water.

4.5. Color Measurement

The color parameters were L^* ($L^* = 100$ indicates white; $L^* = 0$ indicates black), a^* ($+a^*$ indicates redness; $-a^*$ indicates greenness), b^* ($+b^*$ indicates yellow; $-b^*$ indicates blue). Color coordinates were determined in the CIE $L^*a^*b^*$ space for the 10° standard observer and the D 65 standard illuminant. CIE $L^*a^*b^*$ was measured using a Konica Minolta CM-700d spectrophotometer [49].

4.6. Identification of Compounds in Wine

4.6.1. Phenolics

Identification and quantification of the polyphenol values of the extracts were carried out using an ACQUITY Ultra Performance LC system, equipped with a photodiode array detector with a binary solvent manager (Waters Corporation, Milford, MA, USA) and a mass detector equipped with an electrospray ionization (ESI) source operating in negative and positive modes, as described by Oszmiański et al. [51]. Separation of the individual polyphenols was carried out using a UPLC BEH C18 column (1.7 m, 2.1 × 100 mm, Waters Corporation, Milford, MA, USA) at 30 °C. The samples (10 L each) were injected, and then the elution was completed in 15 min with a sequence of linear gradients and isocratic flow rates of 0.45 mL/min The mobile phase consisted of solvent A (2.0% formic acid, v/v) and solvent B (100% acetonitrile). The program began with isocratic elution using 99% of solvent A (0–1 min), then a linear gradient was used until the 12-min point, lowering solvent A to 0%; from 12.5 to 13.5 min, the gradient returned to the initial composition (99% A), then it was maintained at a constant level to re-equilibrate the column. The analysis was carried out using full-scan, data-dependent MS scanning from m/z 100 to 1500. Leucine enkephalin was used as the reference compound at a concentration of 500 pg/L and at a flow rate of 2 L/min; the $[M-H]^-$ ion was detected at 554.2615 Da. The $[M-H]^-$ ion was detected during a 15-min analysis performed within ESI–MS accurate mass experiments, which were permanently introduced via the Lock Spray channel using a Hamilton pump. The lock mass correction was ± 1.00 for the mass window. The mass spectrometer was operated in negative- and positive-ion modes, set to the base peak intensity (BPI) chromatograms, and scaled to 12,400 counts per second (cps) (100%). The optimized MS conditions were as follows: capillary voltage of 2500 V; cone voltage of 30 V, source temperature of 100 °C; desolvation temperature of 300 °C; desolvation gas (nitrogen) flow rate of 300 L/h. Collision-induced fragmentation experiments were performed using argon as the collision gas, with voltage ramping cycles from 0.3 to 2 V. Characterization of the single components was carried out via the retention time and the accurate molecular masses. Each compound was optimized to its estimated molecular mass $[M-H]^-/[M+H]^+$ in the negative and positive modes before and after fragmentation. The data obtained from UPLC–MS were subsequently entered into the MassLynx 4.0ChromaLynx Application Manager software. On the basis of these data, the software is able to scan different samples for the characterized substances. The runs were monitored at the wavelength for flavonol glycosides of 360 nm. The PDA spectra were measured over the wavelength range of 200–800 nm, in steps of 2 nm. The retention times and spectra were compared to those of the pure standard [51].

4.6.2. Glycerol

Analyses were performed using a Shimadzu NEXE-RA XR instrument (Kyoto, Japan) with an RF-20A refractometric detector. Separation was carried out on an Asahipak NH2P-50 250 × 4.6 mm Shodex column (Showa Denko Europe, Munich, Germany). For quantitative measurements, the standard curves prepared for the glycerol standard were used.

4.6.3. Volatile Compound Analysis

Qualitative analysis of the volatile aromas of the two highest-scoring wood-aging wines was carried out using the gas chromatography-mass spectrometer (GC-SPME) technique, using a device equipped with a flame ionization detector (FID) and a DB-WAX capillary column, with helium as a carrier gas. The oven temperature was kept at 40 °C for 7 min, followed by an increase to 230 °C at a rate of 3 °C/min.

4.7. Sensory Evaluation

The wines were subjected to sensory evaluation. A group comprising 35 tasters evaluated the quality of the wine. Before starting the sensory evaluation of the wines, the testers were trained and informed about the purpose of the assessment. The people who made the assessment were not professional testers. Wine samples (30 mL) were evaluated in 100 mL wine glasses. Color, aroma, flavor, acidity, and clarity were assessed on a scale of 1 to 10 (with 10 being the best score for a given characteristic). The arithmetic mean for each trait of wine quality was calculated on the basis of individual assessments and a chart was developed.

4.8. Statistical Analysis

All statistical analyses were performed using the Statistica 12.5 software (StatSoft Polska, Cracow, Poland). The data were then subjected to one-factor variance analysis (ANOVA). Mean comparisons were performed using Tukey's least significant difference (LSD) test; significance was set at $p < 0.05$.

5. Conclusions

Ionizing radiation effectively eliminated the yeast in the wine. The optimal dose was determined to be 2.5 kGy, resulting in a reduction of yeast population by over 90%, except for one strain. Importantly, the quality of the wine was not significantly reduced. Radiation applied at higher doses worsened the organoleptic and chemical properties of the wine. As the radiation dose increased, the yeast in the wine progressively and rapidly declined. The differences in yeast response are related to the extent of radiation exposure.

Among the yeast breeds studied, only the *S. cerevisiae* ES181 strain exhibited rela-tively higher resistance to radiation doses of 1 and 2.5 kGy. In contrast, the other yeast breeds were subject to reduction, starting from 1 kGy, although to varying degrees. Therefore, it is reasonable to conduct a pretest to determine the sensitivity of the yeasts being used in the different levels of reduction.

In general, it can be said that the breed of yeast used has a very strong influence on the quality of the wine. It is justifiable to use commercial yeast breeds to obtain standard-quality wines. However, there are instances where the use of special yeast strains, such as *B. bruxellensis*, can be justified if the aim of vinification is to deliver or produce. *B. bruxellensis* strains rapidly fermented the sugars in the must, resulting in higher alcohol content. This extraction of alcohol contributes to the intensity of the red color in the wine, as it interacts with the anthocyanins.

Supplementary Materials: The following supporting information can be downloaded at: https://www.mdpi.com/article/10.3390/molecules28124867/s1, Table S1: The content of polyphenols [μg/mL].

Author Contributions: Conceptualization, M.B. and I.O.; methodology, M.B., B.J., I.O., S.L.-W., W.M. and U.G.; software, I.O. and M.B.; validation, M.B., I.O., S.L.-W., W.M. and U.G.; formal analysis, M.B. and I.O.; investigation, M.B. and I.O.; resources, M.B. and I.O.; data curation, M.B., I.O., S.L.-W., B.J., W.M. and U.G.; writing—original draft preparation, M.B., I.O., S.L.-W., B.J., W.M. and U.G.; writing—review and editing, M.B., I.O., B.J., S.L.-W.; supervision, I.O.; project administration, I.O.; funding acquisition, I.O. All authors have read and agreed to the published version of the manuscript.

Funding: The research was funded by project No. 00020.DDD.6509.00056.2019.16 carried out under the Rural Development Programme—Action WSPÓŁPRACA M16.

Institutional Review Board Statement: Not applicable.

Informed Consent Statement: Not applicable.

Data Availability Statement: Not applicable.

Conflicts of Interest: The authors declare no conflict of interest.

References

1. Ribéreau-Gayon, P.; Dubourdieu, D.; Donèche, B.; Lonvaud, A. (Eds.) *Handbook of Enology, Volume 1: The Microbiology of Wine and Vinifications*; John Wiley & Sons: Hoboken, NJ, USA, 2006; Volume 1, pp. 73–124.
2. Drtilová, T.; Ďurčanská, K.; Machyňáková, A.; Špánik, I.; Klempová, T.; Furdíková, K. Impact of different pure cultures of *Saccharomyces cerevisiae* on the volatile profile of Cabernet Sauvignonrosé wines. *Czech J. Food Sci.* **2020**, *38*, 94–102. [CrossRef]

3. Loureiro, V.; Malfeito-Ferreira, M. Spoilage activities of *Dekkera/Brettanomyces* spp. In *Food Spoilage Microorganisms*; Blackburn, C., Ed.; Woodhead: Cambridge, UK, 2006; pp. 354–398.
4. Barata, A.; Caldeira, J.; Botellheiro, R.; Pagliara, D.; Malfeito-Ferreira, M.; Loureiro, V. Survival patterns of *Dekkera bruxellensis* in wines and inhibitory effect of sulphur dioxide. *Int. J. Food Microbiol.* **2008**, *121*, 201–207. [CrossRef]
5. Lu, Y.; Sun, F.; Wang, W.; Liu, Y.; Wang, J.; Sun, J.; Mu, J.; Gao, Z. Effects of spontaneous fermentation on the microorganisms diversity and volatile compounds during 'Marselan' from grape to wine. *LWT-Food Sci. Technol.* **2020**, *134*, 110193. [CrossRef]
6. Francesca, N.; Sannino, C.; Settanni, L.; Corona, O.; Barone, E.; Moschetti, G. Microbiological and chemical monitoring of Marsala base wine obtained by spontaneous fermentation during large-scale production. *Ann. Microbiol.* **2014**, *64*, 1643–1657. [CrossRef]
7. Vilanova, M.; Cortés, S.; Santiago, J.L.; Martínez, C.; Fernández, E. Aromatic Compounds in Wines Produced During Fermentation: Effect of Three Red Cultivars. *Int. J. Food Prop.* **2007**, *10*, 867–875. [CrossRef]
8. Malfeito-Ferreira, M. Yeasts and wine off-flavours: A technological perspective. *Ann. Microbiol.* **2011**, *61*, 95–102. [CrossRef]
9. Błaszak, M.; Nowak, A.; Lachowicz, S.; Migdał, W.; Ochmian, I. E-Beam Irradiation and Ozonation as an Alternative to the Sulphuric Method of Wine Preservation. *Molecules* **2019**, *24*, 3406. [CrossRef]
10. Ward, L.R.; Van Schaik, E.; Samuel, J.; Pillai, S.D. Reduction in microbial infection risks from raw milk by Electron Beam Technology. *Radiat. Phys. Chem.* **2020**, *168*, 108567. [CrossRef]
11. Migdał, W.; Owczarczyk, H. Radiation decontamination of meat lyophylized products. *Radiat. Phys. Chem.* **2002**, *63*, 371–373. [CrossRef]
12. Migdał, W.; Owczarczyk, H.; Kędzia, B.; Hołderna-Kędzia, E.; Madajczyk, D. Microbiological decontamination of natural honey by irradiation. *Radiat. Phys. Chem.* **2000**, *57*, 285–288. [CrossRef]
13. Barkai-Golan, R.; Follett, P.A. *Irradiation for Quality Improvement, Microbial Safety and Phytosanitation of Fresh Produce*; Elsevier Science: Amsterdam, The Netherlands, 2017.
14. Smith, B.; Shayanfar, S.; Walzem, R.; Alvarado, C.Z.; Pillai, S.D. Preserving fresh fruit quality by low-dose electron beam processing for vending distribution channels. *Radiat. Phys. Chem.* **2020**, *168*, 108540. [CrossRef]
15. De Bruyn, I. Prospects of Radiation Sterilization of Self-Stable Food. In *Irradiation for Food Safety and Quality*; Loaharanu, P., Thomas, P., Eds.; Technomic Publishing Co. Inc.: Lancaster, PA, USA, 2001.
16. Chmielewski, A.G.; Migdal, W. Radiation decontamination of herbs and spices. *Nukleonika* **2005**, *50*, 179–184.
17. CODEX STAN 106-1983, REV.1-2003. General Standard for Irradiated Foods. 2003. Available online: http://www.codexalimentarius.net/download/standards/16/CXS_106e.pdf (accessed on 1 October 2016).
18. Ehlermann, D.A. The early history of food irradiation. *Radiat. Phys. Chem.* **2016**, *129*, 10–12. [CrossRef]
19. Roberts, P.B. Food irradiation: Standards, regulations and world-wide trade. *Radiat. Phys. Chem.* **2016**, *129*, 30–34. [CrossRef]
20. Buchanan, R.L.; Edelson, S.G.; Snipes, K.; Boyd, G. Inactivation of *Escherichia coli* O157:H7 in Apple Juice by Irradiation. *Appl. Environ. Microbiol.* **1998**, *64*, 4533–4535. [CrossRef] [PubMed]
21. Orlikowski, L.; Migdał, W.; Ptaszek, M.; Gryczka, U. Effectiveness of electron beam irradiation in the control of some soil-borne pathogens. *Nukleonika* **2011**, *56*, 357–362.
22. Nieto-Sandoval, J.M.; Almela, L.; Fernández-López, J.A.; Muñoz, J.A. Effect of Electron Beam Irradiation on Color and Microbial Bioburden of Red Paprika. *J. Food Prot.* **2000**, *63*, 633–637. [CrossRef]
23. Gryczka, U.; Kameya, H.; Kimura, K.; Todoriki, S.; Migdał, W.; Bułka, S. Efficacy of low energy electron beam on microbial decontamination of spices. *Radiat. Phys. Chem.* **2020**, *170*, 108662. [CrossRef]
24. Rio Segade, S.; Orriols, I.; Gerbi, V.; Rolle, L. Phenolic characterization of thirteen red grape cultivars from Galicia by anthocyanin profile and flavanol composition. *J. Int. Sci. Vigne Vin* **2009**, *43*, 189–198. [CrossRef]
25. Morata, A.; Bañuelos, M.A.; Tesfaye, W.; Loira, I.; Palomero, F.; Benito, S.; Callejo, M.J.; Villa, A.; González, M.C.; Suárez-Lepe, J.A. Electron Beam Irradiation of Wine Grapes: Effect on Microbial Populations, Phenol Extraction and Wine Quality. *Food Bioprocess Technol.* **2015**, *8*, 1845–1853. [CrossRef]
26. Breitfellner, F.; Solar, S.; Sontag, G. Effect of γ-Irradiation on Phenolic Acids in Strawberries. *J. Food Sci.* **2002**, *67*, 517–521. [CrossRef]
27. Sajilata, M.; Singhal, R. Effect of irradiation and storage on the antioxidative activity of cashew nuts. *Radiat. Phys. Chem.* **2006**, *75*, 297–300. [CrossRef]
28. Benito, S. The impact of *Torulaspora delbrueckii* yeast in wine-making. *Appl. Microbiol. Biotechnol.* **2018**, *102*, 3081–3094. [CrossRef]
29. Wei, R.; Wang, L.; Ding, Y.; Zhang, L.; Gao, F.; Chen, N.; Song, Y.; Li, H.; Wang, H. Natural and sustainable wine: A review. *Crit. Rev. Food Sci. Nutr.* **2022**, 1–12. [CrossRef] [PubMed]
30. de-la-Fuente-Blanco, A.; Sáenz-Navajas, M.P.; Ferreira, V. Levels of higher alcohols inducing aroma changes and modulating experts' preferences in wine model solutions. *Aust. J. Grape Wine Res.* **2017**, *23*, 162–169. [CrossRef]
31. Cioch, M.; Skotniczny, M.; Kuchta, T.; Satora, P. Characteristics of selected parameters of grape wines obtained from the vineyards of southern Poland. *Postępy Tech. Przetwórstwa Spożywczego* **2018**, *1*, 14–18.
32. Remize, F.; Sablayrolles, J.M.; Dequin, S. Re-assessment of the influence of yeast strain and environmental factors on glycerol production in wine. *J. Appl. Microbiol.* **2000**, *88*, 371–378. [CrossRef]
33. Gawel, R.; VAN Sluyter, S.; Waters, E.J. The effects of ethanol and glycerol on the body and other sensory characteristics of Riesling wines. *Aust. J. Grape Wine Res.* **2007**, *13*, 38–45. [CrossRef]

34. Yalcin, S.K.; Ozbas, Z.Y. Effects of pH and temperature on growth and glycerol production kinetics of two indigenous wine strains of *Saccharomyces cerevisiae* from Turkey. *Braz. J. Microbiol.* **2008**, *39*, 325–332. [CrossRef]
35. Jiang, B.; Zhang, Z. Volatile Compounds of Young Wines from Cabernet Sauvignon, Cabernet Gernischet and Chardonnay Varieties Grown in the Loess Plateau Region of China. *Molecules* **2010**, *15*, 9184–9196. [CrossRef]
36. Wzorek, W.; Pogorzelski, E. *Technology of Grape and Fruit Wine-Making*; SIGMA—NOT: Warsaw, Poland, 1998; pp. 56–71.
37. Cortés-Diéguez, S.; Rodriguez-Solana, R.; Domínguez, J.M.; Díaz, E. Impact odorants and sensory profile of young red wines. *J. Inst. Brew.* **2015**, *121*, 628–635. [CrossRef]
38. Cucciniello, R.; Forino, M.; Picariello, L. How acetaldehyde reacts with low molecular weight phenolics in white and red wines. *Eur. Food Res. Technol.* **2021**, *247*, 2935–2944. [CrossRef]
39. Lambrechts, M.G.; Pretorius, I.S. Yeast and its importance to wine aroma-A review. *S. Afr. J. Enol. Vitic.* **2000**, *21*, 97–129. [CrossRef]
40. Styger, G.; Prior, B.; Bauer, F.F. Wine flavor and aroma. *J. Ind. Microbiol. Biotechnol.* **2011**, *38*, 1145–1159. [CrossRef]
41. Ciani, M.; Comitini, F. Non-Saccharomyces wine yeasts have a promising role in biotechnological approaches to wine-making. *Ann. Microbiol.* **2011**, *61*, 25–32. [CrossRef]
42. Comitini, F.; Capece, A.; Ciani, M.; Romano, P. New insights on the use of wine yeasts. *Curr. Opin. Food Sci.* **2017**, *13*, 44–49. [CrossRef]
43. Spitaels, F.; Wieme, A.; Janssens, M.; Aerts, M.; Daniel, H.-M.; Van Landschoot, A. The Microbial Diversity of Traditional Spontaneously Fermented Lambic Beer. *PLoS ONE* **2014**, *9*, e95384. [CrossRef] [PubMed]
44. Oro, L.; Canonico, L.; Marinelli, V.; Ciani, M.; Comitini, F. Occurrence of *Brettanomyces bruxellensis* on Grape Berries and in Related Winemaking Cellar. *Front. Microbiol.* **2019**, *10*, 415. [CrossRef] [PubMed]
45. Pinto, C.; Pinho, D.; Cardoso, R.; Custódio, V.; Fernandes, J.; Sousa, S.; Pinheiro, M.; Egas, C.; Gomes, A.C. Wine fermentation microbiome: A landscape from different Portuguese wine appellations. *Front. Microbiol.* **2015**, *6*, 905. [CrossRef]
46. Lachowicz, S.; Oszmiański, J.; Rapak, A.; Ochmian, I. Profile and content of phenolic compounds in leaves, flowers, roots, and stalks of *Sanguisorba officinalis* L. determined with the LC-DAD-ESI-QTOF-MS/MS analysis and their in vitro anti-oxidant, antidiabetic, antiproliferative potency. *Pharmaceuticals* **2020**, *13*, 191. [CrossRef] [PubMed]
47. Sterczyńska, M.; Machowski, M.; Jakubowski, M.; Wiśniewski, A. Effectiveness of selected methods of clarification of grape red wines fermented with wild and noble yeast. *Inżynieria Przetwórstwa Spożywczego* **2015**, *4*, 28–33.
48. Mijowska, K.; Ochmian, I.; Oszmiański, J. Rootstock effects on polyphenol content in grapes of 'Regent' cultivated under cool climate conditio. *J. Appl. Bot. Food Qual.* **2017**, *90*, 159–164. [CrossRef]
49. Ochmian, I.; Oszmiański, J.; Lachowicz, S.; Krupa-Małkiewicz, M. Rootstock effect on physicochemical properties and content of bioactive compounds of four cultivars Cornelian cherry fruits. *Sci. Hortic.* **2019**, *256*, 108588. [CrossRef]
50. *ISO 21527-1:2008*; Horizontal Method for the Enumeration of Yeasts and Molds. Part 1: Colony Count Technique in Products with Water Activity Greater than 0.95. International Organization for Standardization: Geneva, Switzerland, 2008. Available online: https://www.iso.org/obp/ui/#!iso:std:iso:21527:-1:ed-1:v1:en (accessed on 27 March 2019).
51. Oszmiański, J.; Kolniak-Ostek, J.; Lachowicz, S.; Gorzelany, J.; Matłok, N. Effect of dried powder preparation process on polyphenolic content and antioxidant capacity of cranberry (*Vaccinium macrocarpon* L.). *Ind. Crop. Prod.* **2015**, *77*, 658–665. [CrossRef]

Disclaimer/Publisher's Note: The statements, opinions and data contained in all publications are solely those of the individual author(s) and contributor(s) and not of MDPI and/or the editor(s). MDPI and/or the editor(s) disclaim responsibility for any injury to people or property resulting from any ideas, methods, instructions or products referred to in the content.

Article

Okara-Enriched Gluten-Free Bread: Nutritional, Antioxidant and Sensory Properties

Mirjana B. Pešić [1,*], Milica M. Pešić [1], Jelena Bezbradica [1], Anđela B. Stanojević [2], Petra Ivković [1], Danijel D. Milinčić [1], Mirjana Demin [1], Aleksandar Ž. Kostić [1], Biljana Dojčinović [3] and Sladjana P. Stanojević [1]

[1] Institute of Food Technology and Biochemistry, Faculty of Agriculture, University of Belgrade, 11080 Belgrade, Serbia; mpesic1801@gmail.com (M.M.P.)
[2] Lund University Center for Sustainable Studies (LUCSUS), Faculty of Social Sciences, 223 62 Lund, Sweden
[3] Institute of Chemistry, Technology and Metallurgy, National Institute of the Republic of Serbia, University of Belgrade, Njegoševa 12, 11000 Belgrade, Serbia
* Correspondence: mpesic@agrif.bg.ac.rs

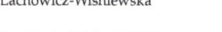

Citation: Pešić, M.B.; Pešić, M.M.; Bezbradica, J.; Stanojević, A.B.; Ivković, P.; Milinčić, D.D.; Demin, M.; Kostić, A.Ž.; Dojčinović, B.; Stanojević, S.P. Okara-Enriched Gluten-Free Bread: Nutritional, Antioxidant and Sensory Properties. *Molecules* **2023**, *28*, 4098. https://doi.org/10.3390/molecules28104098

Academic Editor: Sabina Lachowicz-Wiśniewska

Received: 29 March 2023
Revised: 8 May 2023
Accepted: 9 May 2023
Published: 15 May 2023

Copyright: © 2023 by the authors. Licensee MDPI, Basel, Switzerland. This article is an open access article distributed under the terms and conditions of the Creative Commons Attribution (CC BY) license (https://creativecommons.org/licenses/by/4.0/).

Abstract: The aim of this study was to produce an eco-innovative gluten-free bread with a pleasant taste and a unique formulation that includes the highest quality grains and pseudocereals (buckwheat; rice; and millet); and okara; a by-product of soy milk production. The mixture of pseudocereal and cereal flour contained buckwheat flour 45%, rice flour 33%, and millet flour 22%. Three gluten-free breads; each containing different contents of gluten-free flour (90%, 80%, and 70%, respectively); okara (10%, 20%, and 30%, respectively); and a control sample (without okara); were prepared and subjected to sensory evaluation. The okara-enriched gluten-free bread with the highest sensory score was selected for further analysis of physico-chemical (total proteins; total carbohydrates; insoluble fiber; soluble fiber; sugars; total lipids; saturated fatty acids; and salt) and functional properties (total phenolic content and antioxidant properties). The highest sensory scores were obtained for 30% okara-enriched gluten-free bread including taste; shape; odor; chewiness; and cross-section properties; classifying this bread in the category of very good quality and excellent quality (mean score 4.30 by trained evaluators and 4.59 by consumers). This bread was characterized by a high content of dietary fiber (14%), the absence of sugar; low content of saturated fatty acids (0.8%), rich source of proteins (8.8%) and certain minerals (e.g.,; iron; zinc); and low energy value (136.37 kcal/100g DW). Total phenolic content was 133.75 mgGAE/100g FW; whereas ferric reducing power; ABTS radical cation; and DPPH radical scavenging activity were 119.25 mgAA/100g FW; 86.80 mgTrolox/100g FW; and 49.92 mgTrolox/100g FW; respectively. Okara addition in gluten-free bread production enables the formulation of high-nutritive; good antioxidative; low-energy bread; and better soy milk waste management.

Keywords: okara; sustainable food production; buckwheat; rice; millet; phenolics; antioxidant properties

1. Introduction

Gluten-free bread, according to its composition and method of preparation, belongs to the group of dietary foods [1] and is primarily intended for the nutrition of people suffering from celiac disease, who often have difficulties in applying adequate nutrition [2]. Celiac disease is a lifelong dietary disorder, present worldwide, and defined as "immuno-mediated enteropathy triggered by the ingestion of gluten in susceptible patients" [3,4]. Celiac disease can appear without visible symptoms, but much more often there are symptoms such as an inflammatory disorder of the small intestine (due to difficulties in absorption of many nutrients, such as liposoluble vitamins, folic acid, and iron), abdominal discomfort, weight loss, diarrhea, osteoporosis, fatigue, disorders of multiple organ systems, and an increased risk of some cancers [2–5]. In the case of celiac disease, a gluten-free diet is recommended [2–4].

In the preparation of gluten-free bread, starches from gluten-free grains (corn and rice) and hydrocolloids of natural origin are most often used, as well as buckwheat and rice flour [6,7]. In a lower extent, the flour of millet can also be used [8]. Commonly used commercial buckwheat flour is a nutritious, rich source of starch (~60–73%) and protein (~7–38%) with well-balanced amino acid content (glutamic and aspartic acid, arginine, leucine, glycin, serine, phenylalanine, alanine, and proline) [9–11]. Buckwheat flour also contains a relatively high level of dietary fiber (~3–10%), soluble carbohydrates (~1%), lipids (~1–4%; with dominant fatty acids such as linoleic, oleic, and palmatic acid), minerals (~1–3%, such as Mg, P, K, Ca, Fe, and Zn) vitamins and pseudovitamins (B9, B3, K, and choline), and compounds such as organic acids, tannins, nucleotides, and nucleic acids [8–10]. Buckwheat can also be used as a good source of polyphenols (rutin, catechin, quercetin, and hyperin), and natural antioxidants that are absent in other pseudocereals or grains [11,12]. Consumption of buckwheat in a diet can have several health benefits such as prevention of cardiovascular disease, cancers, and diabetes, as well as probiotic, anti-inflammatory, and antimutagenic activities [11].

Rice flour contains ~90% total carbohydrates (with amylose content about 20%, ~47–90% starch, and ~2.4% dietary fiber of the total carbohydrate composition), ~6–10% protein (with dominant amino acids such as glutamine, aspartate, arginine, leucine, and valine) ~1–2% fat, and about 1% minerals [10,13]. Millet flour contains about 70% carbohydrates [14,15]. Starch is the main carbohydrate of millet grains (50–70% of the total carbohydrate composition), followed by a high content of dietary fiber (13.1% of the total carbohydrate composition) and a low content of monosaccharides (glucose, fructose, and galactose), disaccharide–sucrose, and trisaccharide–raffinose [15–18]. Millet flour contains about 9–12% proteins (with lysine as a dominant amino acid) and 1–2.6% oils (with dominant fatty acids such as palmitic, stearic, oleic, and linoleic) [15]. The importance of millet as a foodstuff, and therefore flour, is reflected in the content of biologically active components (vitamins B1 B2, B3, and E, and minerals K, Ca, P, Mg, Fe and Zn, as well as tannins, flavonoids, phenolic acids, and β-carotene) [15]. Regardless of the nutrition it possesses, the addition of millet flour can negatively affect the sensory quality of gluten-free products (small specific volume and increased hardness), thus the addition of a higher amount is often avoided [8].

However, the production of gluten-free bread is a technological challenge due to several drawbacks compared to gluten-rich bread, such as worst texture, less tasty, and lower nutritional quality due to a higher content in lipids and sugars, and a lower content in protein, dietary fiber, and mineral elements [7]. To overcome these problems, the use of nutritional valuable okara, a soy milk by-product, can be one of the solutions. Okara is characterized by a light-yellow color, mild and neutral flavor [19], and low energy potential (2.78–3.28 kcal/g fresh matter) [19,20]. From 1 kg of soybeans used in the manufacturing of soy milk, about 1.1–1.2 kg of okara is obtained [21,22]. The global production of soybean okara amounts to about 1.4 billion tons per year [23,24], but it is underutilized considering the potential nutritional benefits of okara, causing significant environmental pollution [25]. The composition of okara depends on the genotype of the soybean as well as the method of soy milk production [16,17,26–28]. Approximately 40% of produced okara is used for animal consumption and only 10% for human consumption; 50% of okara ends up as waste. Okara contains ~15–40% proteins (with essential amino acids: phenylalanine, leucine, isoleucine, lysine, valine, threonine, histidine and methionine, and non-essential amino acids: tyrosine, proline, alanine, arginine, glycine, glutamic acid, serine, and asparaginic acid) [29]. The dominant proteins in extracts of okara are subunits of basic 7S globulin ("heavy" (HI,II) and "light" (LI,II) subunits with molecular weight values of 27,000 and 16,000. Basic 7S globulin is desirable because of its nutritional value as it is a cysteine-rich glycoprotein [27]. Okara contains ~32–53% carbohydrates (with ~56–58% total dietary fibers and ~42–55% insoluble dietary fibers of the total carbohydrate composition and monosaccharides 0.17–4.11% (mannose and fructose) and disaccharides 1.61–4.35% (sucrose and maltose) depending on the soybean genotype [20]. In addition, okara contains lipids of a very wide range (~6–22%), depending on the genotype of the soybean and the method used to obtain soy milk (with dominant fatty acids: linoleic, oleic, palmitic, linoleic, and stearic fatty acids) [29]. Okara is rich in bioactive compounds (such as isoflavones and phytoestrogens; vitamins B1, B2, B3, B6, folates (B9), and K; antioxidants polyphenol isoquercetin) and minerals (Na, Mg, N, K, P, Ca, Fe, Zn, and Mn) [20,23,25–31]. The literatures data indicate that okara contains an antioxidant capacity similar to that of beet or pumpkin; much higher than those of tomato, carrot, broccoli, or onion [29]. Furthermore, okara lacks lactose, gluten, and cholesterol, which can have a significant health benefit for the health-compromised consumers [32].

The application of okara can have a significant effect on the structure of the gluten-free matrix and the volume of the bread. The high content of dietary fiber and protein can affect the rheological properties of bakery products [33]. In addition, dietary fibers are now often added to bakery products with the aim of prolonging freshness, which is based on their ability to retain water [34].

It has been shown that okara could be successfully used in the formulation of meat products, biscuits, filled pasta, drinks, candies, gluten bread, and nutritional flour, as well as edible packaging and biodegradable materials [21,25,35,36], due to its good nutritional and functional characteristics [37]. Guimarães et al. [38] formulated gluten-free bread with the addition of okara and corn by-products and obtained a product with improved nutritional characteristics, but poorer sensory/technological characteristics. However, the use of okara in the formulation of gluten-free bread (from different gluten-free grains and pseudocereals) has not yet been thoroughly investigated, whereby a product with good sensory properties has not yet been obtained.

Thus, the aim of this study was to produce soy okara-enriched gluten-free bread based on buckwheat, rice, and millet with high nutritional and low-energy values while achieving maximum sensory quality. Three gluten-free breads, each containing different contents of okara, and a control sample (without okara), were prepared and subjected to sensory evaluation. The bread with the highest sensory score was selected for further analysis of proximate composition and antioxidant properties. The new formulated bread can enable the diversification of gluten-free products on the market and increase a higher utilization of okara, food waste by-product, contributing to environmental protection.

2. Results and Discussion
2.1. Sensory Analyses

Knowing that gluten-free products are considered as products with poor sensory properties [38], sensory analysis of three gluten-free breads enriched with okara, and a control sample (without okara), was performed by first, aiming to select the most acceptable gluten-free bread for consumers. Preparation of sensory-acceptable bread for consumers is very demanding considering that the majority of consumers react negatively to the sensory characteristics of gluten-free bread, in addition to the unfavorable leguminous properties of soy food which are generally unacceptable for most consumers of the Western market [38,39]. The weighted mean scores for the sensory quality of the tested samples were relatively very similar, even for samples with 20% and 30% okara (4.30; Figure 1A), indicating that bread samples belong to the "very good quality" category. Regardless of the potential lipoxygenase activity, the tested breads had high taste scores. The highest score for taste (4.75) was given to the sample of bread with the highest percentage of okara (30%). This agrees with the results of the sensory analysis of soy bread, where it was concluded that beany flavor of soy-food was reduced, and soy flour contributed to a higher general acceptability of the bread [37].

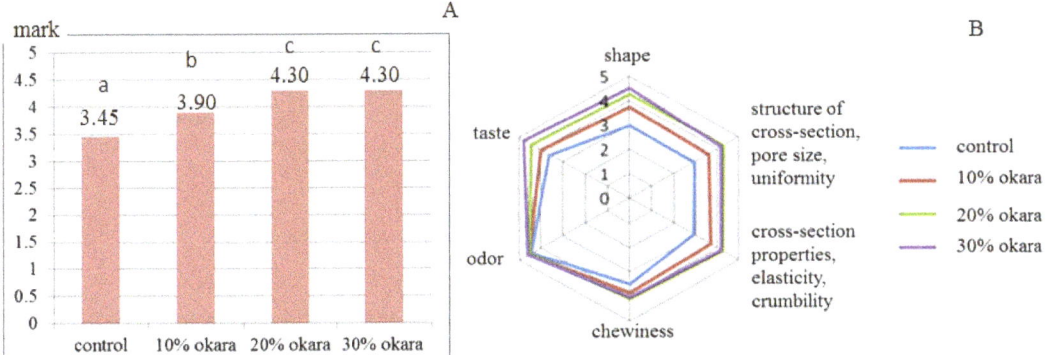

Figure 1. Sensory analysis by trained expert panelists of the bread samples supplemented with 10%, 20%, and 30% okara, respectively, compared with control bread after 24 h. (**A**)—values of weighted scores; (**B**)—scores of individual sensory quality parameters. Means with different small roman letters in Figure 2A are significantly different ($p < 0.05$).

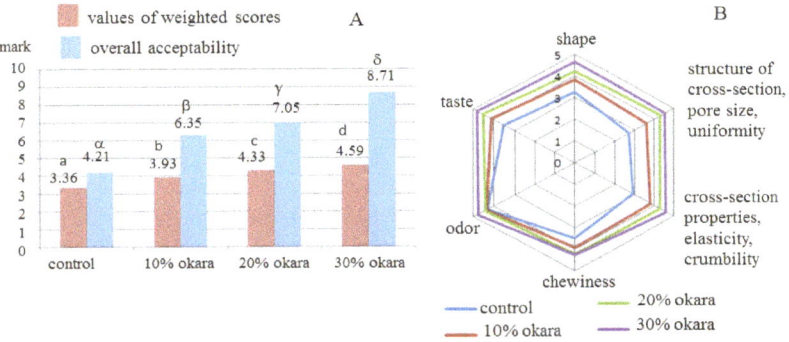

Figure 2. Sensory analysis by consumers of the bread samples enriched with 10%, 20%, and 30% okara, respectively, compared with control bread after 24 h. (**A**)—values of weighted scores (0–5) and overall acceptability obtained by hedonic scale (1–9); (**B**)—scores of individual sensory quality parameters (0–5). Means with different small roman and greek letters in Figure 2A are significantly different ($p < 0.05$).

It is very interesting that the bread containing 30% okara received a very high score for the shape (4.55), as well as for the marks related to the cross-section structure (4.20), considering that it does not contain gluten. Namely, from the technological aspect, the first problem that occurs in the production of gluten-free bread is the possible absence of desirable rheological and textural characteristics of the dough and the final product. Most common breads are made from wheat flour, water, salt, and yeast, with bread making relying on the ability of hydrated gluten to develop a viscoelastic network that traps gas (CO_2) and produces bread with a larger loaf volume and better rheological and textural characteristics [7]. Our results are in contrast with the results of Ostermann-Porcel et al. [19] who identified that the greater presence of okara in gluten-free cookies caused a smaller volume, due to the fibers present in okara flour that interfere with the structure of the matrix, reducing the gas retention capacity of the dough. However, the same authors indicate that the results of scanning electron microscopy of the cross-section of cookies with and without okara showed no noticeable differences. The influence of soybean content on the volume of gluten-free dough/bread is opposite in the literature, too. For example, Melini et al. [40] examined the influence of soybean flour on the volume of gluten-free bread and indicated that there was an increase in the softness of bread with increasing soybean flour content. The rheological properties of the tested gluten-free bread can be influenced by many factors, such as the amount of water added during the preparation of the dough, the content of protein and dietary fiber, and the use of substances with a high affinity for hydrocolloid formation. The literatures data indicate that the presence of dietary fiber and hydrocolloids in optimal amounts can improve the texture of bread [38]. Moreover, the dominant proteins in extracts of okara are subunits of basic 7S globulin [27], which is a cysteine-rich glycoprotein [41] that can affect the structure of the gluten-free matrix by the formation of disulfide bridges.

In addition, bread with 30% okara received a high score for elasticity (4.20). This may be the effect of added guar gum to the dough, since it is known that all gluten-free breads contain hydrocolloids to improve dough behavior [42]. Namely, vegetable gums (such as guar gum) swell and form a gel, which thickens the mass of the dough preventing the loss of gas released during wetting and mixing. Hydrocolloids bind water, stabilize the structure of the crumb, and prevent rapid retrogradation of starch [43]. Additionally, the compatibility of rice flour protein (glutenin fraction in high concentration with albumin and globulin fraction) and soy protein (globulins–β conglycinin and glycinin) could be of particular importance for achieving good dough and bread elasticity. We can assume that the interaction between rice and soybean proteins in the dough was intensified by the direct formation of new, intermolecular covalent bonds, catalyzed by transglutaminases, and by the indirect formation of disulfide bonds. This combination of proteins from different sources and enzymes led to the formation of a network, which improves the structure of gluten-free breads. The interaction between rice and soy proteins in solutions was reported by Wang et al. [44]. The authors reported that protein composites showed significantly improved emulsifying and foaming properties compared to rise protein. Melini et al. [40] identified that the addition of soy flour to gluten-free dough/bread in the amount of 45–60% significantly increased softness.

Statistically significant differences were registered between the ratings of the sensory quality of the control sample and all samples enriched with okara, for all parameters of the sensory analysis, except for the smell. Among gluten-free bread samples, the sensory quality ratings between samples with 10% okara and samples with 20% and 30% okara differed for all quality parameters, except for the smell. Comparing the statistically significant differences for the sensory quality scores between the gluten-free bread enriched with 20% and 30% okara, the shape and taste scores were statistically different between the two samples.

In addition to similar mean quality scores, the samples differed according to the scores for individual parameters of sensory characteristics (Figure 1B). Gluten-free bread prepared with the addition of 30% okara by trained panelists received the highest scores for individual sensory characteristics; taste (4.75), odor (4.55), chewiness (4.15), and shape (4.55).

The obtained ratings (0–5) by consumers (Figure 2) confirmed the results of the sensory analysis obtained by trained expert panelists. After sensory analysis by consumers, the sample with the best evaluation of all quality parameters was the gluten-free bread enriched with 30% okara (Figure 2B). The mean quality rating of gluten-free bread with 30% okara was the highest (4.59) compared with other samples (Figure 2A), which introduced the product in the "excellent quality" category. Furthermore, the results that were obtained using a hedonic scale (1–9) showed a very high score of overall acceptability (8.71), which indicated a very pleasant/acceptable sensory feeling of gluten-free bread enriched with 30% okara.

Based on the obtained results of the sensory analysis, both trained evaluators, and consumers, rated the gluten-free bread enriched with 30% okara as the most acceptable for consumption. For this reason, this sample (with 30% okara) was selected as the most appropriate gluten-free bread enriched with okara for further analyses.

2.2. Proximate Composition

The proximate composition of gluten-free bread containing 30% okara is presented in Table 1. The high content of total carbohydrates was determined, 28.90% with high content of dietary fiber (14.00%, among which 11.11% were insoluble and 2.89% were soluble fibers). Pseudocereals and cereal are considered the most common sources of dietary fiber for bakery products, consisting of cellulose and complex xylans and lignin (in cell walls), arabinoxylans, β-glucan, heteromannans, and esterified phenolic acids (in aleurone and endosperm) [45]. However, legume fibers, including soybean, contain both insoluble (mainly consist of hemicelluloses and cellulose) and soluble dietary fibers (primarily consist of pectin) [46], and are considered to have more advantages than cereal fibers, due to higher content of solubles [45]. Taking this into account, soybean okara can increase the nutritional value of gluten-free bread.

Table 1. Proximate composition of gluten-free bread enriched with 30% okara.

Components	Content (%)
Total proteins	8.80 ± 0.07
Total carbohydrates	28.90 ± 0.04
Insoluble fiber	11.11 ± 0.03
Soluble fiber	2.89 ± 0.06
Sugars (glucose, fructose, sucrose)	0.05 ± 0.006
Total lipids	3.80 ± 0.03
Saturated fatty acids	0.80 ± 0.04
Salt	0.80 ± 0.03
	Content (ppm)
Gluten	14.41 ± 5.33

All results were calculated based on the dry weight of the sample.

Dietary fibers are poorly absorbable or non-absorbable in the human gastrointestinal tract and play a significant physiological/nutritional role in human metabolism [47]. Numerous studies prove their significant effect in preventing/treating various chronic diseases (such as diabetes, obesity, gastrointestinal tract, and cardiovascular diseases) and colorectal cancer [48–52]. Joint WHO and FAO experts recommend an intake of at least 25 g/day of dietary fibers, which can protect against obesity and its consequences [53]. Nevertheless, dietary fiber consumption by humans is usually lower than the recommended value [54]; thus, food technologists and food scientists are trying to develop fiber-enriched products. Bakery products, especially bread, are one of the most widely, and regularly consumed food worldwide [45], and can be an ideal source of dietary fiber and other bioactive compounds in a diet.

In addition to a high content of dietary fiber, gluten-free bread enriched with 30% okara has a very low sugar content (glucose, fructose, and sucrose) (0.05%; Table 1). Okara is characterized by a low content of monosaccharides and disaccharides, which influence the low value of total carbohydrates in a final product. Bearing in mind that these sugars significantly affect the glycemic index of food, it can be assumed that the analyzed bread is not characterized by a high glycemic index.

In addition to being a good source of dietary fiber, okara is also a good source of protein. The analyzed sample of gluten-free bread contained 8.80% of total proteins (Table 1). This is very important considering gluten-free bakery products generally have a lower content of total protein than similar products with gluten [55]. The reason for this is that gluten-free flours generally contain less protein than gluten-rich flours [7]. Therefore, people suffering from celiac disease, by consuming gluten-free bread, are forced to consume proteins from other sources. Segura and Rosell [56] investigated the composition of commercially available gluten-free breads and registered a total protein content of 0.91–2.80% in samples that did not contain added proteins (e.g., casein or proteins of lupine and egg). Therefore, the formulated gluten-free bread, according to legal regulations, can be considered a "product with increased protein content" [1].

Several important things must be considered when proteins are added to gluten-free bread. First, the functionality of added proteins is important since added proteins cannot replace the role of gluten in the dough. Second, the nutritional quality, allergenicity and origin (if it is a product intended for vegans), and price are very important aspects [7]. The addition of soy proteins required indication on the product declaration since they can be allergens in human nutrition. Alternately, it is very important that the allergenic effect of soy proteins can be reduced/neutralized by appropriate technological processes such as pressure techniques (e.g., extrusion and high hydrostatic pressure) and waves (e.g., gamma irradiation, microwave, and ultrasonication) [57].

Checking the possible presence of residual gluten in the formulated bread, 14.41 ppm was obtained. Considering that the FDA [58] prescribed that "any foods that carry the label "gluten-free," "no gluten," "free of gluten," or "without gluten" must contain less than 20 parts per million (ppm) of gluten" this product can be classified in the group of "gluten-free products".

The sample of gluten-free bread with 30% okara was characterized by a low content of total lipids (3.80%) and a very low content of saturated fatty acids (0.08%; Table 1), Such a low content of saturated fatty acids is of great nutritional importance, knowing that they can have a harmful effect on the human body (increase the content of total cholesterol and low-density lipoproteins, which leads to an increased risk of cardiovascular diseases) [59]. Considering that okara, depending on the method of production and soybean genotype, can contain total lipids in a very wide range (0.8–22%) [31], it can be assumed that in products enriched with okara, the content of total lipids will largely depend on their content in okara. For example, Ostermann-Porcel et al. [19] received gluten-free cookies enriched with 30% okara, which contained 16.80% of total lipids, significantly increased the total energy value of the product (387.6 kcal/100g). The total energy value of the formulated gluten-free bread enriched with 30% okara was very low (136.37 kcal/100g or 568.21 kJ/100g; Table 2), which potentially allows for the possibility to use this bread in the diets of obese people. The contribution of digestible carbohydrates to total energy value was the highest, more than 50%, Figure 3.

Table 2. Energy value of gluten-free bread enriched with 30% okara.

Components	kJ/100g	kcal/100g
Energy of proteins	149.60	35.91
Energy of digestible carbohydrates *	302.43	72.58
Energy of lipids	116.18	27.88
Total energy value	568.21	136.37

* The content of digestible carbohydrates was calculated from the difference between total carbohydrates and insoluble dietary fibers.

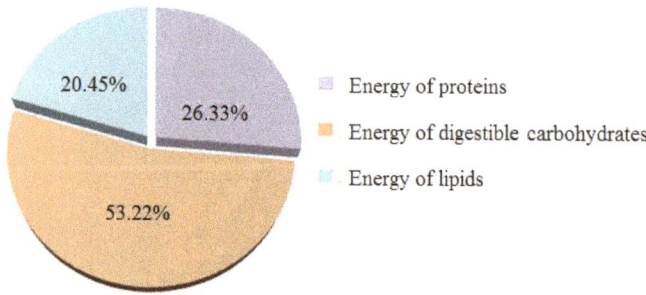

Figure 3. Percentage participation of individual components in the total energy value of gluten-free bread enriched with 30% okara.

The content of table salt in the sample of gluten-free bread enriched with 30% okara was within the limits indicated in the literature (0.80%; Table 1). Šmídová and Rysová [43] reported that the salt content in gluten-free breads can vary widely, from 0.2% to as much as 2.5%. Sodium chloride in the technological sense plays a very important role as it extends the shelf life of the product (by ensuring the microbiological stability of the product), participates in the formation of the taste in food and in the formation of the dough structure of bakery products (where the addition of salt depends on the type of flour) [60]. However, low table salt content is desirable in food since high contents can lead to high blood pressure and cardiovascular problems [61].

Based on the obtained results, it can be concluded that the major drawbacks of commercially available gluten-free bread from nutritional quality aspects can be improved by the addition of 30% okara in the rice/buckwheat/millet flour mixture (Table 3). The results of this study confirm the statement of Melini et al. [40] which states that in the production of gluten-free bread, none of the raw materials can replace gluten, but legumes might "epresent a new forward-looking frontier in gluten-free breadmaking because of their functional and nutritional characteristics."

Table 3. Nutritional quality problems of gluten-free products suggested by Gómez [7] and possible solutions to address them by gluten-free bread enriched with 30% okara.

Nutritional Aspects to Improve	Solution	Outcome
Low protein	Okara proteins are added	High-quality proteins were added and the resulting bread, according to legal regulations, belongs to the group "product with increased protein content". Okara proteins are of high value and contain all essential amino acids, with lysine exceeding the daily requirements. Furthermore, they can reduce the content of cholesterol and triglycerides in the blood. Such properties allow okara proteins to be used for supplementation [19].
Low fiber	Okara fibers are added	Soluble and insoluble fibers are added with okara (dietary fibers of formulated gluten-free bread make up to 50% of total carbohydrates)
Low minerals	Okara minerals are added	Okara contains: Na, Mg, N, K, P, Ca, Fe, Zn, Mn; okara contain Fe^{+2}, which is easily absorbed
High lipids	Okara may contain a low percentage of total lipids [28].	The addition of okara did not increase the lipid content in the formulated gluten-free bread; in addition, the saturated fatty acid content was low. Namely, soy fats, are known as "cardio-healthy fats", they are unsaturated fats (contains an essential omega 3 fatty acids) [62].
High sugars	Okara contain a low percentage of monosaccharides and disaccharides [20].	The content of sugar (glucose, fructose, and sucrose) in the examined bread was very low.

2.3. Total Phenolic Content and Antioxidant Properties

The total phenolic content and antioxidant properties of gluten-free breads enriched with 30% okara are presented in Table 4. TPC in analyzed bread was 133.75 mgGAE/100g. In the literature available to us, there are no data on the content of total phenolics in gluten-free bread. However, in bread made from whole wheat flour, the content of total phenolics was ranged from 50 to 200 mgGAE/100g, depending on the cultivar of wheat [63]. Therefore, it can be concluded that the content of total phenolics in the prepared gluten-free bread was in the range of values for wheat bread.

Table 4. Total phenolic content and antioxidant properties of gluten-free bread enriched with 30% okara.

Total Phenolics	FRP	ABTS	DPPH
mgGAE/100g	mgAA/100g	mgTrolox/100g	
133.75 ± 2.42	119.25 ± 4.29	86.80 ± 2.46	49.42 ± 8.58

All results were calculated based on the fresh weight of the sample. FRP-ferric reducing power. assay; ABTS and DPPH radical scavenging activity assay.

The antioxidant properties of gluten-free bread were evaluated by three assays: FRP, ABTS, and DPPH. The reasoning is that different studies apply different methods and calculations of antioxidant properties; in addition, the combinations of different constituents in the formulation of bread are practically unlimited. In the case of complex food matrices, since there are many different potential antioxidants, of which have different mechanisms of action, it is recommended to use several different antioxidant screening assays [64]. The ABTS radical is soluble in water and organic solvents, enabling determination of hydrophilic and hydrophobic antioxidants, whereas the DPPH radical is soluble in organic solvents and is mainly used for the determination of phenolic antioxidants soluble in organic media [65]. Furthermore, the ABTS assay, unlike the DPPH assay which involves hydrogen atom transfer, are based on both hydrogen atom and electron transfer enabling determination of antioxidants with different mechanisms of action [65]. Antioxidants action involves the reduction of a colored ABTS/DPPH radical, whereas the FRP method monitors the reduction of ferric-iron (Fe^{3+}) to ferrous-iron (Fe^{2+}) by antioxidants [66]. All three methods indicated significant antioxidant activity (Table 4). The DPPH scavenging activity of the analyzed gluten-free bread was 49.42 mgTrolox/100g, whereas the ABTS radical cation scavenging activity was 86.80 mgTrolox/100g (Table 4). It was reported that the DPPH scavenging activity of bread prepared only with millet flour or rice flour was 19.24% and 10.50%, respectively [67]. There are no available data on the FRP of gluten-free bread, however, Shin et al. [37] estimated, using the FRAP method, that soy increases ferric reducing antioxidant power. Turfani et al. [68] indicated that the addition of leguminous vegetables provides bread with an acceptable volume, taste and texture, and increased antioxidant activity.

2.4. Macro- and Micro-Elements

Sodium (4677.08 μg/g) was registered as the most abundant macro-element in the gluten-free bread followed by phosphorus (2821.19 μg/g) and potassium (2002.04 μg/g), as well as sulfur (1374.72 μg/g). Additionally, the presence of calcium and magnesium were registered (849.59 μg/g; 777.04 μg/g; Table 5). Each of the registered macro-elements in the gluten-free bread enriched with 30% okara has an extremely important role in the body. Sodium is the main cation of the extracellular fluid and participates in the maintenance and regulation of the osmotic pressure of the blood plasma and other extracellular fluids, as well as in the synthesis of gastric HCl and in the regulation of the acid-base balance. Potassium, as the main intracellular cation, actively regulates osmotic pressure, affects the synthesis of proteins, ribosomes, and the activity of some enzymes (e.g., pyruvate kinase). In addition, potassium affects muscle activity, especially the activity of the heart muscle, and acts as an antagonist to calcium in its effect on heart rate. Calcium has a structural role in the body (part of solid tissues), neuromuscular (in muscle contraction, neurotransmitter release, and excitability control), enzymatic (as a coenzyme of blood coagulation factors), and hormonal (as an intracellular secondary messenger). Phosphorus participates in the process of ossification and deposition/transfer of energy as an integral part of macro-energetic compounds. Phosphorus is also involved in the reactions of phosphorylation and biosynthesis of some coenzymes. Magnesium ion is a cofactor and activator of many enzymes, participates in the activation of amino acids in protein synthesis, and reduces neuromuscular excitability. Sulfur is included in the structure of some amino acids and proteins, as well as in the thioester macro-energetic compound–coenzyme A and mucopolysaccharides. In addition, sulfur is involved in oxidation-reduction and detoxification processes in the body [69].

Table 5. Content of mineral elements of gluten-free bread enriched with 30% okara.

Macro-Elements (µg/g)			
Ca	849.59	Na	4677.08
K	2002.04	P	3831.19
Mg	777.04	S	1374.72
Pooled std		4.11	
Micro-elements (µg/g)			
Co	n.d.	Mn	8.91
Cr	0.19	Ni	0.87
Cu	2.96	Sr	2.42
Fe	24.71	Zn	17.42
Pooled std		0.25	
Toxic elements (µg/g)			
Al	4.15	Cd	0.02
As	n.d.	Li	0.04
B	5.82	Pb	n.d.
Ba	0.59		
Pooled std	0.09	Pooled std	0.00

Data are expressed as mean and pooled standard deviation (Pooled std) of two replicates. n.d. <0.005 µg/g.

The most dominant micro-elements were iron (27.71 µg/g) and zinc (17.42 µg/g), but a significant content of manganese was also recorded (8.91). Iron as an integral part of specific proteins (heme proteins, e.g., hemoglobin, myoglobin, cytochromes, and proteins that do not contain heme but bind iron: ferritin, flavoproteins, transferrin) plays a central role in the transport of oxygen, as well as in the transport of electrons in the respiratory chain and participates in energy metabolism [66]. In addition, iron participates in the synthesis of steroid hormones and bile acids as well as in the detoxification of the organism [70]. Zinc is part of many enzymes, plays a role in the body's immune system and significantly affects the preservation of cell integrity (by stabilizing the molecular structures of the components of cell membranes and cells). Manganese is an activator of many enzymes [70].

These results indicated a very favorable composition of macro- and micro-elements of the gluten-free bread. Ibidapo et al. [71] analyzed bread made from wheat flour and malted millet with 15% added okara and reported the highest content of calcium (1278.0 µg/g) among macro-elements, and iron (14.7 µg/g) and zinc (8.1 µg/g) among micro-elements. Maggio et al. [72] presented an overview of several traditional gluten-free products. Among other products, they demonstrated the mineral composition of gluten-free breadsticks that contained the highest content of sodium (3209.0 µg/g) and magnesium (1207.0 µg/g) among macro-elements, whereas iron (62.0 µg/g) was dominant among microelements. However, it is difficult to compare these results with the results of this study since the composition of the analyzed breadsticks is not known. Rybicka and Gliszczyńska-Świgło [73], after comparing the mineral composition of different gluten-free products concluded, that gluten-free products containing millet and buckwheat have a better composition of macro- and micro-elements than products containing rice.

The presence of potentially toxic elements was registered in analyzed samples: boron (5.82 µg/g) and aluminum (4.15 µg/g). Namely, boron plays a role in the synthesis of uridine-diphosphate-glucose, by affecting the reaction of glucose-1-phosphate into uridine-triphosphate and affects the activity of dehydrogenase and oxidoreductase enzymes. In larger amounts, boron will damage the brain. Aluminum is important in the activity of the nervous system, participates in cellular respiration, and activates B complex vitamins. However, aluminum can have a harmful effect on the central nervous system, kidneys, and bone marrow. For these reasons, food should always be checked for the presence of these potentially toxic elements [74]. The presence of aluminum was registered in gluten-free biscuits (18 µg/g) and breadsticks (1.5 µg/g) [72]. The presence of trace amounts of lead [72] which should not be present in food was also registered in these products. The presence of lead was not registered in the analyzed gluten-free bread.

2.5. Dietary Reference Intakes

The value of dietary reference intake showed the nutritional contribution of individual components of the analyzed gluten-free bread. A portion of 50 g of the analyzed gluten-free bread would meet as much as ~34% of the daily protein needs of children aged 1–3 years and ~23% of the needs of children aged 4–8 years. This is very important, considering that children are in the growing phase of life. Martos and López [74] obtained similar values (20%) for covering daily protein needs in children aged 4–8 years by a portion of gluten-free bread (50 g) enriched with *Prosopis nigra* flour. However, the same authors obtained lower values for the coverage of daily protein needs by examining gluten-free bread (portion 50 g) for men and females, which were 7–8%. The daily need for protein between men and females would be met by a portion of gluten-free bread (50 g) enriched with okara in the value of 9 to approximately 13% depending on age (Table 6).

Table 6. Dietary reference intakes (%DRI) for carbohydrate and protein covered by a portion of 50g of gluten-free bread enriched with 30% okara.

Life Stage Group	Nutrients								
	Carbohydrate			Total Fiber			Proteins		
	DRI (g/d)	CBS (g/50g)	%DRI	DRI (g/d)	CBS (g/50g)	%DRI	DRI (g/d)	CBS (g/50g)	%DRI
Children									
1–3 y				16		36.84	13		33.85
4–8 y				25		28.00	19		23.16
Man									
9–13 y				31		22.58	34		12.94
14–18 y				38		18.42	52		8.46
19–50 y	130	14.45	11.12		7.00			4.40	
51–70 y				30		23.33	56		7.86
>70 y									
Females									
9–13 y				26		26.93	34		12.94
14–18 y									
19–70 y				21		33.33	46		9.57
>70 y									
Pregnancy	175		8.26	28		25.00	71		6.20
Lactation	210		7.19	29		24.14			

DRI-Dietary Reference Intakes [75]. No values for fat were detected, thus %RDI values for a portion of the tested bread are not presented in this study. CBS-Contribution of a bread serving (50 g). d—day; y—year.

A serving of gluten-free bread enriched with okara would meet the daily carbohydrate needs in the value of 11% for children, men, and females of all ages. However, the portion of analyzed gluten-free bread with okara would meet the daily needs in dietary fiber in a significantly higher percentage: for children aged 1–3 years ~37%, for children aged 4–8 years 28%, for men and females from 18 years to approximately 27%, depending on gender and age (Table 6). Such a high daily intake of dietary fiber is a consequence of the high presence of fiber in okara (~56–58% total dietary fibers of the total carbohydrate composition) [26]. The literature data show that a portion of gluten-free bread enriched with *Prosopis nigra* flour can contribute to the daily needs for dietary fiber in a significantly lower value (%DR = 10–16%) [74].

A portion of 50 g of analyzed gluten-free bread enriched with 30% okara would satisfy significant daily needs for macro- and micro-elements (Tables 7 and 8).

Table 7. Dietary reference intakes (%DRI) for macro-elements covered by a portion of 50 g of gluten-free bread enriched with 30% okara.

Life Stage Group	Macro-Elements					
	Ca	K	Mg	Na	P	S
Children						
1–3 y	93.04	3.34	48.57	33.29	41.64	68.74
4–8 y	58.15	2.63	29.89	19.52	38.31	
Men						
9–13 y	35.78	2.22	16.19		15.32	68.74
14–18 y			9.48	15.53		
19–30 y			9.71			
31–50 y	46.52	2.13			27.37	13.75
51–70 y			9.25	17.92		
>70 y	38.77			19.42		
Females						
9–13 y	35.78	2.22	16.19		15.32	68.74
14–18 y			10.79	17.92		
19–30 y			12.53			
31–50 y	46.52	2.13			27.37	13.75
51–70 y			12.14			
>70 y	38.77			19,42		
Pregnancy	46.52	2.13	11.10	15.53	27.37	/
Lactation		1.96	12.34			

y—year.

Table 8. Dietary reference intakes (%DRI) for micro-elements covered by a portion of 50 g of gluten-free bread enriched with 30% okara.

Life Stage Group	Micro-Elements					
	Fe	Zn	Mn	Cu	Cr	Ni
Children						
1–3 y	17.65	28.67	37.13	43.53	86.36	21.75
4–8 y	12.36	17.20	29.70	33.64	63.33	14.50
Men						
9–13 y	15.44	10.75	23.45	21.14	38.00	7.25
14–18 y	11.23		20.25	16.63	27.14	
19–50 y		7.82				4.35
51–70 y	15.44		19.37	16.44	31.67	
>70 y						
Females						
9–13 y	15.44	10.75	27.84	21.14	45.24	7.25
14–18 y	8.24	9.56	24.75	16.63	39.58	
19–50 y	6.86				38.00	
51–70 y	15.44	10.75	22.28	16.44	47.50	4.35
>70 y						
Pregnancy	4.58	7.82	22.28	14.80	31.67	
Lactation	13.73	7.17	17.13	11.38	21.11	

y—year.

2.6. Potential Benefits and Future Trends

In conclusion, there are several benefits of producing and consuming the formulated gluten-free bread enriched with 30% okara:

This bread does not contain gluten; thus, it is a suitable food for people who are intolerant to gluten and suffer from celiac disease. Studies have shown that 1% of the world's population suffers from celiac disease and the only solution is to consume a gluten-free diet [76].

The high content of dietary fiber and the absence of sugar allow people with insulin resistance and diabetes to consume this bread. The range of products with these characteristics is significant since 442 million people in the world live with diabetes and that number is rapidly increasing [77]. By 2030, it is predicted that 7.8% of the world's population will suffer from this disease [77]. While the prevalence of insulin resistance (metabolic syndrome) ranges from 15.5 to 46.5%, among adults, worldwide [78].

This bread is a source of nutritionally valuable proteins, which distinguish it from other gluten-free breads on the market and makes it suitable for athletes' nutrition. For example, prolonged daily training can increase the need for proteins, not only because they encourage good muscle work but also regenerate damaged tissues [79].

A very low content of saturated fatty acids in the tested bread is desirable. It is known that different saturated fatty acids (short-chain, medium-chain, and long-chain) contribute differently to increasing the level of LDL cholesterol in the blood, and that a complex food system (which can significantly affect the digestion of lipids and their absorption) has a major impact [59,80]. Today, there are epidemiological data that saturated fatty acids can negatively affect coronary heart disease and the development of some types of cancer [81,82]. It is recommended to reduce the intake of saturated fatty acids to less than 10% of the total daily energy consumption and to replace them with unsaturated fatty acids [61].

The low energy value makes this bread suitable for use in diets for treating obesity. Worldwide, more than 1 billion people are obese—39 million children, 340 million adolescents, and 650 million adults—and these numbers are increasing [83]. WHO [83] estimates that " . . . by 2025, approximately 167 million people–adults and children–will become less healthy because they are overweight or obese".

The presence of phenolics in the tested bread contributes to the antioxidant properties of this product. Thus, gluten-free bread enriched with 30% okara is a source of natural antioxidants, which is increasingly important in food preparation and nutrition. Namely, since about 1980, natural antioxidants have been used as an alternative to synthetic antioxidants. The toxicity of synthetic antioxidants was extensively studied indicating caution in their use. Therefore, in this sense, natural antioxidants represent nutritionally healthier and safer compounds compared to synthetic antioxidants [84,85].

Good sensory characteristics, above all good taste, provide the possibility of using this food by a wide group of consumers, who do not have metabolic and health problems, and want to eat healthy. Considering that bread is practically the most common food item in the diets of many people, and the low price of this product, it can be available to many consumers.

A potentially wide range of biologically active components (e.g., phenolics), minerals, and dietary fibers give this bread the properties of a "functional food". The functional food sector in the food industry, with an annual growth rate of 7.4%, has experienced significant growth in recent years. The reason for such large growth in this area is not only due to technological progress in the food industry, but also due to the formulation of new products that will meet the needs of consumers who today are increasingly aware of the need to consume nutritionally/bioactive compound rich food to act preventively and preserve health [86].

Finally, but not in the least, the production of this bread is in accordance with the principles of a sustainable and circular economy, as well as procedures for proper waste handling in the food industry.

Regardless of the efforts made in the food industry, the production of gluten-free bread still presents significant technological problems. This confirms the poor quality of gluten-free bread currently available on the market [87,88]. The absence of gluten has a huge, negative impact on product characteristics and dough rheology, as the dough is less elastic, and the bread has a poorer texture and smaller volume compared to products containing wheat flour [87–89]. This is why it is a big challenge to obtain gluten-free bread by selecting and using raw materials that can replace/imitate the role of gluten in a good way, in order to obtain a product of satisfactory quality [90]. The results of this study indicate that using soy okara as an alternative source of protein can achieve these goals.

3. Material and Methods

3.1. Material

Buckwheat, rice, millet flours, sunflour oil, yeast, guar gum, and salt were purchased in the local markets. The soybean variety used in the production of okara was "Olga" selected by the Maize research Institute Zemin Polje (Belgrade, Serbia). Transglutaminase was provided by the Purtos Group in Serbia (Bijgaarden, Belgum).

3.2. Okara Production

Okara was obtained according to the procedure proposed by Stanojevic et al. [20], with modification. Soaked soybeans (grain:water = 1:5, 14 h, at 15 °C) were ground and cooked (grain:water = 1:6, 30 min, at 100 °C) in a soy milk maker (Mester, D1158-W11QG, Dongfeng Town, Zhongshan City, Guangdong Province, China). After filtering of the soybean suspension (through a muslin cloth) and hand squeezing, soy milk and okara were obtained. The okara was then dried (to a moisture content of 15%) on a heated surface with constant manual stirring.

3.3. Gluten-Free Bread Production

Gluten-free flour composed of buckwheat flour 45%, rice flour 33%, and millet flour 22% was mixed with okara in the ratios of 90%:10%; 80%:20%, 70%:30%, respectively. The ingredients necessary for quality dough were therefore added to the mixture of gluten-free flour and okara: vegetable oil (4%), yeast (4%), guar gum (0.03%), transglutaminase (0.01%), salt (1.8%), and water (160%). Sucrose was not added during the preparation of the dough. After preparing the dough and pouring it into molds, fermentation (at 25 °C; 30 min) and baking (at 175–180 °C; 55 min) of gluten-free bread were followed. The control bread was prepared in the same procedure but without adding okara.

3.4. Sensory Analyses

Sensory evaluation of bread was carried out 24 h after baking by 8 trained expert panelists, according to the procedure by Purić et al. [91] as well as by 150 consumers (79 women and 71 men). Evaluation of sensory qualities was performed with a point-system (0–5) by both groups of evaluators and with a "hedonic scale" by consumers. In the process of scoring with a point-system, weighted quality scores were made with the following importance of coefficients: for taste–6, for smell, chewiness, and all parameters of the structure of cross-section–3 and for shape–2. The maximum value of the mark during the evaluation was 5, so the quality category had five levels: excellent quality (quality score > 4.5), very good quality (3.5 < score $\leq$ 4.5), good quality (2.5 < score $\leq$ 3.5), poor/unsatisfactory quality (1.5 < score $\leq$ 2.5), and very poor quality (score $\leq$ 1.5) [92].

Using the "hedonic scale" in evaluating the overall acceptability of gluten-free bread enriched with 30% okara, consumers were required to give their opinion by choosing one of the 9 offered sensory feelings: (1 = extremely dislike, 2 = very much dislike, 3 = moderately dislike, 4 = slightly dislike, 5 = neither like/nor dislike, 6 = slightly like, 7 = moderately like, 8 = very much like, and 9 = extremely like) [92].

Evaluation with both the point system and the "hedonic scale" was performed in two repetitions. The consumers were non-smokers, and the sensory analysis was performed anonymously. None of the evaluators was informed about the composition of the samples before the evaluation. Samples were served on glass trays, with random number labels.

3.5. Proximate Composition

Total protein content was determined by Kjeldahl method [93] with a 6.25 as conversion factor. The content of insoluble and soluble dietary fiber was determined according to the AOAC 991.43 method [94]. The content of total carbohydrates was determined according by difference [95] and the sugar content (glucose, fructose, and sucrose) was determined using the SRPS E.L8.007:1980 and SRPS E.L8.011:1980 method [96,97]. Total lipid content was determined by Soxhlet method [98] and the content of saturated fatty acids was determined by SRPS EN ISO 12966-2:2017 [99]. The salt content was determined using the method of Julshamn and Lea [100]. For the quantification of gluten, immunological ELISA assay was used [101]. The content of macro- and micro-elements was determined using ICP-OES analysis according by Kostić et al. [102]; (instrument model Thermo Scientific iCAP 6500 Duo ICP, Thermo Fisher Scientific, Cambridge, United Kingdom; with iTEVA operating software). Total phenolic content (TPC) was determined according to the procedure described by Pešić et al. [103] and results were expressed as milligrams of gallic acid equivalents per 100 g of fresh weight (mgGAE/100g).

3.6. Total Energy Value

The total energy value was calculated from the proximate composition using the content of protein, lipid, and carbohydrate, multiplied by their combustion equivalents in the body (17 kJ/g for protein, 37 kJ/g for lipid, 17 kJ/g for available carbohydrate and 8 kJ/g for available dietary fiber) [97]. The results were expressed on a fresh weight basis.

3.7. Determination of Antioxidant Properties

The antioxidant properties of gluten-free bread enriched with okara was tested using three different tests: ferric reducing power assay (FRP), ABTS$^{\bullet+}$ and DPPH$^{\bullet}$ scavenging activity (ABTS/DPPH) according to Milinčić et al. [104]. The results of FRP assay were expressed as milligrams of ascorbic acid equivalents per 100 g of fresh weight (mg AA/100g). Antioxidant activity determined by the ABTS and DPPH assays were expressed as milligrams of Trolox equivalent per 100 g of fresh weight (mgTrolox/100g).

3.8. Dietary Reference Intakes

Percentage of the dietary reference intakes of nutrients (%DčRI) was calculated according to Martos and López [74] for children (from 1–8 years), men and women (from 9 to more than 70 years), and for pregnant and lactating. The calculation was made for a portion of bread of 50 g, based on the guidelines for Dietary Reference Intakes for energy, carbohydrate, fiber, fat, fatty acids, protein, and amino acids (macronutrients) by the National Academy of Science Institute of Medicine [75].

3.9. Statistical Analysis

Statistica software version 8.0 (StatSoft Co., Tulsa, OK, USA) was applied for statistical analysis. Data are expressed as mean and standard deviation of three replicates, or as mean and pooled standard deviation (*Pooled std*) of two replicates. In statistical data processing, Pearson's correlation coefficients and Tukey's test at $p < 0.05$ were applied.

4. Conclusions

Dried okara can be a great alternative as a new food ingredient. Okara is a suitable nutritional supplement in the production of gluten-free bread. According to current regulations, the formulated bread based on buckwheat, rice, and millet enriched with 30% okara belongs to the group of "gluten-free products" and "products with increased protein content". The present research has shown that eco-innovative gluten-free bread with the addition (30%) of soy okara has the potential to be a source of carbohydrates, especially dietary fiber, high-value proteins, phenolic compounds, and macro- and micro-elements that are valuable in human health. The analyzed bread showed good antioxidant properties and high sensory scores, obtained for taste, shape, odor, chewiness, and cross-section properties. Gluten-free bread enriched with 30% okara can potentially meet significant values of the daily needs for carbohydrates, dietary fiber, protein, macro- and microelements of children, men, and females. In addition to its high nutritional value, this bread was distinguished by its low energy value, thus it can be suitable in diets intended for the treatment of obesity.

In addition, the production of this eco-innovative bread was in accordance with the sustainable method of production, circular economy, and proper waste management in food production, which is one of the important goals of modern life from an economic and ecological aspect.

Author Contributions: Conceptualization, M.B.P. and S.P.S.; formal analysis, M.B.P. and S.P.S.; funding acquisition, S.P.S.; investigation, M.M.P., J.B., A.B.S., P.I., D.D.M., M.D., A.Ž.K. and B.D.; methodology, M.B.P., M.D. and S.P.S.; supervision, M.B.P. and S.P.S.; writing—original draft, M.B.P. and S.P.S.; writing—review and editing, M.B.P. and S.P.S. All authors have read and agreed to the published version of the manuscript.

Funding: This work was supported by the Ministry of Education, Science and Technological Development of Republic of Serbia, Grant No. 451-03-9/2021-14/200116 and the Ministry of Science, Technological Development and Innovation of Republic of Serbia Grant No. 451-03-47/2023-01/200116.

Institutional Review Board Statement: Not applicable.

Informed Consent Statement: Not applicable.

Data Availability Statement: Data are contained within the article.

Conflicts of Interest: The authors declare no conflict of interest.

References

1. Official Gazette of the RS (2018): Rulebook on the Healthiness of Dietary Products. Available online: https://www.pravno-informacioni-sistem.rs/SlGlasnikPortal/eli/rep/sgrs/ministarstva/pravilnik/2010/45/2/reg (accessed on 28 March 2023). (In Serbian).
2. Egea, B.M.; De Sousa, L.T.; Dos Santos, C.D.; De Oliveira Filho, G.J.; Guimarães, M.R.; Yoshiara, Y.L. Application of soy, corn, and bean by-products in the gluten-free baking process: A Review. *Food Bioprocess Technol.* **2023**. [CrossRef]
3. Kennedy, N.P.; Feighery, C. Clinical features of coeliac disease today. *Biomed. Pharmacother.* **2000**, *54*, 373–380. [CrossRef] [PubMed]
4. Anderson, R.P. Coeliac disease. Review. *Aust. Fam. Physician* **2005**, *34*, 239–242. [PubMed]
5. Presutti, R.J.; Cangemi, J.R.; Cassidy, H.D.; Hill, D.A. Celiac Disease. *Am. Fam. Physician* **2007**, *76*, 1795–1802.
6. Torbica, A.; Hadnadev, M.; Dapčević, T. Rheological, textural and sensory properties of gluten-free bread formulations based on rice and buckwheat flour. *Food Hydrocolloid* **2010**, *24*, 626–632. [CrossRef]
7. Gómez, M. Gluten-free bakery products: Ingredients and processes. *Adv. Food Nutr. Res.* **2022**, *99*, 189–237. [CrossRef]
8. Demin, M. *Gluten-Free Grains and Cereals, New Technologies in Processing*; University of Belgrade, Faculty of Agriculture: Belgrade, Serbia, 2017; ISBN 978-86-7834-283-7. Available online: https://agris.fao.org/agris-search/search.do?recordID=RS2020000110 (accessed on 28 March 2023) (In Serbian)
9. Bhinder, S.; Kaur, A.; Singhb, B.; Yadav, P.M.; Singh, N. Proximate composition, amino acid profile, pasting and process characteristics of flour from different Tartary buckwheat varieties. *Food Res. Int.* **2020**, *130*, 108946. [CrossRef]
10. Espinoza-Herrera, J.; Martínez, L.M.; Serna-Saldívar, O.S.; Chuck-Hernández, C. Review methods for the modification and evaluation of cereal proteins for the substitution of wheat gluten in dough systems. *Foods* **2021**, *10*, 118. [CrossRef]
11. Przybylski, R.; Gruezynska, E. A review of nutritional and nutraceutical components of buckwheat. *Eur. J. Plant Sci. Biotechnol.* **2009**, *3*, 10–22.
12. Inglett, E.G.; Chen, D.; Berhow, M.; Lee, S. Antioxidant activity of commercial buckwheat flours and their free and bound phenolic compositions. *Food Chem.* **2011**, *125*, 923–929. [CrossRef]
13. Oppong, D.; Panpipat, W.; Chaijan, M. Chemical, physical, and functional properties of Thai indigenous brown rice flours. *PLoS ONE* **2021**, *16*, e0255694. [CrossRef]
14. Parameswaran, K.P.; Sadasivam, S. Changes in the carbohydrates and nitrogenous components during germination of proso millet, *Panicum miliaceum*. *Plant Food Hum Nutr.* **1994**, *45*, 97–102. [CrossRef]
15. Taylor, R.N.J. Millets: Their unique nutritional and health-promoting attributes, Chapter 4. In *Gluten-Free Ancient Grains Cereals, Pseudocereals, and Legumes: Sustainable, Nutritious, and Health-Promoting Foods for the 21st Century*; Taylor, R.N.J., Awika, M.J., Eds.; Elsevier: Duxford, UK, 2017; pp. 55–103. [CrossRef]
16. Becker, R.; Lorenz, K. Saccharides in proso and foxtail millets. *J. Food Sci.* **1978**, *43*, 1412–1414. [CrossRef]

17. Casey, P.; Lorenz, K. Millet–functional and nutritional properties. *Bakers Digest.* **1977**, *51*, 45–57.
18. Ferriola, D.; Stone, M. Sweetener effects on flaked millet breakfast cereals. *J. Food Sci.* **1998**, *63*, 726–729. [CrossRef]
19. Ostermann-Porcel, M.V.; Quiroga-Panelo, N.; Rinaldoni, N.A.; Campderrós, E.M. Incorporation of okara into gluten-free cookies with high quality and nutritional value. *J. Food Qual.* **2017**, *2017*, 4071585. [CrossRef]
20. Stanojevic, P.S.; Barac, B.M.; Pesic, B.M.; Jankovic, S.V.; Vucelic-Radovic, V.B. Bioactive proteins and energy value of okara as a byproduct in hydrothermal processing of soy milk. *J. Agric. Food Chem.* **2013**, *61*, 9210–9219. [CrossRef]
21. Guimarãesa, R.M.; Silvaa, T.E.; Lemesb, A.C.; Boldrina, M.C.F.; da Silvaa, M.A.P.; Guimarães, S.F.; Egeaa, M.B. A soybean by-product as an alternative to enrich vegetable paste. *LWT Food Sci. Technol.* **2018**, *92*, 593–599. [CrossRef]
22. Khare, S.K.; Jha, K.; Gandhi, A.P. Citric acid production from okara (soyresidue) by solid-state fermentation. *Bioresour. Technol.* **1995**, *54*, e323–e325. [CrossRef]
23. Kamble, D.B.; Rani, S. Bioactive components, in vitro digestibility, microstructure and application of soybean residue (okara): A review. *Legume Sci.* **2020**, *2*, e32. [CrossRef]
24. Tao, X.; Cai, Y.; Liu, T.; Long, Z.; Huang, L.; Deng, X.; Zhao, Q.; Zhao, M. Effects of pretreatments on the structure and functional prop-erties of okara protein. *Food Hydrocolloid* **2019**, *90*, 394–402. [CrossRef]
25. Colletti, A.; Attrovio, A.; Boffa, L.; Mantegna, S.; Cravotto, G. Valorisation of by-products from soybean (*Glycine max* (L.) Merr.) processing. *Molecules* **2020**, *25*, 2129. [CrossRef] [PubMed]
26. Van der Riet, W.B.; Wight, A.W.; Cilliers, J.J.L.; Datel, J.M. Food chemical investigation of tofu and its byproduct okara. *Food Chem.* **1989**, *34*, 193–202. [CrossRef]
27. Stanojevic, P.S.; Barac, B.M.; Pesic, B.M.; Vucelic- Radovic, V.B. Composition of proteins in okara as a by-product in hydrothermal processing of soymilk. *J. Agric. Food Chem.* **2012**, *60*, 9221–9228. [CrossRef] [PubMed]
28. Stanojevic, P.S.; Barac, M.B.; Pesic, M.B.; Zilic, M.S.; Kresovic, M.M.; Vucelic-Radovic, V.B. Mineral elements, lipoxygenase activity, and antioxidant capacity of okara as a byproduct in hydrothermal processing of soy milk. *J. Agric. Food Chem.* **2014**, *62*, 9017–9023. [CrossRef]
29. Kumar, V.; Rani, A.; Husain, L. Investigations of amino acids profile, fatty acids composition, isoflavones content and antioxidative properties in soy okara. *Asian J. Chem.* **2016**, *28*, 903–906. [CrossRef]
30. Lu, F.; Liu, Y.; Li, B. Okara dietary fiber and hypoglycemic effect of okara foods. *Bioact. Carbohydr. Diet. Fibre.* **2013**, *2*, 126–132. [CrossRef]
31. Li, S.; Zhu, D.; Li, K.; Yang, Y.; Lei, Z.; Zhang, Z. Soybean curd residue: Composition, utilization, and related limiting factors. *ISRN Ind. Eng.* **2013**, *2013*, 1–8. [CrossRef]
32. Salgado, J.M.; Donado-Pestana, M.C. Soy as a functional food. In *Soybean and Nutrition*; El-Shemy, H., Ed.; InTech: Changzhou, China, 2011; pp. 21–44.
33. Ostermann Porcel, V.M.; Campderrós, E.M.; Rinaldoni, N.A. Effect of Okara flour addition on the physical and sensory quality of wheat bread. *MOJ Food Process. Technol.* **2017**, *4*, 184–190. [CrossRef]
34. Fendri, L.; Chaari, F.; Maaloul, M.; Kallel, F.; Abdelkafi, L.; Ellouz Chaabouni, S.; Ghribi-Aydi, D. Wheat bread enrichment by pea and broad bean pods fibers: Effect on dough rheology and bread quality. *LWT-Food Sci. Technol.* **2016**, *73*, 584–591. [CrossRef]
35. Garrido, T.; Etxabide, A.; Leceta, I.; Cabezudo, S.; de la Caba, K.; Guerrero, P. Valorization of soya by-products for sustainable packaging. *J. Clean. Prod.* **2014**, *64*, e228–e233. [CrossRef]
36. Santos, D.C.D.; Oliveira, F.J.G.D.; Silva, J.D.S.; Sousa, M.F.D.; Vilela, M.D.S.; Silva, M.A.P.D.; Egea, M.B. Okara flour: Its physicochemical, microscopical and functional properties. *Nutr. Food Sci.* **2019**, *49*, 1252–1264. [CrossRef]
37. Shin, D.-J.; Kim, W.; Kim, Y. Physicochemical and sensory properties of soy bread made with germinated, steamed, and roasted soy flour. *Food Chem.* **2013**, *141*, 517–523. [CrossRef]
38. Guimarães, R.M.; Pimentel, T.C.; de Rezende, T.A.M.; de Santana Silva, J.; Falcão, H.G.; Ida, E.I.; Egea, M.B. Gluten-free bread: Effect of soy and corn co-products on the quality parameters. *Eur. Food Res. Technol.* **2019**, *245*, 1365–1376. [CrossRef]
39. Sandri, L.T.B.; Santos, F.G.; Fratelli, C.; Capriles, V.D. Development of gluten-free bread formulations containing whole chia flour with acceptable sensory properties. *Food Sci. Nutr.* **2017**, *5*, 1021–1028. [CrossRef]
40. Melini, F.; Melini, V.; Luziatelli, F.; Ruzzi, M. Current and forward-looking approaches to technological and nutritional im-provements of gluten-free bread with legume flours: A critical review. *Compr. Rev. Food Sci. Food Saf.* **2017**, *16*, 1101–1121. [CrossRef]
41. Yoshizawa, T.; Shimizu, T.; Yamabe, M.; Taichi, M.; Nishiuchi, Y.; Shichijo, N.; Unzai, S.; Hirano, H.; Sato, M.; Hashimoto, H. Crystal structure of basic 7S globulin, a xyloglucan-specific endo-β-1,4-glucanase inhibitor protein-like protein from soybean lacking inhibitory activity against endo-β-glucanase. *FEBS J.* **2011**, *278*, 1944–1954. [CrossRef]
42. Miñarro, B.; Albanell, E.; Aguilar, N.; Guamis, B.; Capellas, M. Effect of legume flours on baking characteristics of gluten-free bread. *J. Cer. Sci.* **2012**, *56*, e476–e481. [CrossRef]
43. Šmídová, Z.; Rysová, J. Gluten-free bread and bakery products technology. *Foods* **2022**, *11*, 480. [CrossRef]
44. Wang, T.; Xu, P.; Chen, Z.; Zhou, X.; Wang, R. Alteration of the structure of rice proteins by their interaction with soy protein isolates to design novel protein composites. *Food Funct.* **2018**, *9*, 4282–4291. [CrossRef]
45. Lin, S. Dietary fiber in bakery products: Source, processing, and function. *Adv. Food Nutr. Res.* **2022**, *99*, 37–100. [CrossRef] [PubMed]

46. Rehinan, Z.; Rashid, M.; Shah, H.W. Insoluble dietary fibre components of food legumes as affected by soaking and cooking processes. *Food Chem.* **2004**, *85*, 245–249. [CrossRef]
47. Poutanen, K.S.; Fiszman, S.; Marsaux, C.F.M.; Pentikäinen, S.P.; Steinert, R.E.; Mela, D.J. Recommendations for characterization and reporting of dietary fibres in nutrition research. *Am. J. Clin. Nutr.* **2018**, *108*, 437–444. [CrossRef] [PubMed]
48. Eastwood, M.A.; Morris, E.R. Physical properties of dietary fibre that influence physiological function: A model for polymers along the gastrointestinal tract. *Am. J. Clin. Nutr.* **1992**, *55*, 436–442. [CrossRef]
49. Du, H.; van der A, D.L.; Boshuizen, C.H.; Forouhi, N.G.; Wareham, N.J.; Halkjær, J.; Tjønneland, A.; Overvad, K.; Jakobsen, M.U.; Boeing, H.; et al. Dietary fibre and subsequent changes in body weight and waist circumference in European men and women. *Am. J. Clin. Nutr.* **2010**, *91*, 329–336. [CrossRef]
50. Dahm, C.C.; Keogh, R.H.; Spencer, E.A.; Greenwood, D.C.; Key, T.J.; Fentiman, I.S.; Shipley, M.J.; Brunner, E.J.; Cade, J.E.; Burley, V.J.; et al. Dietary fibre and colorectal cancer risk: A nested case–control study using food diaries. *JNCI J. Natl. Cancer I* **2010**, *102*, 614–626. [CrossRef]
51. Chuang, S.-C.; Norat, T.; Murphy, N.; Olsen, A.; Tjønneland, A.; Overvad, K.; Boutron-Ruault, M.C.; Perquier, F.; Dartois, L.; Kaaks, R.; et al. Fibre intake and total and cause-specific mortality in the european prospective investigation into cancer and nutrition cohort. *Am. J. Clin. Nutr.* **2012**, *96*, 164–174. [CrossRef]
52. Satija, A.; Hu, F.B. Cardiovascular benefits of dietary fibre. *Curr. Atheroscler. Rep.* **2012**, *14*, 505–514. [CrossRef]
53. WHO/FAO Expert Consultation on Diet NatPoCD. Diet, Nutrition, and the Prevention of Chronic Diseases. WHO Technical Report Series 91634-63. 2003. Available online: https://apps.who.int/iris/bitstream/handle/10665/42665/WHO_TRS_916.pdf;jsessionid=49613CE7E4E7BF52F2A6E5FD7F2CD55F?sequence=1 (accessed on 12 December 2022).
54. Stephen, A.M.; Champ, M.M.J.; Cloran, S.J.; Fleith, M.; van Lieshout, L.; Mejborn, H.; Burley, V.J. Dietary fibre in Europe: Current state of knowledge on definitions, sources, recommendations, intakes and relationships to health. *Nutr. Res. Rev.* **2017**, *30*, 149–190. [CrossRef]
55. Conte, P.; Fadda, C.; Piga, A.; Collar, C. Techno-functional and nutritional performance of commercial breads available in Europe. *Food Sci. Technol. Int.* **2016**, *22*, 621–633. [CrossRef]
56. Segura, M.E.M.; Rosell, M.C. Chemical composition and starch digestibility of different gluten-free breads. *Plant. Foods Hum. Nutr.* **2011**, *66*, 224–230. [CrossRef]
57. Kerezsi, A.D.; Jacquet, N.; Blecker, C. Advances on physical treatments for soy allergens reduction–A review. *Trends Food Sci. Technol.* **2022**, *122*, 24–39. [CrossRef]
58. FDA–U.S. Food and Drug Administration. Gluten and Food Labeling. 2018. Available online: https://www.fda.gov/food/nutrition-education-resources-materials/gluten-and-food-labeling (accessed on 12 December 2022).
59. Wahrburg, U. What are the health effects of fat? *Eur. J. Nutr.* **2004**, *43*, I/6–I/11. [CrossRef]
60. McCann, T.H.; Day, L. Effect of sodium chloride on gluten network formation, dough microstructure and rheology in relation to breadmaking. *J. Cer. Sci.* **2013**, *57*, 444e452. [CrossRef]
61. WHO, World Health Organization. Fat intake. 2013. Available online: https://www.who.int/data/gho/indicator-metadata-registry/imr-details/3418 (accessed on 14 December 2022).
62. Ma, C.-Y.; Liu, W.-S.; Kwokb, K.C.; Kwokb, F. Isolation and characterization of proteins from soymilk residue (okara). *Food Res. Int.* **1997**, *29*, 199–805. [CrossRef]
63. Falcinelli, B.; Calzuola, I.; Gigliarelli, L.; Torricelli, R.; Polegri, L.; Vizioli, V.; Benincasa, P.; Marsili, V. Phenolic content and antioxidant activity of wholegrain breads from modern and old wheat (*Triticum aestivum* L.) cultivars and ancestors enriched with wheat sprout powder. *Ital. J. Agron.* **2018**, *13*, 1220. [CrossRef]
64. Power, O.; Jakeman, P.; Fitzgerald, R.J. Antioxidative peptides: Enzymatic production, in vitro and in vivo antioxidant activity and potential applications of milk-derived antioxidative peptides. *Amino Acids* **2013**, *44*, 797–820. [CrossRef]
65. Gülçin, İ. Antioxidant activity of food constituents: An overview. *Arch. Toxicol.* **2012**, *86*, 345–391. [CrossRef]
66. Floegel, A.; Kim, D.-O.; Chung, S.-J.; Koo, S.I.; Chun, O.K. Comparison of ABTS/DPPH assays to measure antioxidant capacity in popular antioxidant-rich US foods. *J. Food Comp. Anal.* **2011**, *24*, 1043–1048. [CrossRef]
67. Banu, I.; Aprodu, I. Assessing the performance of different grains in gluten-free bread applications. *Appl. Sci.* **2020**, *10*, 8772. [CrossRef]
68. Turfani, V.; Narducci, V.; Durazzo, A.; Galli, V.; Carcea, M. Technological, nutritional and functional properties of wheat bread enriched with lentil or carob flours. *LWT–Food Sci. Technol.* **2017**, *78*, 361–366. [CrossRef]
69. Pavlović, D.D. Metabolism of water and minerals. Chapter XI. In *Biochemistry*; Busarčević, V., Ed.; "Savremena Administracija"a.d.: Belgrade, Serbia, 1995; pp. 817–881, ISBN 86-387-0750-9. (In Serbian)
70. Stanojević, S.; Pešić, M. Mineral elements. In *Food Biochemistry*; Radivojević, D., Ed.; University of Belgrade: Belgrade, Serbia, 2017; pp. 189–205. (In Serbian)
71. Ibidapo, O.P.; Henshaw, F.O.; Shittu, T.A.; Afolabi, W.A. Quality evaluation of functional bread developed from wheat, malted millet (*Pennisetum Glaucum*) and 'Okara' flour blends. *Sci. Afr.* **2020**, *10*, e00622. [CrossRef]
72. Maggio, A.; Orecchio, S.; Barreca, S. Review on chemical composition of gluten-free food for celiac people. Review Article. *Integr. Food Nutr. Metab.* **2019**, *6*, 1–11. [CrossRef]
73. Rybicka, I.; Gliszczyńska-Świgło, A. Minerals in grain gluten-free products. The content of calcium, potassium, magnesium, sodium, copper, iron, manganese, and zinc. *J. Food Comp. Anal.* **2017**, *59*, 61–67. [CrossRef]

74. Martos, A.G.T.; López, E.P. Chemical composition, percent of dietary reference intake, and acceptability of gluten free bread made from Prosopis nigra flour, added with hydrocolloids. *Food Sci. Technol.* **2018**, *38*, 619–624. [CrossRef]
75. National Academy of Science; Institute of Medicine. Dietary Reference Intakes for Energy, carbohydrate, Fiber, Fat, Fatty Acids, Protein, and Amino Acids (Macronutrients). Washington D.C. NAP. 2005. Available online: https://nap.nationalacademies.org/catalog/10490/dietary-reference-intakes-for-energy-carbohydrate-fiber-fat-fatty-acids-cholesterol-protein-and-amino-acids (accessed on 10 January 2023).
76. King, A.J.; Jeong, J.; Underwood, F.E.; Quan, J.; Panaccione, N.; Windsor, W.J.; Coward, S.; deBruyn, J.; Ronksley, P.E.; Shaheen, A.-A.; et al. Incidence of celiac disease is increasing over time: A systematic review and meta-analysis. *Am. J. Gastroenterol.* **2020**, *115*, 507–525. [CrossRef]
77. WHO. World Health Organization Global Report on Diabetes. 2016. Available online: https://www.who.int/health-topics/diabetes#tab=tab (accessed on 28 March 2023).
78. Saklayen, M.G. The global epidemic of the metabolic syndrome. *Curr. Hypertens. Rep.* **2018**, *20*, 12. [CrossRef]
79. Burke, L. *Practical Sports Nutrition*; Bahrke, M.S., Ed.; Human Kinetics: Champaign, IL, USA, 2007; pp. 1–26.
80. Astrup, A.; Bertram, C.S.H.; Bonjour, J.-P.; de Groot, C.P.L.; de Oliveira Otto, C.M.; Feeney, L.E.; Garg, M.L.; Givens, I.; Kok, F.J.; Krauss, R.M.; et al. WHO draft guidelines on dietary saturated and trans fatty acids: Time for a new approach? *BMJ Brit. Med. J.* **2019**, *366*, l4137. [CrossRef]
81. Glade, M.J. Food, nutrition and the prevention of cancer: A global perspective. American Institute for Cancer Research/World Cancer Research Fund, American Institute for Cancer Research, 1997. *Nutrition* **1999**, *15*, 523–526. [CrossRef]
82. Hu, F.B.; Manson, J.E.; Willett, W.C. Types of dietary fat and risk of coronary heart disease: A Critical Review. *J. Am. Coll. Nutr.* **2001**, *20*, 5–19. [CrossRef]
83. WHO, World Health Organization. World Obesity Day 2022—Accelerating Action to Stop Obesity. 2022. Available online: https://www.who.int/news/item/04-03-2022-world-obesity-day-2022-accelerating-action-to-stop-obesity (accessed on 16 January 2023).
84. Gupta, D. Methods for determination of antioxidant capacity: A review. *Int. J. Pharm. Sci. Res.* **2015**, *6*, 546–566. [CrossRef]
85. Augustyniak, A.; Bartosz, G.; Čipak, A.; Duburs, G.; Horáková, L.; Łuczaj, W. Natural and synthetic antioxidants: An updated overview. *Free Radic. Res.* **2010**, *44*, 1216–1262. [CrossRef]
86. Arshad, M.S.; Khalid, W.; Ahmad, S.R.; Khan, K.M.; Ahmad, H.M.; Safdar, S.; Kousar, S.; Munir, H.; Shabbir, U.; Zafarullah, M.; et al. Functional foods and human health: An overview. In *Functional Foods–Phytochemicals and Health Promoting Potential*; Arshad, M.S., Ahmad, H.M., Eds.; IntechOpen: London, UK, 2021.
87. Matos, M.E.; Rosell, C.M. Understanding gluten-free dough for reaching breads with physical quality and nutritional balance. *J. Sci. Food Agric.* **2015**, *95*, 653–661. [CrossRef]
88. Ren, Y.; Linter, B.R.; Linforth, R.; Foster, T.J. A comprehensive investigation of gluten free bread dough rheology, proving and baking performance and bread qualities by response surface design and principal component analysis. *Food Funct.* **2020**, *11*, 5333–5345. [CrossRef]
89. Duodu, K.G.; Taylor, J.R.N. The quality of breads made with non-wheat flours. In *Breadmaking*; Woodhead Publishing: Cambridge, UK, 2012; pp. 754–782.
90. Torbica, A.; Belović, M.; Tomić, J. Novel breads of non-wheat flours. *Food Chem.* **2019**, *282*, 134–140. [CrossRef]
91. Purić, M.; Rabrenović, B.; Rac, V.; Pezo, L.; Tomašević, I.; Demin, M. Application of defatted apple seed cakes as a by-product for the enrichment of wheat bread. *LWT Food Sci. Technol.* **2020**, *130*, 109391. [CrossRef]
92. Stanojevic, P.S.; Barać, B.M.; Pešić, B.M.; Vucelic-Radovic, V.B. Protein composition and textural properties of inulin-enriched tofu produced by hydrothermal process. *LWT Food Sci. Technol.* **2020**, *126*, 109309. [CrossRef]
93. AACC method 46-13 Crude protein-micro Kjeldahl method. In *Proceedings of the 10th in Approved Methods of the AACC: Vol. II AACC International*; American Association of Cereal Chemist. Approved Methods Committee: St Paul, MC, USA, 2000; ISBN 1891127128.
94. AOAC Official Method 991.43. ANNEX G: Total, Soluble, and Insoluble Dietary Fibre in Foods. 1995. Available online: https://acnfp.food.gov.uk/sites/default/files/mnt/drupal_data/sources/files/multimedia/pdfs/annexg.pdf (accessed on 28 March 2023).
95. FAO-Food and Agriculture Organization of the United Nations. Food energy–methods of analysis and conversion factors. *Food Nutr.* **2002**, *77*, 57–60.
96. SRPS E.L8.007:1980; Determination of Glucose and Fructose by Enzymatic Method (UV-Test). Institute of Standardization of Serbia: Belgrade, Serbia, 1980. Available online: https://iss.rs/sr_Cyrl/project/show/iss:proj:3372 (accessed on 28 March 2023).
97. SRPS E.L8.011:1980; Determination of Sucrose and Glucose by Enzymatic Method (UV-Test). Institute of Standardization of Serbia: Belgrade, Serbia, 1980. Available online: https://iss.rs/sr_Cyrl/project/show/iss:proj:3376 (accessed on 28 March 2023).
98. AOAC Method 16th. In *Official Methods of Analysis*, 16th ed.; Association of Official Analytical Chemists: Gaithersburg, MD, USA, 2006.
99. SRPS EN ISO 12966-2:2017; Animal and Vegetable Fats and Oils–Gas Chromatography of Fatty Acid Methyl Esters–Part 2: Preparation of Methyl Esters of Fatty Acids (ISO 12966-2:2017). Institute of Standardization of Serbia: Belgrade, Serbia, 2017. Available online: https://iss.rs/sr_Cyrl/project/show/iss:proj:63206 (accessed on 18 January 2023).

100. Julshamn, K.; Lea, P. Determination of sodium in foods by flame atomic absorption spectrometry after microwave digestion: NMKL interlaboratory study. *J. AOAC Int.* **2005**, *88*, 1212–1216. [CrossRef]
101. AOAC Official Method 2012.01. Gliadin as a Measure of Gluten in Foods Containing Wheat, Rye, and Barley. 2012. Available online: http://www.aoacofficialmethod.org/index.php?main_page=product_info&cPath=1&products_id=2965 (accessed on 21 January 2023).
102. Kostić, Ž.A.; Pešić, B.M.; Mosić, D.M.; Dojčinović, P.B.; Natić, N.M.; Trifković, Đ.J. Mineral content of some bee-collected pollen from Serbia. *Arch. Ind. Hyg. Toxicol.* **2015**, *66*, 251–258. [CrossRef]
103. Pešić, B.M.; Milinčić, D.D.; Kostić, Ž.A.; Stanisavljević, S.N.; Vukotić, N.G.; Kojić, O.M.; Gašić, M.U.; Barać, B.M.; Stanojević, P.S.; Popović, A.D.; et al. In vitro digestion of meat-and cereal-based food matrix enriched with grape extracts: How are polyphenol composition, bioaccessibility and antioxidant activity affected? *Food Chem.* **2019**, *284*, 28–44. [CrossRef] [PubMed]
104. Milinčić, D.D.; Stanisavljević, S.N.; Kostić, Ž.A.; Gašić, M.U.; Stanojević, P.S.; Tešić, L.Ž.; Pešić, B.M. Bioaccessibility of phenolic compounds and antioxidant properties of goat-milk powder fortified with grape-pomace-seed extract after in vitro gastrointestinal digestion. *Antioxidants* **2022**, *11*, 2164. [CrossRef] [PubMed]

Disclaimer/Publisher's Note: The statements, opinions and data contained in all publications are solely those of the individual author(s) and contributor(s) and not of MDPI and/or the editor(s). MDPI and/or the editor(s) disclaim responsibility for any injury to people or property resulting from any ideas, methods, instructions or products referred to in the content.

Article

The Effect of the Addition of Ozonated and Non-Ozonated Fruits of the Saskatoon Berry (*Amelanchier alnifolia* Nutt.) on the Quality and Pro-Healthy Profile of Craft Wheat Beers

Józef Gorzelany [1], Michał Patyna [1], Stanisław Pluta [2], Ireneusz Kapusta [3], Maciej Balawejder [4] and Justyna Belcar [1,*]

[1] Department of Food and Agriculture Production Engineering, University of Rzeszow, 4 Zelwerowicza Street, 35-601 Rzeszów, Poland; gorzelan@ur.edu.pl (J.G.); mpatyna95@gmail.com (M.P.)
[2] Department of Horticultural Crop Breeding, The National Institute of Horticultural Research, Konstytucji 3 Maja 1/3 Street, 96-100 Skierniewice, Poland; stanislaw.pluta@inhort.pl
[3] Department of Food Technology and Human Nutrition, University of Rzeszow, 4 Zelwerowicza Street, 35-601 Rzeszów, Poland; ikapusta@ur.edu.pl
[4] Department of Food Chemistry and Toxicology, University of Rzeszow, Ćwiklińskiej 1A Street, 35-601 Rzeszów, Poland; mbalawejder@ur.edu.pl
* Correspondence: justyna.belcar@op.pl

Citation: Gorzelany, J.; Patyna, M.; Pluta, S.; Kapusta, I.; Balawejder, M.; Belcar, J. The Effect of the Addition of Ozonated and Non-Ozonated Fruits of the Saskatoon Berry (*Amelanchier alnifolia* Nutt.) on the Quality and Pro-Healthy Profile of Craft Wheat Beers. *Molecules* **2022**, *27*, 4544. https://doi.org/10.3390/molecules27144544

Academic Editor: Adele Papetti

Received: 27 June 2022
Accepted: 14 July 2022
Published: 16 July 2022

Publisher's Note: MDPI stays neutral with regard to jurisdictional claims in published maps and institutional affiliations.

Copyright: © 2022 by the authors. Licensee MDPI, Basel, Switzerland. This article is an open access article distributed under the terms and conditions of the Creative Commons Attribution (CC BY) license (https://creativecommons.org/licenses/by/4.0/).

Abstract: Research into the suitability of domestic raw materials, including, for example, new wheat cultivars and fruit additives for the production of flavoured beers, is increasingly being undertaken by minibreweries and craft breweries. The fruits of the Saskatoon berry are an important source of bioactive compounds, mainly polyphenols, but also macro- and microelements. The fruits of two Canadian cultivars of this species, 'Honeywood' and 'Thiessen', were used in this study. Physicochemical analysis showed that wheat beers with the addition of non-ozonated fruit were characterised by a higher ethanol content by 7.73% on average. On the other hand, enrichment of the beer product with fruit pulp obtained from the cv. 'Thiessen' had a positive effect on the degree of real attenuation and the polyphenol profile. Sensory evaluation of the beer product showed that wheat beers with the addition of 'Honeywood' fruit were characterised by the most balanced taste and aroma. On the basis of the conducted research, it can be concluded that fruits of both cvs. 'Honeywood' and 'Thiessen' can be used in the production of wheat beers, but the fermentation process has to be modified in order to obtain a higher yield of the fruit beer product.

Keywords: Saskatoon berry fruit; ozonation; wheat beers; quality of fruity wheat beers

1. Introduction

Wheat beers are beverages for which several raw materials are used, such as: unmalted wheat (e.g., Witbier-style beers), wheat malt (generally 40% to 60% of the raw material charge), barley malt, hops, water and yeast [1,2]. A higher degree of turbidity and a more persistent yet delicate beer head and low bitterness sensation compared to barley beers are characteristics of wheat beers [3–6]. A typical wheat beer is a beverage fermented using a strain of yeast which is most commonly *Saccharomyces cerevisae*. The colour of wheat beer is light golden and often opaque. Wheat beers are characterised by their original palatability due to the wide range of compounds produced during the fermentation process (including phenols, aldehydes, esters and their derivatives) giving the sensation of vanilla, cloves, banana or fresh fruit, among others [1,3–5]. Wheat beers are also characterised by a high content of antioxidant compounds, including polyphenols [7].

Fruit beers have become a summer trend among beverages in recent years, mainly in the form of so-called radlers, that is, beer drinks characterised by the addition of fruit juice or fruit flavour [8,9]. The enrichment of beers with fruit increases the content of bioactive compounds and antioxidant activity of beer beverages, and also determines their sensory qualities, e.g., the taste and aroma of beers [10,11]. The most common beers on the world markets are cherry, raspberry, banana, strawberry or beers with the addition of exotic fruits. Due to the nutritional and health-promoting qualities of the Saskatoon berry (*Amelanchier alnifolia* Nutt.), it can be a valuable addition to the production of fruit beers.

The Saskatoon berry is a shrub belonging to the rose family found in Europe, North America, Africa and the eastern part of Asia [12,13]. It is most widely cultivated in Canada and recently also on a small scale in Finland, the Czech Republic, Lithuania, Latvia and Poland [13,14]. The fruits of the Saskatoon berry are a source of many health-promoting nutrients and can be used as functional food ingredients. They are particularly rich in soluble and insoluble fibre, vitamins such as tocopherol, riboflavin, ascorbic acid, pyridoxine, thiamine and riboflavin, minerals, i.e., manganese, magnesium, iron, calcium and potassium, sugars including sucrose, glucose, fructose and sorbitol, organic acids, protein and pectin. The caloric value of the fruit is averaged at 85 kcal/100 g [12]. The main groups of polyphenols found in the fruit of the Saskatoon berry include flavanols, anthocyanins, flavonols and phenolic acids [15]. The fruit peel of the Saskatoon berry is rich in anthocyanins, including cyanidin derivatives, flavonols and quercetin derivatives. The skin and pulp of the Saskatoon berry are rich in phenolic acids, including chlorogenic acid and neochlorogenic acid. Compared to other fruits, the Saskatoon berry has a 20% higher antioxidant content than cranberries [16], while a 40% higher content compared to aronia (*Aronia melanocaroa* L.) [17]. Other health-promoting properties of the fruit are related to the content of carotenoids and triterpenoids, which show anti-inflammatory effects [18,19]. Consumption of the Saskatoon berry has positive effects on vision, the cardiovascular system and contributes to a lower blood pressure [8].

A factor that can positively influence the production process and the fruit beer's quality is ozonation fruit and then their being added to the fermentation wort. Ozone is a chemical with strong oxidising properties that causes the disinfection of plant raw material subjected to the process, thus extending its technological shelf life [20]. The antimicrobial action of ozone influences the reduction of microbiological infections during the fermentation of beers with ozone-treated fruit, which has a significant impact both on the fermentation process itself and on the quality of the finished product (taste and aroma). Ozonation of fruits can be performed both before harvest (reducing the occurrence of diseases; e.g., grey mould—*Botrytis cinerea*) and at particular stages of raw material processing (improving processing properties [8]). The use of ozone treatment has a positive effect in the reduction of water losses during fruit storage, increasing antioxidant activity or reducing the release of ethylene by treated fruit. Ozone can be used in two forms: aqueous or gaseous, but studies on fruit ozonation have shown a better efficiency of the process with the latter form [21–23].

The purpose of this study was to determine the physicochemical properties, sensory properties and antioxidant activity of wheat beers with the addition of ozonated and non-ozonated Saskatoon berry fruits, and to determine the possibility of the practical application of the research results to expand the range of fruit beers and to use these fruits in a new food industry.

2. Results and Discussion
2.1. Physicochemical Characteristics of Fruit Wheat Beers

Fruit beers should be characterised by the colour of the finished product coming from the added fruit and show good sensory and health-promoting qualities. The results on the evaluation of the physical and chemical parameters of wheat beers are presented in Table 1.

Table 1. Physicochemical analysis of the beers produced with an addition of the Saskatoon berry fruit.

Type of Beer	CB	HB0	HB1	TB0	TB1
Apparent extract [%; m/m]	3.31 [a] ± 0.09	4.53 [c] ± 0.03	4.09 [b] ± 0.04	4.71 [d] ± 0.01	4.42 [c] ± 0.02
Real extract [%; m/m]	4.85 [a] ± 0.05	5.35 [c] ± 0.05	5.41 [c] ± 0.01	5.05 [b] ± 0.05	4.82 [a] ± 0.04
Original extract [%; m/m]	14.88 [e] ± 0.06	13.21 [b] ± 0.06	13.70 [d] ± 0.10	12.84 [a] ± 0.04	13.43 [c] ± 0.03
Degree of final apparent attenuation [%]	77.64 [e] ± 0.06	65.71 [b] ± 0.07	70.14 [d] ± 0.06	63.32 [a] ± 0.02	67.09 [c] ± 0.09
Degree of final real attenuation [%]	67.71 [d] ± 0.06	59.50 [a] ± 0.50	60.51 [b] ± 0.07	60.67 [b] ± 0.03	64.11 [c] ± 0.03
Content of alcohol [%; m/m]	5.24 [d] ± 0.04	4.08 [a] ± 0.07	4.32 [b] ± 0.02	4.04 [a] ± 0.04	4.48 [c] ± 0.04
Content of alcohol [%; v/v]	4.18 [d] ± 0.04	3.24 [a] ± 0.04	3.44 [b] ± 0.04	3.21 [a] ± 0.01	3.56 [c] ± 0.10
Colour [EBC units]	20.1 [a] ± 0.3	23.1 [b] ± 0.0	25.2 [d] ± 0.2	24.0 [c] ± 0.0	26.9 [e] ± 0.1
Titratable acidity [0.1M NaOH/100 mL]	3.46 [a] ± 0.06	3.55 [b] ± 0.05	3.64 [c] ± 0.03	3.71 [c] ± 0.02	4.22 [d] ± 0.03
pH	4.54 [b] ± 0.06	4.41 [a] ± 0.08	4.42 [a] ± 0.02	4.40 [a] ± 0.10	4.47 [b] ± 0.03
Content of carbon dioxide [%]	15.4 [d] ± 0.2	13.9 [c] ± 0.4	14.2 [c] ± 0.1	12.5 [a] ± 0.1	13.4 [b] ± 0.0
Bitter substances [IBU]	0.46 [a] ± 0.06	0.44 [a] ± 0.03	0.46 [a] ± 0.02	0.42 [a] ± 0.02	0.47 [a] ± 0.00
Energy value [kcal/100 mL]	57.22 [e] ± 0.02	50.41 [b] ± 0.06	52.34 [d] ± 0.61	48.88 [a] ± 0.13	51.01 [c] ± 0.08

Data are expressed as mean values ($n = 3$) ± SD; SD—standard deviation. Mean values within rows with different letters are significantly different ($p < 0.05$). [a,b,c,d,e]—statistically significant differences for the effect: physicochemical properties of beer × type of beer. CB—control wheat beer; HB—cv. 'Honeywood'; TB—cv. 'Thiessen'; 0—wheat beer with treated Saskatoon berry fruit; 1—wheat beer with non-treated Saskatoon berry fruit.

Wheat beers enriched with Saskatoon berry fruit pulp were characterised by an apparent extract of 4.09–4.71% m/m and a real extract of 4.82–5.41% m/m and were, on average, 24.94% and 6.01% higher than wheat beers without these fruit (CB; Table 1). Compared to barley beers enriched with Saskatoon berry fruit [8], wheat beers were characterised by a higher apparent extract by 33.3% on average and a real extract by 7.36% on average, which affected the attenuation of beer and a lower ethanol content.

As reported by Mascia et al. [24], beer attenuation significantly influenced the ethanol content, which was a determining factor in the beer beverages (content of alcohol) and in the taste and aroma profile of the finished product. In our study, the highest apparent attenuation among beers enriched with Saskatoon berry fruit was characterised by HB0 beer, while the real attenuation was characterised by TB1 beer. All fruit wheat beers were characterised by lower values of the assessed parameters (14.26% and by 9.62%, respectively) in relation to the control wheat beer—CB (Table 1). The lower attenuation of the fruit beers affected the ethanol content, which ranged from 4.04% v/v to 4.48% v/v. Beers with the addition of non-ozonated fruits of the Saskatoon berry were characterised by a higher ethanol content of 7.73% on average in comparison with beers enriched with ozonated fruits of this species (Table 1). Fruity barley beers (enriched with the pulp of the Saskatoon berry) studied by Gorzelany et al. [8] were characterised by a higher ethanol content of 5.03% v/v on average. In cherry and blueberry fruit-enriched beers, Yang et al. [11] investigated apple beer and cranberry beer and reported ethanol contents of 3.5% v/v and 3.6% v/v, respectively. The ethanol content in raspberry fruit beer was between 2.8–3.5% v/v [9]. In the study by Baigts-Allende et al. [10], cherry beers were characterised by an alcohol content in the range of 3.2–8.0% v/v, with raspberries being 2.5–5.7% v/v and blackcurrants being 7.1% v/v. In contrast, Nedyalkov et al. [25] obtained an ethanol content of 5.13% v/v in the barley beer with bilberry, whereas beers produced with the addition of mango juice and pulp were found with a calorific value of 34.13–36.73 kcal and alcohol contents of 4.13–4.27% v/v [26]. The fruit wheat beers' characterised caloric content of the finished product was at the level of 48.88–52.34 kcal/100 mL (Table 1). The caloric content of the barley beers enriched with the pulp of the ozonated and non-ozonated fruits of the Saskatoon berry was lower and averaged 44.83 kcal/100 mL [8].

The addition of pulp from Saskatoon berry fruits to wheat beers significantly affected the colour of the finished product and the process of the ozonating fruits reduced the colour intensity of the colour of the fruit beers by 8.01% on average, compared to wheat beers enriched with non-ozonated fruits (Table 1; Figures 1 and 2). The most intensive colour among the fruit wheat beers was characterised by beer enriched with the non-ozonated fruits of the Saskatoon berry cultivar 'Thiessen' (TB1). In the study by Gorzelany et al. [8], the average colour intensity of the barley beers enriched with fruits of the Saskatoon berry (ozonated and non-ozonated fruits) was 22.18 EBC units. Beers with added blackcurrants were characterised by a colour of 14.97 EBC units [10]. In the study by Patrașcu et al. [9], beers with raspberry fruit were marked by a colour of 21.16 EBC units.

Figure 1. The appearance of the obtained wheat beers with addition of the Saskatoon berry fruits (from left to right)—CB—control beer, HB0—cv. 'Honeywood' with ozone-treated fruit and HB1—untreated fruit.

Figure 2. The appearance of the obtained wheat beers with addition of the Saskatoon berry fruits (from left to right)—CB—control beer, TB0—cv. 'Thiessen' with ozone-treated fruit and TB1—untreated fruit.

Compared to the control wheat beer (CB), the fruit beers were characterised by a slightly higher acidity of the finished product, especially those enriched with non-ozonated fruits of the Saskatoon berry, cultivar 'Thiessen', designated as TB1 (Table 1). In addition, the fruits of the Saskatoon berry cultivar 'Thiessen' not subjected to the ozonation process were characterised by the highest acidity, both for fruits subjected to the ozonation process and fruits of the cultivar 'Honeywood' (ozonated and non-ozonated fruit). The pH value of all fruit beers and the control beer (CB) was at a similar level, from 4.40 to 4.54 (Table 1). In the study by Gorzelany et al. [8], the pH of barley beers (with the addition of Saskatoon berry fruits) averaged 4.53, and the acidity of the finished beer product was 2.2–2.3. In the study by Patrașcu et al. [9], beers with raspberry fruit were characterised by acidity and pH, respectively: 2.84–3.50 and 4.24. Nardini and Garaguso [27], when analysing fruit beers, found that the pH was in the range of 3.56–4.86. Adadi et al. [28] reported the pH value and acidity of beer enriched with sea-buckthorn berries amounting to 3.9 and 2.2, respectively. It is worth noting that the lower the pH of the finished beer product, the lower the risk of infection and development of undesirable microflora [26]. Similarly, fruits subjected to the ozonation process, which destroys or inactivates the microflora present in the fruit peel, may constitute a safer batch added on the seventh day of beer wort fermentation from the point of view of reducing microbiological risk.

In our studies, the addition of Saskatoon berry fruit reduced the perception of bitterness in wheat beers. The finished product enriched with ozonated fruits was characterised by a lower bitterness content of 4.35% in comparison with wheat beers with the addition of non-ozonated fruits of the Saskatoon berry (Table 1). The main factor that affects the bitterness sensation in beers is the used cultivar, its dose and the content of chemical compounds, including α-acids. The boiling time with hops is also an important factor, on which the rate of the protein–polyphenol reaction also depends [3,29]. The reduction in the bitterness sensation in fruity wheat beers is also related to the addition of pulp from the fruit of the Saskatoon berry, which is characterised by a relatively high sugar content; on average it was 14.78 g/100 g d.m., depending on the Saskatoon berry cultivar [18]. In our study, the carbon dioxide content of wheat beers, with or without the addition of Saskatoon berry fruit, ranged from 0.42% to 0.47% (Table 1). Gorzelany et al. [8] obtained a similar carbon dioxide content from barley beers with an addition of these fruits. Patraşcu et al. [9] reported contents of carbon dioxide in lemon beer samples in the range of 0.48–0.55%, in grapefruit beer amounting to 0.52% and in cranberry beer amounting to 0.55%.

2.2. Content of Bioactive Compounds in Fruit Beers

Beers are beverages, mostly alcoholic, but at the same time, they contain in their composition compounds of an antioxidant nature, the main representatives of which are polyphenols, but also vitamins, melanoids or bitter acids [30,31]. The antioxidant activity (determined by three methods: DPPH·, FRAP and ABTS$^+$) of wheat beers enriched with pulp from non-ozonated and ozonated fruits of the Saskatoon berry was presented in Table 2.

Table 2. Antioxidant potential of fruit beers with Saskatoon berry fruit pulp added.

Type of Beer	CB	HB0	HB1	TB0	TB1
DPPH· [mM TE/L]	2.27 [a] ± 0.07	2.34 [b] ± 0.04	2.94 [d] ± 0.01	2.71 [c] ± 0.07	2.42 [b] ± 0.02
FRAP [mM Fe^{2+}/L]	2.19 [d] ± 0.04	1.46 [a] ± 0.06	1.97 [c] ± 0.03	1.52 [a] ± 0.02	1.63 [b] ± 0.03
ABTS$^+$ [mM TE/L]	1.81 [a] ± 0.05	2.02 [b] ± 0.02	2.18 [c] ± 0.02	1.96 [b] ± 0.04	2.22 [c] ± 0.08

Data are expressed as mean values (n = 3) ± SD; SD—standard deviation. Mean values within rows with different letters are significantly different ($p < 0.05$). [a,b,c,d]—statistically significant differences for the effect: antioxidant activity of beer × type of beer. CB—control wheat beer; HB—cv. 'Honeywood'; TB—cv. 'Thiessen'; 0—wheat beer with treated Saskatoon berry fruit; 1—wheat beer with non-treated Saskatoon berry fruit; TE—expressed as Trolox equivalent (mM TE/L).

The higher antioxidant activity of wheat beers determined by the DPPH and ABTS methods was found in fruit beers, compared to beer without added fruit (control CB). The final products of the fermentation process enriched with non-ozonated Saskatoon berry showed on average 5.78% higher antioxidant activity determined by the DPPH method and 9.55% higher activity determined by the ABTS method compared to beers enriched with pulp from ozonated fruit pulp (Table 2.). Wheat beer without added fruit (CB) showed the highest reducing capacity of the beers (FRAP method). Similarly, as in the case of the antioxidant activity determined by the DPPH and ABTS methods of the analysed beers, finished products with added fruit pulp without ozonation were also characterised by higher activity, on average by 17.22% (Table 2). The antioxidant activity of the used cultivars of Saskatoon berry fruit (determined by the DPPH method) showed that the fruit subjected to the ozonation process was characterised by a slightly higher antioxidant activity in relation to the non-ozonated fruit. Barley beers enriched with Saskatoon berry fruit pulp had slightly lower antioxidant activity, determined by the DPPH and ABTS methods, while higher activity was determined by the FRAP method in relation to wheat fruit beers [8]. At the same time, barley beers showed a positive effect of the addition of ozonated fruits on antioxidant activity, in contrast to wheat beers, whose antioxidant activity was higher when pulp from non-ozonated fruits was added [8]. Deng et al. [32] enhanced beer with omija fruit added during the fermentation process and reported antioxidant activity, measured by a DPPH assay, amounting to 1.68 mM TE/L, and reducing capacity, assessed with FRAP, at a level of 2.4 mM Fe^{2+}/L. Portuguese commercial fruit beers with lemon flavour were reported to have an antioxidant capacity in the range of 0.035–0.037 mM TE/L, according to the DPPH assay, and at a level of 0.008 mM TE/L, according to the ABTS assay [33]. The Saskatoon berry fruit, as an addition to the analysed beers, showed very high antioxidant activity, such as for the cultivar 'Honeywood'—21 mM/100 g d.m. (by the FRAP method) and 31.06 mM/100 g d.m. (by $ABTS^+$ method), and for the cultivar 'Thiessen'—32.32 mM/100 g d.m. (determined by the $ABTS^+$ method [15,18].

Polyphenolic compounds present in beers are mainly derived from the malt (70–80%) and the hops used [34]. The degree of fineness of the malt, as well as the conditions of the mashing and boiling process with hops, significantly affect the total polyphenol content [29]. Polyphenolic compounds are diverse substances with different biologically active effects, including antioxidant and antiradical activity [35]. We confirmed that the addition of Saskatoon berry fruits to wheat beer increased the content of the total polyphenols, on average by 37.37% compared to beer without the addition of fruits (CB). The differences in the content of polyphenolic compounds in fruit beers (addition of non-ozonated and ozonated fruit) were statistically significant (Table 3). The degree of the transfer of phenolic compounds contained in the fruit to wheat beer depends on the degree of grinding of the fruit. The use of fruit pulp as an input to the fermenting wort increases the contact with the solution, which leaches and transfers the chemical compounds found in the fruit through the disrupted cell wall, thus enriching the finished beer product [26]. According to Gorzelany et al. [8], fruity barley beers with the addition of the Saskatoon berry had an average total polyphenol content of 381 mg GAE/L for beers enriched with non-ozonated fruit and 388 mg GAE/L for beers with the addition of ozonated fruit. The data from the literature showed that the total polyphenol content of the beers enriched with Cornelian cherry was 350 mg GAE/L [36] and with goji berries was 415 mg GAE/L [37]. The addition of persimmon juice led to a decrease in the total polyphenol content in the beer samples from 433.32 mg GAE/L (25% juice addition) to 290.34 mg GAE/L (75% juice addition; [34]). Portuguese commercial fruit beers with lemon flavour were found with total polyphenol contents in the range of 240–304 mg GAE/L [33].

Table 3. Contents of polyphenols and polyphenolic profile identified by UPLC-PDA-TQD-MS in wheat beer.

No.	Compound [mg/L]	Rt [min]	[M-H]⁻ (m/z)	Fragment ions (m/z)	Absorbance maxima (nm)	Type of Beer				
						CB	HB0	HB1	TB0	TB1
	Contents of polyphenols [mg GAE/L]					243.90 ^{a} ± 1.85	382.83 ^{c} ± 0.92	413.43 ^{d} ± 0.76	383.65 ^{c} ± 0.43	377.86 ^{b} ± 0.46
1.	K-3-O-sophoroside	3.97	609	285	264, 324	0.95 ± 0.02	n.d.	n.d.	n.d.	n.d.
2.	K-3-O-rut-7-O-glc	4.09	755	593, 285	264, 324	0.92 ± 0.02	n.d.	n.d.	n.d.	n.d.
3.	K-3-O-glc-7-O-glc	4.20	609	447, 285	264, 324	1.35 ± 0.04	n.d.	n.d.	n.d.	n.d.
4.	Neo-chlorogenic acid	2.88	353	191	299sh, 324	n.d.	0.71 ^{a} ± 0.08	1.21 ^{c} ± 0.12	0.82 ^{a} ± 0.14	1.07 ^{b} ± 0.09
5.	Chlorogenic acid	3.54	353	191	299sh, 324	n.d.	0.83 ^{a} ± 0.08	2.17 ^{c} ± 0.36	0.51 ^{a} ± 0.05	1.46 ^{b} ± 0.01
6.	Sinapic acid glucoside	4.04	385	223	299sh, 326	n.d.	1.63 ^{b} ± 0.04	1.05 ^{a} ± 0.09	2.18 ^{c} ± 0.01	2.23 ^{c} ± 0.26
7.	Caffeic acid	4.13	179	161	299sh, 327	n.d.	0.57 ^{a} ± 0.00	0.96 ^{b} ± 0.01	0.90 ^{b} ± 0.10	0.87 ^{b} ± 0.09
8.	K-3-O-glc-pent	4.40	579	285	264, 350	n.d.	0.76 ^{b} ± 0.02	0.66 ^{a} ± 0.08	0.73 ^{b} ± 0.07	0.80 ^{b} ± 0.02
9.	K-3-O-rut	4.51	593	285	264, 350	n.d.	0.79 ^{a} ± 0.02	0.78 ^{a} ± 0.08	0.92 ^{a} ± 0.02	0.81 ^{a} ± 0.00
10.	K-3-O-rha-7-O-pent	4.63	563	431, 285	264, 344	n.d.	1.00 ^{b} ± 0.03	0.97 ^{ab} ± 0.02	1.03 ^{bc} ± 0.00	0.94 ^{a} ± 0.02
11.	Ferulic acid derivative	5.42	610	193	299sh, 327	n.d.	1.09 ^{c} ± 0.05	0.79 ^{a} ± 0.00	1.02 ^{b} ± 0.02	1.00 ^{b} ± 0.03
	Total					3.22 ^{a} ± 0.08	7.37 ^{b} ± 0.05	8.57 ^{c} ± 0.69	8.10 ^{bc} ± 0.34	9.16 ^{cd} ± 0.40

Data are expressed as mean values ($n = 3$) ± SD; SD—standard deviation. Mean values within rows with different letters are significantly different ($p < 0.05$), ^{a,b,c,d}—statistically significant differences for the effect: contents of polyphenols and polyphenolic profile of beer × type of beer. CB—control wheat beer; HB—cv. 'Honeywood'; TB—cv. 'Thiessen'; 0—wheat beer with treated Saskatoon berry fruit; 1—wheat beer with non-treated Saskatoon berry fruit. K—kaempferol; glc—glucoside; rut—rutinoside; pent—pentoside; rha—rhamnoside; GAE—equivalent of gallic acid (mg GAE/L).

The identification of polyphenolic compounds in fruity wheat beers was based on the analysis of characteristic spectral data: the mass-to-charge ratio m/z and the maximum of radiation. In our study, a total of 11 polyphenolic compounds were identified, whose spectral properties are presented in Table 3. In control wheat beer (CB), three polyphenolic compounds belonging to the flavonol group (compounds 1–3) were identified. Their representatives were kaempferol derivatives, of which the highest concentration (1.35 mg/L) was determined for K-3-O-glucoside-7-O-glucoside (Table 3). In wheat beers enriched with fruits of the Saskatoon berry, eight compounds belonging to the group of hydroxycinnamic acid derivatives (compounds 4–7; 11) and to the group of flavonols (compounds 8–10) were identified. The content of the polyphenolic compounds in wheat beers enriched with ozonated Saskatoon berry fruits was, on average, 7.74 mg/L, while with the addition of non-ozonated fruits, it was slightly higher and amounted, on average, to 8.87 mg/L (Table 3). The results of the present study confirmed the effect of ozonation on the reduction of the content of polyphenolic compounds in beers with the addition of fruits of the Saskatoon berry, which was previously obtained by Gorzelany et al. [8]. The lower concentration of polyphenolic compounds in ozonated fruit-enriched beers is most likely related to the interaction between the ozone remaining on the fruit and the products of the fermentation process, but this hypothesis is not fully understood. However, there are results in the international literature that confirm the positive effect of ozone on the polyphenolic profile of different fruits [20,38–41].

Among the phenolic acids present in wheat beers, the content of the chlorogenic acid content was on average 63.09% higher in beers enriched with non-ozonated fruits of the Saskatoon berry. The highest concentration of this acid (2.17 mg/L) was determined for beer with the addition of non-ozonated fruits of the 'Honeywood' cultivar, designated as HB1 (Table 3). Chlorogenic acid was present in ozonated fruit wheat beers, in contrast to ozonated fruit barley beers in which its presence was not detected. In contrast, the main representative of the polyphenolic compounds in barley beers with added fruit was caffeic acid, whose concentration averaged 4.01 mg/L [8]. In the wheat beers with the addition of Saskatoon berry fruit, the caffeic acid was between 0.57 and 0.96 mg/L (Table 3). From the health point of view, caffeic acid is responsible for blocking some substances with carcinogenic effect, e.g., nitrosamines, and it also affects the oxidation process of lipoproteins and LDL cholesterol fraction [42]. The caffeic acid in beers is generally in the range of 0.00–23.50 mg/L [43]. In barley beers with bilberry (fruit addition in the amount of 167 g/L), the content of caffeic acid was 13.01 mg/L, chlorogenic acid was 90.19 mg/L and neochlorogenic acid was 52.24 mg/L [25]. Wheat beers enriched with fruits of the cv. 'Thiessen' of the Saskatoon berry were further characterised by a high concentration of sinapic acid derivative; on average, it was 50.0% more than in fruits of the cv. 'Honeywood' (Table 3). Kaempferol glycosides have also been identified in fruit wheat beers, which have strong antioxidant, anticancer and supportive properties in cardiovascular diseases. In addition, they can be supportive substances in autoimmune diseases and for transplant patients [44]. In barley beers, kaempferol compounds are most commonly found at 0.10—1.64 mg/L [43]. Flavonoid glycosides (including kaempferol derivatives), as well as chlorogenic acid, caffeic acid or sinapic acid contained in beers impart astringency and acidity sensations in the mouth, as well as, although to a much lower extent, also a bitterness sensation which affects the sensory experience of the finished beer product [44].

2.3. Sensory Analysis of Fruit Wheat Beers

The sensory characteristics of a finished fruit beer product have a significant impact on its attractiveness and acceptance by consumers. The taste and aroma qualities of fruity wheat beers can influence consumers' preference to purchase a particular beer, or this purchase will only be a one-off. The results of the sensory evaluation of fruit wheat beers carried out by a 13-member panel are presented in Table 4 and Figures 3 and 4.

Table 4. Sensory analysis of fruit wheat beer.

Type of Beer	CB	HB0	HB1	TB0	TB1
Aroma	4.20 [a] ± 0.38	4.23 [a] ± 0.83	4.15 [a] ± 0.99	4.23 [a] ± 0.73	4.00 [a] ± 1.00
Taste	3.79 [a] ± 0.27	4.08 [ab] ± 1.34	4.69 [b] ± 0.48	4.54 [b] ± 0.66	4.08 [ab] ± 0.76
Foam stability	3.51 [b] ± 0.17	2.46 [a] ± 0.97	2.69 [a] ± 0.48	3.00 [ab] ± 0.58	3.08 [ab] ± 0.86
Bitterness	4.06 [a] ± 0.11	3.62 [a] ± 0.87	3.92 [a] ± 0.86	3.77 [a] ± 0.93	3.46 [a] ± 0.52
Saturation	3.71 [a] ± 0.32	4.31 [ab] ± 0.63	4.84 [c] ± 0.38	4.15 [ab] ± 0.80	4.38 [bc] ± 0.65
Overall impression	3.91 [a] ± 0.47	3.82 [a] ± 0.75	4.18 [a] ± 0.48	4.11 [a] ± 0.49	3.83 [a] ± 0.57

Data are expressed as mean values ($n = 3$) ± SD; SD—standard deviation. Mean values within rows with different letters are significantly different ($p < 0.05$).). [a,b,c]—statistically significant differences for the effect: sensory analysis of beer × type of beer. CB—control wheat beer; HB—cv. 'Honeywood'; TB—cv. 'Thiessen'; 0—wheat beer with treated Saskatoon berry fruit; 1—wheat beer with non-treated Saskatoon fruit.

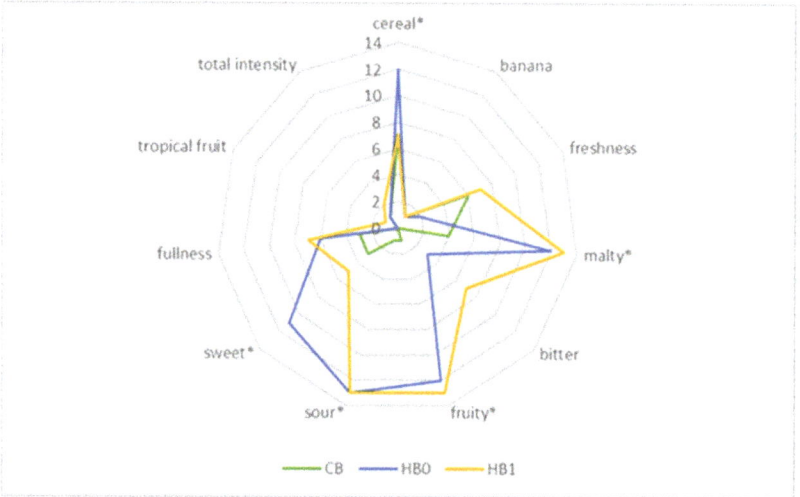

Figure 3. Sensory profile of wheat beers—control (CB) and sample with addition of the cv. 'Honeywood' fruit untreated (HB1) and treated with ozone (HB0) (* marks the attributes which were statistically different at $p \leq 0.05$).

Fruity wheat beers were characterised by a similar smell sensation of the finished product (4.00–4.23 points, on a 5-point rating scale), but with a statistically different taste. The highest taste sensory ratings were obtained for wheat beer enriched with non-ozonated fruit of the cv. 'Honeywood' (Table 4, Figure 3) and with ozonated fruit of the cv. 'Thiessen' (Table 4., Figure 4). The taste and smell of beer are influenced not only by the raw materials used, but also by the products of the fermentation process (such as aldehydes, phenols, or esters) affecting the taste profile of a given beer. Among the quality attributes of fruity wheat beers, the stability of the beer head was assessed the lowest, especially for beers enriched with the cv. 'Honeywood' fruit irrespective of the applied ozonation process (Table 4). Sensory evaluation confirmed the results of the physicochemical analysis regarding the lower bitterness sensation in fruity wheat beers compared to the control beer (CB).

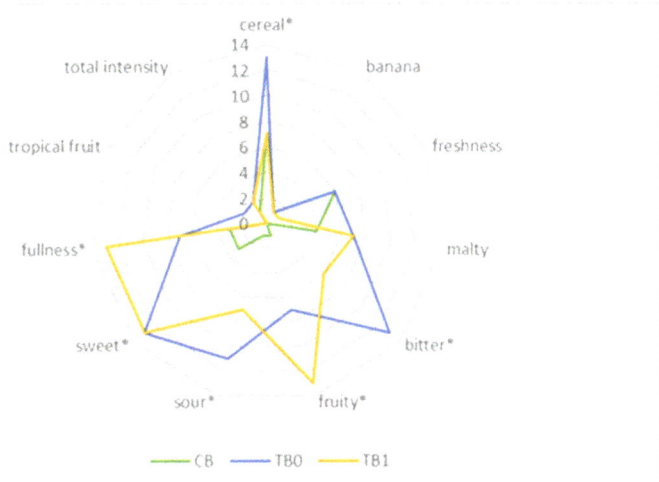

Figure 4. Sensory profile of wheat beers—control (CB) and sample with addition of the cv. 'Thiessen' fruit untreated (TB1) and treated with ozone (TB0) (* marks the attributes which were statistically different at $p \leq 0.05$).

The sensory profile of the wheat fruit beers varied. Sensory evaluation showed that the most balanced flavour profile was characterised by wheat beers enriched with unfermented the cv. 'Honeywood' fruit irrespective of the cultivar used (HB1 and TB1; Figures 3 and 4). The significant quality attributes of the fruited wheat beers with the cv. 'Honeywood' fruit were cereal and malty, fruity, sweet and sour tastes and aroma (Figure 3.). On the other hand, wheat beers enriched with fruit from the cv. 'Thiessen' additionally had a bitter aftertaste, especially for beers with ozonated fruit added (Figure 4). The aftertaste of the sour, astringent or bitter fruit of wheat beers is related to the varying content of the polyphenols responsible, including caffeic acid and chlorogenic acid [29]. In a study on the possibility of enriching barley beers with the Saskatoon berry, Gorzelany et al. [8] obtained similar results for the flavour and aroma profile and a bitter aftertaste was also clearly perceived in the finished product, especially in the finished product enriched with non-ozonated fruit. Chemical compounds important for beer flavour are formed by interactions between carbonyl compounds, esters, sulphur compounds, alcohols, phenolic compounds or organic acids [45]. Beers characterised by fruity notes with a sweet aftertaste and pleasant aroma are more preferred and desired by consumers compared to traditional types of beers [28,46].

3. Materials and Methods

3.1. Material

Common wheat (*Triticum aestivum* L.), the winter variety 'Elixer', was used to produce wheat beers. The grain came from a field experiment collected in the year 2021, in Przeworsk (50°03'31" N 22°29'37" E), Podkarpackie Province (south-east Poland). After full maturity, the grain was harvested and a 5-day wheat malt was prepared (the malting process methodology is described by Belcar et al. [47]). The wheat malt had the following characteristics: extract potential—85.7% d.m. (d.m.—dry matter), total protein content—11.6% d.m., content of soluble protein—4.67% d.m., diastatic power—324 WK, and degree of final attenuation—82.14%.

Commercial barley malt from the Viking Malt malting plant in Strzegom (Poland) was also used to brew the beers. The barley malt had the following characteristics: extract potential—80.0% d.m., total protein content—11.4% d.m., content of soluble protein—3.75% d.m., diastatic power—324 WK, and degree of final attenuation—82.1%. The wheat and barley malts were refined in a Cemotec disc mill (FOSS). The brewing stock consisted of 40% wheat malt and 60% commercial barley malt.

Fruits of two Canadian cultivars of Saskatoon berry 'Honeywood' and 'Thiessen' were used to enrich the wheat beers. Ripe fruits weighing 2 kg each were harvested by hand from 6-year-old bushes grown in an implementation experiment in a field at the Experimental Orchard in Dąbrowice (51.9163° N/20.1009° E) of the National Institute of Horticultural Research (InHort) in Skierniewice, central Poland, at the beginning of July 2021. In the laboratory of the Department of Agriculture and Food Production Engineering of the University of Rzeszów, the Saskatoon berry fruits were divided into two samples of 1 kg each (one part was left without ozonation and the other part was ozonated). Until beers were produced, ozonated and non-ozonated fruits of Saskatoon berry were directly frozen and stored in a freezer (temp. -18 °C). The non-ozonated Saskatoon fruit cultivar 'Honeywood' had the following chemical parameters: total polyphenol content— 3.67 g GAE·1000 g^{-1} d.m., antioxidant activity (DPPH test)—17.38 mM TE·100g^{-1} d.m., and total acidity—0.503 g·100 g^{-1}, whereas cultivar 'Thiessen' had the following chemical parameters: total polyphenol content—5.16 g GAE·1000 g^{-1} d.m., antioxidant activity (DPPH test)—20.97 mM TE·100g^{-1} d.m., and total acidity—0.963 g·100 g^{-1}.

3.2. Ozonation Process

The Saskatoon berry fruits of both cultivars were placed on a metal grid inside a plastic container with dimensions L × W × H—0.6 × 0.4 × 0.4 m—and ozonated for 22 min—ozone concentration 10 ppm, flow time 4 $m^3 \cdot h^{-1}$, temperature 20 °C. The TS 30 ozone generator (Ozone Solution, Hull, MA, USA) with a 106 M UV Ozone Solution detector (Ozone Solution, Hull, MA, USA) was used to generate ozone. The ozone-treated Saskatoon fruit cultivar 'Honeywood' had the following chemical parameters: total polyphenol content— 3.81 g GAE·1000 g^{-1} d.m., antioxidant activity (DPPH test)—17.40 mM TE·100g^{-1} d.m., and total acidity—0.352 g 100 g^{-1}, whereas cultivar 'Thiessen' had the following chemical parameters: total polyphenol content—3.23 g GAE·1000 g^{-1} d.m., antioxidant activity (DPPH test)—21.12 mM TE·100g^{-1} d.m., and total acidity—0.691 g·100 g^{-1}.

3.3. Beer Production

The production process was carried out using the infusion method in the laboratory of the Department of Agriculture and Food Production Engineering, University of Rzeszów. Total of 3.0 kg of barley malt and 2.0 kg of wheat malt were grated and placed in a ROYAL RCBM-40N mash kettle (Expondo; Poland; assuming a process efficiency of 80%) and 15.0 L of water (3 L of water for each kilogram of malt). The mashing, boiling process with hops and cooling of the beer wort were carried out according to the methodology described by Gorzelany et al. [8].

Each of the five beer worts produced was characterised by an extract of 12.0 °P. The cooled worts were transferred to fermentation containers with a capacity of 30 L each and inoculated with *Saccharomyces cerevisae* Fermentis Safale US-05 yeast (6×10^9/g), which had previously undergone a rehydration process, according to the manufacturer's instructions (0.58 g d.m./L of wort). The fermentation process was carried out at 21 °C. After 7 days of fermentation, 1 kg of Saskatoon berry fruit was added to the fermenting beer in the form of pulp and left to ferment for another 14 days. After 21 days, the beers were bottled, with a solution of sucrose (0.3%) added to water for refermentation and to obtain an appropriate degree of beer saturation. The resulting beers were kept at 20 °C. Sensory and physicochemical analyses were performed one month after bottling.

Wheat beers enriched with the cv. 'Honeywood' fruit were designated as HB0 (ozonated fruit) and as HB1 (non-ozonated fruit), while wheat beers with the cv. 'Thiessen' fruit were designated as TB0 (ozonated fruit) and as TB1 (non-ozonated fruit). The wheat beer without added fruit as a control was designated CB. A total of 5 wheat beers were produced.

3.4. Analysis of Quality Indicators of Beers

Alcohol content [% m/m and % v/v], apparent extract [% m/m], real extract [% m/m] and original extract [% m/m] of beer, and apparent [%] and real [%] attenuation degree was marked according to method 9.4 EBC [48]. The titratable acidity of fruit wheat beers was determined by subjecting beer samples to titration with 0,1 M NaOH, with end point at pH = 8.2. The energy value of wheat beers was calculated following the formula: [kcal/100 mL] = (7 × A (% v/v) + (4 × Er (% v/v) × ρ). pH, colour [EBC units], carbon dioxide content [%] and bitterness content [IBU units] of beer were determined according to the methodology described by Belcar et al. [1]. The analyses were performed in three replications.

3.5. Content of Bioactive Compounds in Fruit Beers

The total polyphenol content [mg GAE/L], by using the Folin–Ciocalteu method, and the polyphenol profile [mg/L] in the beers were determined according to the methodology described by Gorzelany et al. [8]. Determination of polyphenolic compounds [mg/L] was carried out using the UPLC equipped with a binary pump, column and sample manager, photodiode array detector (PDA), and tandem quadrupole mass spectrometer (TQD) with electrospray ionisation (ESI) source working in negative mode (Waters, Milford, MA, USA), according to the method of Żurek et al. [49]. Separation was performed using the UPLC BEH C18 column (1.7 µm, 100 mm × 2.1 mm, Waters) at 50 °C, at flow rate of 0.35 mL/min. The injection volume of the samples was 5 µL. The mobile phase consisted of water (solvent A) and 40% acetonitrile in water, v/v (solvent B). The following TQD parameters were used: capillary voltage of 3500 V, con voltage of 30 V, con gas flow 100 L/h, source temperature 120 °C, desolvation temperature 350 °C and desolvation gas flow rate of 800 L/h. Polyphenolic identification and quantitative analyses were performed on the basis of the mass-to-charge ratio, retention time, specific PDA spectra, fragment ions and comparison of data obtained with commercial standards and literature findings. The analyses were performed in three replications.

3.6. Antioxidant Activity

The antioxidant activity of fruit beers (by DPPH· [mM TE/L], FRAP [mM Fe^{2+}/L] and $ABTS^+$ [mM TE/L]) was determined according to the methodology described by Gorzelany et al. [8].

3.6.1. DPPH Test

A 0.05 mM/L solution of DPPH (2,2-diphenyl-1-picrylhydrazyl) in ethanol was prepared for this purpose. A 7.8 mL sample of the solution was placed in a test tube with 0.2 mL of diluted (2×) beer and incubated in darkness for 60 min at 37 °C; subsequently, the absorbance at the wavelength λ = 517 nm was examined using a UV-Vis V-5000 spectrophotometer (Shanghai Metash Instruments Co. Ltd., Shanghai, China). The control contained distilled water rather than beer. The results were expressed as trolox equivalent (mM TE/L).

3.6.2. FRAP Test

The materials prepared for this purpose included a 10 mM/L TPTZ (2,4,6-tripyridyl-s-triazine) solution, a 20 mM/L $FeCl_3·6H_2O$ solution, an acetate buffer with pH = 3.6, as well as a 40 mM/L HCl solution. Subsequently, the FRAP reagent was prepared by mixing 25 mL of the acetate buffer with 2.5 mL of the TPTZ dissolved in HCl and 2.5 mL of $FeCl_3·6H_2O$. A 6 mL sample of FRAP solution was placed in a test tube with 0.2 mL of the beer and incubated at a temperature of 37 °C for 10 min; subsequently, the absorbance at the wavelength of λ = 593 nm was examined using a UV-Vis V-5000 spectrophotometer (Shanghai Metash Instruments Co. Ltd., Shanghai, China). The control contained distilled water instead of beer. The results of the FRAP test were expressed as mM Fe^{2+}/L.

3.6.3. ABTS Test

A 7 mM/L ABTS (2,2'-azinobis(3-ethylbenzothiazoline-6-sulfonic acid) solution and a 2.45 mM/L potassium persulphate solution were prepared for this purpose. The solutions were combined at 1:0.5 ratio, and stored for 12–16 h in darkness to enable development of ABTS cation. The ABTS$^+$ solution was diluted with distilled water to achieve absorbance of 0.700 ± 0.002 (at wavelength λ = 734 nm). A 3 mL portion of the diluted ABTS$^+$ solution was placed in the test tube with 0.3 mL of the beer and after 6 min, the absorbance value at the wavelength λ = 734 nm was determined using a UV-Vis V-5000 spectrophotometer (Shanghai Metash Instruments Co. Ltd., Shanghai, China). The results were corrected to account for dilution and expressed as trolox equivalent (mM TE/L).

All the analyses were performed in three replications.

3.7. Sensory Analysis

The sensory analysis was performed by an expert team of 13 persons (6 women and 7 men, 25–40 years old) in the sensory evaluation laboratory according to the EBC method 13.13 [50]. The beer samples (of 200 mL each) were served after cooling to 12 °C, coded in random order in 250 mL transparent plastic cups. Oral rinse water was administered between each evaluation. The sensory analysis of the beers was performed using a 5-grade rating scale for individual quality attributes: aroma (5—very intense, distinct, pleasant; 1—undetectable/unpleasant aroma), flavour (5—very tasty; 1—unpalatable); beer foam stability (5—very stable; 1—unstable), bitterness (5—weakly intense; 1—very intense) and saturation (5—high; 1—low or none). The average score obtained described the overall impression of the beer evaluated (5—very good; 1—bad) of the wheat beers analysed. Furthermore, a sensory profile was used to assess the taste and aroma of the beers' analyses, in which quality characteristics were determined (malty, fruity, sweet, cereal, intense, fullness, fresh, phenolic, bitter and sour) according to the EBC 13.12 method [51]. The sensory profile of fruit beers produced with the addition of non-ozonated and ozonated fruits of the Saskatoon berry was compared with beers without the addition of fruit (control).

3.8. Statistical Analysis

The results of the fruit beers were presented as a mean value with standard deviation. Statistical analysis of the results was performed using Statistica 13.3 statistical software (TIBCO Software Inc., Tulsa, OK, USA). Two-factor ANOVA of variance ANOVA was used in the analyses in a complete randomised design with a significance level of α = 0.05 for the individual results of physical and chemical analysis, polyphenol content and antioxidant activity of the fruit beers. Comparisons of mean values were done using the HSD-Tukey test.

4. Conclusions

Fruity wheat beers enriched with Saskatoon berry fruit pulp are characterised by a higher colour intensity and lower bitterness sensation, but at the same time insufficient attenuation, affecting the ethanol content. The antioxidant activity of wheat beers and the content of the total polyphenols for both tested cultivars of the Saskatoon berry were at a similar level. However, analysis of the polyphenol profile showed a significantly higher content of polyphenolic compounds in wheat beers enriched with non-ozonated fruits of the cv. 'Thiessen'. The results of the sensory evaluation show that wheat beers with the addition of the cv. 'Honeywood' fruit are characterised by the most balanced taste and aroma. On the basis of the research results obtained, we can conclude that fruits of both cvs. 'Honeywood' and 'Thiessen' can be used in the production of wheat beers. However, unlike barley beers, the fermentation process has to be modified in order to obtain a higher yield of the fruit beer product, and the ozonation process had a positive effect on improving the quality of fruit wheat beers.

Author Contributions: Conceptualisation, J.B. and J.G.; methodology, J.B., I.K. and M.B.; validation, J.B. and J.G.; formal analysis, J.B.; investigation, J.B.; writing—original draft preparation, J.B. and M.P.; writing—review and editing, J.G., S.P., M.B. and I.K.; visualisation, J.B.; supervision, J.G. and S.P.; project administration, J.B.; funding acquisition, J.G. All authors have read and agreed to the published version of the manuscript.

Funding: This research was funded by the programme of the Minister of Science and Higher Education named "Regional Initiative of Excellence" in the years 2019–2022, project number 026/RID/2018/19, the amount of financing PLN 9 542 500.00.

Institutional Review Board Statement: Not applicable.

Informed Consent Statement: Not applicable.

Data Availability Statement: Not applicable.

Conflicts of Interest: The authors declare no conflict of interest.

Sample Availability: Samples of the compounds used in research are available from the authors.

References

1. Belcar, J.; Buczek, J.; Kapusta, I.; Gorzelany, J. Quality and Pro-Healthy Properties of Belgian Witbier-Style Beers Relative to the Cultivar of Winter Wheat and Raw Materials Used. *Foods* **2022**, *11*, 1150. [CrossRef] [PubMed]
2. Kunze, W. *Technology Brewing and Malting*, 4th ed.; VLB: Berlin, Germany, 2010; pp. 108, 843.
3. Byeon, Y.S.; Lim, S.-T.; Kim, H.-J.; Kwak, H.S.; Kim, S.S. Quality Characteristcs of Wheat Malts with Different Country of Origin and Their Effect on Beer Brewing. *J. Food Quality* **2021**, *2021*, 2146620. [CrossRef]
4. Hu, X.; Jin, Y.; Du, J. Differences in protein content and foaming properties of cloudy beers based on wheat malt content. *J. Inst. Brew.* **2019**, *125*, 235–241. [CrossRef]
5. Wu, X.; Du, J.; Zhang, K.; Ju, Y.; Jin, Y. Changes in protein molecular weight during cloudy wheat beer brewing. *J. Inst. Brew.* **2015**, *121*, 137–144. [CrossRef]
6. Depraetere, S.; Delvaux, F.; Coghe, S.; Delvaux, F.R. Wheat Variety and Barley Malt Properties: Influence on Haze Intensity and Foam Stability of Wheat Beer. *J. Inst. Brew.* **2004**, *110*, 200–206. [CrossRef]
7. He, G.; Du, J.; Zhang, K.; Wei, G.; Wang, W. Antioxidant capability and potableness of fresh cloudy wheat beer stored at different temperatures. *J. Inst. Brew.* **2012**, *118*, 386–392. [CrossRef]
8. Gorzelany, J.; Michałowska, D.; Pluta, S.; Kapusta, I.; Belcar, J. Effect of Ozone-Treated or Untreated Saskatoon Fruits (*Amelanchier alnifolia* Nutt.) Applied as an Additive on the Quality and Antioxidant Activity of Fruit Beers. *Molecules* **2022**, *27*, 1976. [CrossRef]
9. Patraşcu, L.; Banu, I.; Bejan, M.; Aprodu, I. Quality parameters of fruit beers available on Romanian market. *St. Cerc. St. CICBIA* **2018**, *19*, 323–335.
10. Baigts-Allende, D.K.; Pérez-Alva, A.; Ramírez-Rodrigues, M.A.; Palacios, A.; Ramírez-Rodrigues, M.M. A comparative study of polyphenolic and amino acids profiles of commercial fruit beers. *J. Food Compos. Anal.* **2021**, *100*, 103921. [CrossRef]
11. Yang, Q.; Tu, J.; Chen, M.; Gong, X. Discrimination of Fruit Beer Based on Fingerprints by Static Headspace-Gas Chromatography-Ion Mobility Spectrometry. *J. Am. Soc. Brew. Chem.* **2022**, *80*, 298–304. [CrossRef]
12. Lavola, A.; Karjalainen, R.; Julkunen-Tiitto, R. Bioactive polyphenols in leaves, stems, and berries of Saskatoon (*Amelanchier alnifolia* Nutt) cultivars. *J. Agric. Food Chem.* **2012**, *60*, 427–433. [CrossRef] [PubMed]
13. Ozga, J.A.; Saeed, A.; Reinecke, D.M. Anthocyanins and nutrient components of Saskatoon fruits (*Amelanchier alnifolia* Nutt.). *Can. J. Plant. Sci.* **2006**, *86*, 193–197. [CrossRef]
14. Mazza, G.; Cottrell, T. Carotenoids and cyanogenic glucosides in Saskatoon berries (*Amelanchier alnifolia* Nutt.). *J. Food Compos. Anal.* **2008**, *21*, 249–254. [CrossRef]
15. Lachowicz, S.; Oszmiański, J.; Wiśniewski, R.; Seliga, Ł.; Pluta, S. Chemical parameters profile analysis by liquid chromatography and antioxidative activity of the Saskatoon berry fruits and their components. *Eur. Food Res. Technol.* **2019**, *245*, 2007–2015. [CrossRef]
16. Oszmiański, J.; Kolniak-Ostek, J.; Lachowicz, S.; Gorzelany, J.; Matłok, N. Effect of dried powder preparation process on polyphenolic content and antioxidant capacity of cranberry (*Vaccinium macrocarpon* L.). *Ind. Crops Prod.* **2015**, *77*, 658–665. [CrossRef]
17. Lachowicz, S.; Oszmiański, J. Saskatoon—A valuable raw material for processing. *Ferm Fruit Veget Ind.* **2016**, *6*, 25–27. (In Polish)
18. Lachowicz, S.; Oszmiański, J.; Pluta, S. The composition of bioactive compounds and antioxidant activity of Saskatoon berry (*Amelanchier alnifolia* Nutt.) genotypes grown in central Poland. *Food Chem.* **2017**, *235*, 234–243. [CrossRef]
19. Lachowicz, S.; Oszmiański, J.; Seliga, Ł.; Pluta, S. Phytochemical composition and antioxidant capacity of seven Saskatoon berry (*Amelanchier alnifolia* Nutt.) genotypes grown in Poland. *Molecules* **2017**, *22*, 853. [CrossRef]
20. Zardzewiały, M.; Matłok, N.; Piechowiak, T.; Gorzelany, J.; Balawejder, M. Ozone Treatment as a Process of Quality Improvement Method of Rhubarb (*Rheum rhaponticum* L.) Petioles during Storage. *Appl. Sci.* **2020**, *10*, 8282. [CrossRef]

21. Jaramillo-Sánchez, G.; Contigiani, E.V.; Castro, M.; Hodara, K.; Alzamora, S.; Loredo, A.; Nieto, A. Freshness maintenance of blueberries (*Vaccinium corymbosum* L.) during postharvest using ozone in aqueous phase: Microbiological, structure, and mechanical issues. *Food Bioprocess Technol.* **2019**, *12*, 2136–2147. [CrossRef]
22. Piechowiak, T.; Antos, P.; Józefczyk, R.; Kosowski, P.; Skrobacz, K.; Balawejder, M. Impact of ozonation process on the microbiological contamination and antioxidant capacity of highbush blueberry *Vaccinum corymbosum* L. fruit during cold storage. *Ozone Sci. Eng.* **2019**, *41*, 1540922. [CrossRef]
23. Contigiani, E.V.; Jaramillo-Sánchez, G.; Castro, M.A.; Gomez, P.L.; Alzamora, S.M. Postharvest quality of strawberry fruit (*Fragaria x Ananassa* Duch cv. Albion) as affected by ozone washing: Fungal spoilage, mechanical properties, and structure. *Food Bioprocess Technol.* **2018**, *11*, 1639–1650. [CrossRef]
24. Mascia, I.; Fadda, C.; Dostálek, P.; Olšovská, J.; Del Caro, A. Preliminary characterization of an Italian craft durum wheat beer. *J. Inst. Brew.* **2014**, *120*, 495–499. [CrossRef]
25. Nedyalkov, P.; Bakardzhiyski, I.; Dinkova, R.; Shopska, V.; Kaneva, M. Influence of the time of bilberry (*Vaccinium myrtillus* L.) addition on the phenolic and protein profile of beer. *Acta Sci. Pol. Technol. Aliment.* **2022**, *21*, 5–15. [CrossRef]
26. Gasiński, A.; Kawa-Rygielska, J.; Szumny, A.; Czubaszek, A.; Gąsior, J.; Pietrzak, W. Volatile Compounds Content Physicochemical Parameters and Antioxidant Activity of Beers with Addition of Mango Fruit (*Mangifera Indica*). *Molecules* **2020**, *25*, 3033. [CrossRef] [PubMed]
27. Nardini, M.; Garaguso, I. Characterization of bioactive compounds and antioxidant activity of fruit beers. *Food Chem.* **2020**, *305*, 125437. [CrossRef]
28. Adadi, P.; Kovaleva, E.G.; Glukhareva, T.V.; Shatunova, S.A.; Petrov, A.S. Production and analysis of non-traditional beer supplemented with sea buckthorn. *Agron. Res.* **2017**, *15*, 1831–1845. [CrossRef]
29. Habschied, K.; Košir, I.J.; Krstanović, V.; Kumrić, G.; Mastanjević, K. Beer Polyphenols—Bitterness, Astrigency, and Off-Flavors. *Beverages* **2021**, *7*, 38. [CrossRef]
30. Bogdan, P.; Kordialik-Bogacka, E. Antioxidant activity of beers produced with the addition of unmalted quinoa and amaranth. *Food Sci. Technol. Qual.* **2016**, *3*, 118–126. (In Polish) [CrossRef]
31. Ditrych, M.; Kordialik-Bogacka, E.; Czyżowska, A. Antiradical and Reducing Potential of Commercial Beer. *Czech J. Food Sci.* **2015**, *33*, 261–266. [CrossRef]
32. Deng, Y.; Lim, J.; Nguyen, T.T.H.; Mok, I.-K.; Piao, M.; Kim, D. Composition and biochemical properties of ale beer enriched with lignans from Schisandra chinensis Baillon (omija) fruits. *Food Sci. Biotechnol.* **2020**, *29*, 609–617. [CrossRef] [PubMed]
33. Gouvintas, I.; Breda, C.; Barros, A.I. Characterization and Discrimination of Commercial Portuguese Beers Based on Phenolic Composition and Antioxidant Capacity. *Foods* **2021**, *10*, 1144. [CrossRef] [PubMed]
34. Martínez, A.; Vegara, S.; Martí, N.; Valero, M.; Saura, D. Physicochemical characterization of special persimmon fruit beers using bohemian pilsner malt as a base. *J. Inst. Brew.* **2017**, *123*, 319–327. [CrossRef]
35. Mikyška, A.; Dušek, M.; Slabý, M. How does fermentation, filtration and stabilization of beer affect polyphenols with health benefits. *Kvasny Prumysl.* **2019**, *65*, 120–126. [CrossRef]
36. Adamenko, K.; Kawa-Rygielska, J.; Kucharska, A.Z. Characteristics of Cornelian cherry sour non-alcoholic beers brewed with the special yeast *Saccharomycodes ludwigii*. *Food Chem.* **2019**, *312*, 125968. [CrossRef] [PubMed]
37. Ducruet, J.; Rébénaque, P.; Diserens, S.; Kosińska-Cagnazzo, A.; Héritier, I.; Andlauer, W. Amber ale beer enriched with goji berries—the effect on bioactive compound content and sensorial properties. *Food Chem.* **2017**, *226*, 109–118. [CrossRef]
38. Piechowiak, T.; Grzelak-Błaszczyk, K.; Sójka, M.; Balawejder, M. Changes in phenolic compounds profile and glutathione status in raspberry fruit during storage in ozone-enriched atmosphere. *Postharvest Biol. Technol.* **2020**, *168*, 111277. [CrossRef]
39. Piechowiak, T.; Antos, P.; Kosowski, P.; Skrobacz, K.; Józefczyk, R.; Balawejder, M. Impact of ozonation process on the microbiological and antioxidant status of raspberry (*Rubus ideaeus* L.) fruit during storage at room temperature. *Agric. Food Sci.* **2019**, *28*, 35–44. [CrossRef]
40. Lv, Y.; Tahir, I.I.; Olsson, M.E. Effect of ozone application on bioactive compounds of apple fruit during short-term cold storage. *Sci. Hortic.* **2019**, *253*, 49–60. [CrossRef]
41. Alothman, M.; Kaur, B.; Fazilah, A.; Bhat, R.; Karim, A.A. Ozone—induced changes of antioxidant capacity of fresh-cut tropical fruits. *Innov. Food Sci. Emerg. Technol.* **2010**, *11*, 666–671. [CrossRef]
42. Gawlik-Dziki, U. Phenolic acids as bioactive food ingredients. *Food Sci. Technol. Qual.* **2004**, *4*, 30–40. (In Polish)
43. Radonjič, S.; Maraš, V.; Raičević, J.; Košmerl, T. Wine or Beer? Comparison, Changes and Improvement of Polyphenolic Compounds during Technological Phases. *Molecules* **2020**, *25*, 4960. [CrossRef]
44. Ma, X.; Yang, W.; Kallio, H.; Yang, B. Health promoting properties and sensory characteristics of phytochemicals in berries and leaves of sea buckthorn (*Hippophaë rhamnoides*). *Crit. Rev. Food Sci. Nutr.* **2022**, *62*, 3798–3816. [CrossRef] [PubMed]
45. Faltermaier, A.; Waters, D.; Becker, T.; Arendt, E.; Gastl, M. Common wheat (*Triticum aestivum* L.) and its use as a brewing cereal—A review. *J. Inst. Brew.* **2014**, *120*, 1–15. [CrossRef]
46. Viejo, G.; Sigfredo, C.; Sigfredo, F.; Damir, T.; Amruta, G.; Frank, D. Chemical characterization of aromas in beer and their effect on consumers liking. *Food Chem.* **2019**, *293*, 479–485. [CrossRef]
47. Belcar, J.; Sekutowski, T.R.; Zardzewiały, M.; Gorzelany, J. Effect of malting process duration on malting losses and quality of wheat malts. *Acta Univ. Cibin. Ser. E Food Technol.* **2021**, *25*, 221–232. [CrossRef]

48. Analytica EBC. 9.4—Original, Real and Apparent Extract and Original Gravity of Beer. In *European Brewery Convention*; Hans Carl Getränke-Fachverlag: Nürnberg, Germany, 2004.
49. Żurek, N.; Karatsai, O.; Rędowicz, M.J.; Kapusta, I. Polyphenolic Compounds of Crataegus Berry, Leaf, and Flower Extracts Affect Viability and Invasive Potential of Human Glioblastoma Cells. *Molecules* **2021**, *26*, 2656. [CrossRef]
50. Analytica EBC. 13.13—Sensory Analysis: Routine Descriptive Test Guideline. In *European Brewery Convention*; Hans Carl Getränke-Fachverlag: Nürnberg, Germany, 2004.
51. Analytica EBC. 13.12—Sensory Analysis: Flavour Terminology and Reference Standards. In *European Brewery Convention*; Hans Carl Getränke-Fachverlag: Nürnberg, Germany, 2004.

Article

Feasibility of Defatted Juice from Sea-Buckthorn Berries (*Hippophae rhamnoides* L.) as a Wheat Beer Enhancer

Justyna Belcar * and Józef Gorzelany

Department of Food and Agriculture Production Engineering, University of Rzeszow, 4 Zelwerowicza Street, 35-601 Rzeszów, Poland; gorzelan@ur.edu.pl
* Correspondence: justyna.belcar@op.pl

Abstract: Juice made from sea-buckthorn berries (*Hippophae rhamnoides* L.) is a valuable source of bioactive compounds, vitamins, as well as micro- and macronutrients. By applying defatted sea-buckthorn juice, it is possible to enhance wheat beer and change its sensory properties and the contents of bioactive compounds in the finished product. A sensory assessment showed that wheat beers with a 5% v/v addition of sea-buckthorn juice were characterised by a balanced taste and aroma (overall impression). Physicochemical analyses showed that, compared to the control samples, wheat beers enhanced with defatted sea-buckthorn juice at a rate of 5% v/v or 10% v/v had high total acidity with respective mean values of 5.30 and 6.88 (0.1 M NaOH/100 mL), energy values lower on average by 4.04% and 8.35%, respective polyphenol contents of 274.1 mg GAE/L and 249.7 mg GAE/L, as well as higher antioxidant activity (measured using DPPH, FRAP, and ABTS assays). The findings show that the samples of wheat beer enhanced with sea-buckthorn juice had average ascorbic acid contents of 2.5 and 4.5 mg/100 mL (in samples with 5% v/v and 10% v/v additions, respectively) and contained flavone glycosides, e.g., kaempferol-3-O-glucuronide-7-O-hexoside. Based on the current findings, it can be concluded that wheat beer enhanced with sea-buckthorn juice could emerge as a new trend in the brewing industry.

Keywords: sea-buckthorn; defatted juice; wheat beer; beer quality; bioactive compounds; antioxidant potential of beer

Citation: Belcar, J.; Gorzelany, J. Feasibility of Defatted Juice from Sea-Buckthorn Berries (*Hippophae rhamnoides* L.) as a Wheat Beer Enhancer. *Molecules* **2022**, *27*, 3916. https://doi.org/10.3390/molecules27123916

Academic Editor: Francesco Cacciola

Received: 30 May 2022
Accepted: 16 June 2022
Published: 18 June 2022

Publisher's Note: MDPI stays neutral with regard to jurisdictional claims in published maps and institutional affiliations.

Copyright: © 2022 by the authors. Licensee MDPI, Basel, Switzerland. This article is an open access article distributed under the terms and conditions of the Creative Commons Attribution (CC BY) license (https://creativecommons.org/licenses/by/4.0/).

1. Introduction

Beer is a type of beverage which contains four main ingredients: malt, hops, water, and yeast. In wheat beers, some of the barley malt (most commonly from 40 to 60% of the total input material) is replaced with wheat malt or unmalted wheat grain [1]. Wheat beers are characterised by an original flavour owing to the wide range of chemical compounds (e.g., phenols, aldehydes, and esters and their derivatives) produced in the process of top fermentation that contribute to the flavour, which can resemble vanilla, cloves, bananas, or fresh fruit; the effect is produced by the interaction of the two types of malt (barley and wheat) as well as the addition of hops. The final product of the brewing process is characterised by delicate and stable frothy foam, a slightly bitter taste, and haziness [2–4]. Wheat beers also have high contents of antioxidant compounds including polyphenols [5].

In recent years, a trend has been observed in consumers' increasing preference for fruit beers, which mainly include morello cherry, raspberry, banana, and strawberry, as well as other exotic fruit beers. The fruit component may be introduced by adding pulp, juice, concentrate, or aroma, most commonly during the fermentation process. By enhancing beer products with fruit, it is possible to improve their sensory qualities (such as colour, aroma, and taste), and to increase the health-promoting properties of the beverage, which are linked to higher contents of bioactive compounds (e.g., polyphenols), resulting in the higher antioxidant activity of fruit beers compared to traditional beer products [6,7]. In global markets we can encounter the very popular Radler-style beverages, i.e., a combination of beer and flavoured sugar syrup or fruit juice [8,9], as well as Belgian specialty beers produced as a result of spontaneous fermentation with the addition of raspberries or morello cherries, respectively, known as 'Framboise' and 'Kriek' Lambic beer [6,10,11].

The fruit of sea-buckthorn (*Hippophae rhamnoides* L.), depending on the variety, are yellow to orange in colour, which results from the high concentrations of carotenoids (including lutein, carotene, and zeaxanthin). Sea-buckthorn berries have also been reported to have high contents of health-promoting compounds such as polyphenols (e.g., flavanols and chlorogenic acid), organic acids, micro- and macronutrients, and vitamins [12–15], including very high levels of ascorbic acid (on average from 53 up to 1550 mg·100 g^{-1} [14,16]), and they do not contain ascorbinase enzyme responsible for the decomposition of ascorbic acid [17]. Owing to their contents of terpenes, alcohols, tannins, and aldehydes, sea-buckthorn berries have a characteristic aroma [12].

The extraction of juice from sea-buckthorn berries is a complex process that may produce changes in the chemical composition and bioavailability of nutrients in the final product. Depending on the preservation process applied (e.g., high-temperature short-term method; HTST), the changes that take place affect the sensory properties, mainly the taste of the juice; on the other hand, a short-term thermal treatment process does not lead to the degradation of ascorbic acid, the content of which decreases during the production of the juice by about 5–11% in relation to the amount of this compound in fresh fruit. Additional technological processes, such as filtration and clarification, contribute to a decrease in the ascorbic acid content. The high-pressure processing (200–600 MPa) that is applied to preserve juice does not produce changes in the quality of the final product; however, it results in a reduced size of particles contained in the juice, enhancing the yellow-orange colour of sea-buckthorn juice [18,19]. Both sea-buckthorn berries and sea-buckthorn juice contain fatty acids, including oleic and palmitoleic acids, phospholipids, and phytosterols [17]. Because these compounds are present, sea-buckthorn juice must be separated into an aqueous fraction and an oil fraction. Lipids present in the oil fraction bind into protein–lipid complexes with soluble low-molecular-weight proteins originating from the malts and they reduce the stability of the structure of the beer head, which comprises carbon dioxide molecules [3,4,20]. The high total acidity of sea-buckthorn juice, which is on average in the range of 2.1–9.1 g·100 mL^{-1} depending on the variety of the raw material, is significantly related to the content of malic and quinic acid (90% share in organic acids; [19]). The bitter and pungent taste of sea-buckthorn juice can be balanced in combination with other beverages, including tea, coffee, wine, and beer [21]. The addition of defatted sea-buckthorn juice to wheat beer, which is characterised by a delicate, slightly sweet taste, may be an interesting and original option acceptable for consumers.

The purpose of this study was to identify the physicochemical, sensory, and antioxidant properties of wheat beers produced with the addition of defatted sea-buckthorn juice. The study also assessed the applicability of the findings to expand the assortment of fruit beers and to make use of sea-buckthorn berries in a new sector of the food industry.

2. Results and Discussion

2.1. Physicochemical Characteristics of the Wheat Beers

The findings describing the physicochemical parameters of wheat beers enhanced with defatted sea-buckthorn juice are shown in Table 1.

The contents of the apparent extract in the wheat beer samples were in the range of 3.33–4.06% m/m; significantly higher contents of the apparent extract were found in the beer samples with defatted sea-buckthorn juice added at a rate of 10% v/v (E10 and L10, Table 1). The highest contents of real extract and original extract were identified in the control samples (E0 and L0; Table 1).

Table 1. Results of physicochemical analysis of wheat beers with defatted sea-buckthorn juice added.

Type of Beer	E0	E5	E10	L0	L5	L10
Apparent extract [%; m/m]	3.33 [a] ± 0.06	3.58 [b] ± 0.02	4.06 [d] ± 0.04	3.52 [b] ± 0.02	3.85 [c] ± 0.05	4.01 [d] ± 0.01
Real extract [%; m/m]	4.81 [d] ± 0.03	4.61 [c] ± 0.04	4.52 [b] ± 0.02	4.51 [b] ± 0.01	4.49 [b] ± 0.04	4.33 [a] ± 0.03
Original extract [%; m/m]	14.88 [e] ± 0.10	14.04 [b] ± 0.04	13.48 [a] ± 0.06	14.38 [d] ± 0.08	14.25 [c] ± 0.05	13.59 [a] ± 0.04
Degree of final apparent attenuation [%]	77.62 [f] ± 0.03	74.50 [d] ± 0.10	69.88 [a] ± 0.02	75.52 [e] ± 0.02	72.98 [c] ± 0.02	70.49 [b] ± 0.01
Degree of final real attenuation [%]	67.67 [c] ± 0.05	67.17 [b] ± 0.02	66.47 [a] ± 0.02	68.64 [f] ± 0.04	68.49 [e] ± 0.05	68.14 [d] ± 0.04
Content of alcohol [%; m/m]	5.28 [d] ± 0.05	4.92 [b] ± 0.02	4.66 [a] ± 0.06	5.16 [c] ± 0.06	5.10 [c] ± 0.10	4.82 [b] ± 0.02
Content of alcohol [%; v/v]	4.20 [d] ± 0.10	3.92 [b] ± 0.00	3.71 [a] ± 0.01	4.11 [b] ± 0.01	4.06 [b] ± 0.06	3.84 [b] ± 0.02
Colour [EBC units]	25.1 [d] ± 0.2	24.1 [c] ± 0.1	23.1 [b] ± 0.0	25.0 [d] ± 0.2	22.9 [b] ± 0.3	22.3 [a] ± 0.2
Titratable acidity [0.1 M NaOH/100 mL]	3.46 [b] ± 0.04	5.44 [d] ± 0.03	7.55 [f] ± 0.04	3.05 [a] ± 0.04	5.15 [c] ± 0.05	6.21 [e] ± 0.01
pH	4.54 [c] ± 0.03	3.95 [b] ± 0.05	3.73 [a] ± 0.03	4.64 [d] ± 0.04	3.99 [b] ± 0.05	3.73 [a] ± 0.03
Bitter substances [IBU]	15.4 [b] ± 0.10	18.1 [d] ± 0.10	19.7 [f] ± 0.00	14.7 [a] ± 0.10	17.5 [c] ± 0.00	19.1 [e] ± 0.20
Content of carbon dioxide [%]	0.46 [b] ± 0.00	0.47 [b] ± 0.02	0.44 [a] ± 0.02	0.47 [b] ± 0.03	0.48 [b] ± 0.01	0.44 [a] ± 0.02
Energy value [kcal/100 mL]	57.22 [f] ± 0.10	53.21 [c] ± 0.10	51.02 [a] ± 0.06	54.48 [e] ± 0.10	53.98 [d] ± 0.02	51.35 [b] ± 0.10

Data are expressed as mean value (n = 3) ± SD; SD—standard deviation. Mean values within a row with different letters are significantly different ($p < 0.05$). E—'Elixer' cultivar; L—'Lawina' cultivar; 0—wheat beer without defatted sea buckthorn juice; 5—wheat beer with 5% v/v defatted sea buckthorn juice; 10—wheat beer with 10% v/v defatted sea buckthorn juice.

The course of the fermentation process and the degree of the final fermentation affect the content of ethyl alcohol, the basic component of beer-type beverages, which is responsible for the sensory characteristics of beer that are perceived by consumers [22]. The degree of the final apparent fermentation in the wheat beer samples ranged from 69.88 to 77.62%, with the highest values identified in the control samples (E0 and L0). An increase in the concentration of defatted sea-buckthorn juice led to a significant decrease in the final apparent fermentation by an average of 3.69% in the samples with a 5% v/v addition of sea-buckthorn juice and by 8.33% in the samples with sea-buckthorn juice added at a rate of 10% v/v (Table 1). The values of the final true fermentation identified in the wheat beer samples were less varied but statistically different, independent from the beer samples acquired from wheat malt produced from grains of 'Lawina' and 'Elixer' wheat varieties. Similar to the degree of the final fermentation, the highest alcohol contents were identified in the control samples (E0 and L0), whereas the beer samples with defatted sea-buckthorn juice added at a rate of 5% v/v and 10% v/v were found with ethanol contents that were lower on average by 4.02% and by 9.19%, respectively (Table 1). According to Gasiński et al. [23], fruit beer should have a higher ethyl alcohol content compared to beer that is not enhanced with fruit. The lower contents of ethyl alcohol in the investigated wheat beer samples could be linked to the addition of defatted sea-buckthorn juice, which contains relatively low concentrations of total sugars (on average 4.94–5.72% relative to the variety) and reducing sugars (on average 1.59–1.83% relative to the variety; [13]); these are processed by the yeast in the fermentation process only to a small degree. Furthermore, the addition of sea-buckthorn juice led to an increase in the volume of the finished beer product while decreasing the concentration of the ethanol in the investigated wheat beer samples. A study by Nordini and Garaguso [10] showed that apple beer had an alcohol content of 5.2% v/v, whereas beer samples enriched with orange peel were found with an ethanol content of 6.0% v/v. On the other hand, Baigts-Allende et al. [6] reported an alcohol content of 4.0–8.2% v/v in citrus beer and 2.5–3.5% v/v in apple beer. Yang et al. [7] investigated apple beer and cranberry beer and reported ethanol contents of 3.5% v/v and 3.6% v/v, respectively. Patraşcu et al. [9] reported ethanol contents in lemon beer samples in the range of 1.9–4.0% v/v, in grapefruit beer samples of 1.9–2.5% v/v, and in cranberry beer samples of 4.0% v/v. The high ethanol content of the investigated wheat beers, in particular in the control samples (E0 and L0), corresponded to a relatively high calorific value of the finished product, which ranged from 54.48 to 57.22 kcal/100 mL; on the other hand, the addition of defatted sea-buckthorn juice led to a decrease in the calorific value of the wheat beer samples by an average of 6.2% (Table 1).

Wheat beer as a rule is darker in colour compared to barley beer (depending on the beer style). The wheat beer samples produced without the addition of defatted sea-buckthorn juice (E0 and L0) were found with a slightly darker colour; however, an increased concentration of sea-buckthorn juice added to the beer led to a lighter colour in the wheat beer samples (Figures 1 and 2; Table 1). Baigts-Allende et al. [6] reported that barley beer produced with the addition of citric fruit was found with a colour of 5.8 EBC units, whereas apple beers were characterised by a slightly stronger colour in the range of 6.34–9.81 EBC units. Patraşcu et al. [9] assessed the colour of lemon, grapefruit, and cranberry beers and reported respective values of 6.75–6.83 EBC units, 16.98–17.36 EBC units, and 5.55 EBC units.

Figure 1. The appearance of the obtained wheat beers (from left)—control (E0); with 5% v/v addition of defatted sea-buckthorn juice (E5); and with 10% v/v addition of defatted sea-buckthorn juice (E10).

Figure 2. The appearance of the obtained wheat beers (from left)—control (L0); with 5% v/v addition of defatted sea-buckthorn juice (L5); and with 10% v/v addition of defatted sea-buckthorn juice (L10).

The addition of defatted sea-buckthorn juice to wheat beer led to a decrease in the pH value by an average of 13.51% in the samples with a 5% v/v addition of sea-buckthorn juice and by an average of 18.74% in the samples with a 10% v/v addition of sea-buckthorn juice relative to the control samples (E0 and L0), which corresponded to a significant increase in the acidity (even twofold in the case of the samples with sea-buckthorn juice added at a rate of 10% v/v; Table 1). Adadi et al. [24] reported that the pH value and acidity of beer enriched with sea-buckthorn berries amounted to 3.9 and 2.2, respectively. A study by Nordini and Garaguso [10] reported that apple beer samples had a pH of 4.42, whereas beer samples enhanced with orange peel were found with a pH value of 4.86. Patraşcu et al. [9] investigated lemon, grapefruit, and cranberry beers, which were found with an acidity of 4.0–4.64; 4.0–4.4; and 3.76, respectively, whereas the pH values in these types of beer were found to be 2.85–3.09, 3.27–3.49, and 2.91, respectively. A lower pH in beer results in a reduced growth of undesirable microflora, and consequently leads to the greater microbiological stability of the finished beer product [23]. The addition of sea-buckthorn juice, characterised by high acidity and a low pH, on the seventh day during the fermentation process further reduces the risk of microbiological contamination, which is of great importance in the production of fruit beer.

All the wheat beer samples were found with similar contents of carbon dioxide (0.44–0.48%; Table 1). Patraşcu et al. [9] reported contents of carbon dioxide in lemon beer samples in the range of 0.48–0.55%, in grapefruit beer samples of 0.52%, and in cranberry beer samples of 0.55%. The contents of bitter substances in the wheat beer samples enhanced with defatted sea-buckthorn juice were at a similar level (17.5–19.7 IBU; Table 1) and the value increased with a higher addition of sea-buckthorn juice and was significantly greater compared to the wheat beer control samples (E0 and L0; Table 1). The bitter taste in the investigated beer samples originates not only from the basic raw material used in the production of the beer (i.e., hops) but also from the added sea-buckthorn juice (polyphenolic compounds present in juice). The bitter taste and contents of bitter substances in beer are significantly impacted by the variety and dose of the hops applied, the degree of isomerisation of α-acids during the process of boiling the wort with hops, and the reaction of proteins with polyphenols contained in the malt [2,25]. Sea-buckthorn berries contain terpenoids, tannins, as well as aldehydes and alcohols, which are responsible for the bitter taste and characteristic aroma in products enhanced with sea-buckthorn berries [12].

2.2. Contents of Bioactive Compounds in Wheat Beers

Ascorbic acid is a chemical compound known for its antioxidant properties. The effects produced by ascorbic acid include the strengthening of defence mechanisms; it is also involved in the synthesis of collagen and the absorption of iron in the human body and it promotes the transcription of mRNA and the treatment of scurvy [18,26]. Sea-buckthorn berries were reported to have high contents of ascorbic acid, ranging from 53–131 mg·100^{-1} g [14] to 114–1550 mg·100 g^{-1} [16], depending on the sea-buckthorn variety and the timing of the harvest; on the other hand, they do not contain the ascorbinase enzyme involved in the decomposition of ascorbic acid in fruit and vegetables. Wheat beer samples enhanced with defatted sea-buckthorn juice had low contents of ascorbic acid, and this compound was not identified in the E0 and L0 beer samples (Table 2). Because of the very rapid decomposition of ascorbic acid, this compound is generally not found in fruit beers (e.g., those with the addition of lemon, grapefruit, black currant, and strawberry), although the fruit added to the beer typically has high contents of ascorbic acid (the respective reported values being 25–53 mg·100 g^{-1}; 4–34.4 mg·100 g^{-1}; 181–215 mg·100 g^{-1}; and 41.2–60 mg·100 g^{-1} [27]). A study by Pimentel et al. [28] used camu-camu fruit (ascorbic acid contents in the range of 2.4–3 g·100 g^{-1} of fruit) to enhance Witbier-type beer. The beer was found with an ascorbic acid content of 15.8 mg·100 mL^{-1}.

The contents of polyphenols, bitter substances, vitamins, and melanoidins impact the antioxidant potential of beer [29,30]. The addition of sea-buckthorn juice to wheat beer significantly increased the antioxidant activity of the beer samples assessed using three methods (DPPH·, FRAP, ABTS⁺·), and the increase was associated with a higher concentration of defatted sea-buckthorn juice in the wheat beer samples (Table 2). In comparison, Nordini and Garaguso [10] reported that beer samples produced with the addition of orange peel were found with an antioxidant capacity, assessed using ABTS, of 2.67 mM TE/L, and a reducing capacity, shown by a FRAP assay, of 5.65 mM Fe^{2+}/L. They also applied FRAP and ABTS assays to measure the antioxidant capacity of apple beer and reported the respective values of 3.08 mM Fe^{2+}/L and 1.62 mM TE/L. The addition of juice from persimmons (kaki) to barley beer led to a decrease in the antioxidant capacity, measured using an ABTS assay, from 6.36 mM TE/L (in a control sample made from 100% barley wort) to 1.65 mM TE/L (in samples made from 25% wort and 75% kaki fruit juice [11]). Deng et al. [31] enhanced beer with omija fruit added during the fermentation process and reported antioxidant activity, measured by DPPH assay, of 1.68 mM TE/L, and a reducing capacity, assessed with FRAP, of 2.4 mM Fe^{2+}/L. Portuguese commercial fruit beers flavoured with lemon were reported to have antioxidant capacity in the range of 0.035–0.037 mM TE/L, according to DPPH assay, and a level of 0.008 mM TE/L, according to an ABTS assay [32].

Table 2. Content of ascorbic acid and antioxidant activity of wheat beers.

Type of beer	E0	E5	E10	L0	L5	L10
Content of ascorbic acid [mg/100 mL]	n.d.	2.5 [a] ± 0.4	4.5 [b] ± 0.0	n.d.	2.5 [a] ± 0.1	4.5 [b] ± 0.3
DPPH [mM TE/L]	2.27 [b] ± 0.03	2.71 [d] ± 0.02	2.76 [d] ± 0.06	2.19 [a] ± 0.04	2.39 [c] ± 0.01	2.74 [d] ± 0.04
FRAP [mM Fe^{2+}/L]	2.79 [b] ± 0.04	3.62 [d] ± 0.02	3.89 [e] ± 0.07	2.53 [a] ± 0.03	3.25 [c] ± 0.05	4.09 [f] ± 0.04
ABTS⁺· [mM TE/L]	1.81 [a] ± 0.01	2.18 [c] ± 0.02	2.47 [e] ± 0.01	1.97 [b] ± 0.01	2.34 [d] ± 0.04	2.62 [f] ± 0.02

Data are expressed as mean value ($n = 3$) ± SD; SD—standard deviation. Mean values within a row with different letters are significantly different ($p < 0.05$). E—'Elixer' cultivar; L—'Lawina' cultivar; 0—wheat beer without defatted sea buckthorn juice; 5—wheat beer with 5% v/v defatted sea buckthorn juice; 10—wheat beer with 10% v/v defatted sea buckthorn juice; n.d.—not detected; TE—expressed as Trolox equivalent (mM TE/L).

Polyphenolic compounds occurring in beer mainly originate from malt (70–80%) and hops [11]. The way raw materials are prepared (refinement of malt), as well as conditions during the processes of mashing and boiling with hops, significantly affect the total polyphenol contents and the degree of the isomerisation of polyphenols in the finished beer product [25]. Polyphenolic compounds have varied chemical structures, which is associated with their diverse capacity for biological activity, including antioxidant activity [33]. The polyphenols contained in beer significantly affect the sensory perceptions of consumers, such as a sense of thickness, a bitter or sour taste, as well as a fullness of flavour. By adding wheat malt (as a part of the input material) it is possible to increase the total polyphenol content in the finished beer product [34]. The total polyphenol contents in wheat beer samples enhanced with sea-buckthorn juice were on average 14.78% (5% v/v addition) and 6.45% (10% v/v addition) higher compared to control samples E0 and L0 (Table 3). Sea-buckthorn berries are found with total polyphenol contents in the range of 128–490 mg·100 g^{-1} depending on the variety and timing of the harvest [13,18]. The content of polyphenolic compounds is significantly affected by the methods applied during juice production, or more specifically by the process of fruit crushing (breaking cell walls), as well as the thermal processes used to preserve the finished product [23]. Nardini and Garaguso [10] reported total polyphenol contents of 639 mg GAE/L and 399 mg GAE/L, respectively, in beer produced with the addition of orange peel and in apple beer (GAE—equivalent of gallic acid). Gasiński et al. [23] investigated beer with the addition of mangoes and reported slightly lower polyphenol contents in the range of 218.6–267.6 mg GAE/L. The addition of persimmon juice led to a decrease in the total polyphenol contents in the beer samples from 433.32 mg GAE/L (25% juice addition) to 290.34 mg GAE/L (75% juice addition [11]). Portuguese commercial fruit beers flavoured with lemon were found with total polyphenol contents in the range of 240–304 mg GAE/L [32].

Table 3. Contents of polyphenols and polyphenolic profiles identified by UPLC-PDA-TQD-MS in wheat beer.

Compound [mg/L]	Rt [min]	[M-H]⁻ (m/z)	Fragment ions(m/z)	Absorbance maxima (nm)	E0	E5	E10	L0	L5	L10
Q-3-O-rut-7-O-glc	3.35	771	609, 301	255, 350	t.c.	0.61 ^{b} ± 0.00	0.55 ^{a} ± 0.05	t.c.	0.53 ^{a} ± 0.05	0.66 ^{b} ± 0.06
K-3-O-glc	3.73	447	285	264, 324	t.c.	0.68 ^{a} ± 0.08	0.95 ^{b} ± 0.02	t.c.	0.72 ^{a} ± 0.02	0.90 ^{b} ± 0.15
K-3-O-sophoroside	3.97	609	285	264, 324	0.92 ^{b} ± 0.02	t.c.	t.c.	0.65 ^{a} ± 0.06	t.c.	t.c.
K-3-O-rut-7-O-glc	4.09	755	593, 285	264, 324	0.92 ^{b} ± 0.02	t.c.	t.c.	0.73 ^{a} ± 0.01	t.c.	t.c.
K-3-O-glc-7-O-glc	4.20	609	447, 285	264, 324	1.31 ^{b} ± 0.04	t.c.	t.c.	0.81 ^{a} ± 0.09	t.c.	t.c.
K-3-O-gluc-7-O-glc	4.61	623	447, 285	264, 324	t.c.	1.39 ^{a} ± 0.02	2.41 ^{b} ± 0.07	t.c.	1.34 ^{a} ± 0.04	2.20 ^{b} ± 0.22
K-3-O-gluc-7-O-glc	5.37	623	447, 285	264, 324	t.c.	0.90 ^{a} ± 0.03	1.42 ^{c} ± 0.06	t.c.	0.88 ^{a} ± 0.04	1.34 ^{b} ± 0.03
Total					3.15 ^{b} ± 0.08	3.57 ^{c} ± 0.07	5.33 ^{c} ± 0.10	2.18 ^{a} ± 0.14	3.46 ^{c} ± 0.04	5.09 ^{c} ± 0.34
					243.9 ^{c} ± 0.8	277.0 ^{f} ± 0.7	264.6 ^{d} ± 0.6	223.3 ^{a} ± 0.3	271.2 ^{e} ± 0.8	234.8 ^{b} ± 0.5

Data are expressed as mean value (n = 15) ± SD; SD—standard deviation. Mean values within a row with different letters are significantly different ($p < 0.05$). E—'Elixer' cultivar; L—'Lawina' cultivar; 0—wheat beer without defatted sea buckthorn juice; 5—wheat beer with 5% v/v defatted sea buckthorn juice; 10—wheat beer with 10% v/v defatted sea buckthorn juice. Q-quercetin; K—kaempferol; glc—glucoside; rut—rutinoside; gluc—glucuronide t.c.—trace content below LOQ; GAE—equivalent of gallic acid (mg GAE/L).

Polyphenolic compounds in samples of wheat beer enhanced with sea-buckthorn juice were identified based on an analysis of characteristic spectral data: the mass-to-charge ratio m/z and the maximum absorption of radiation. The characteristics of six polyphenolic compounds that were identified are shown in Table 3. All the identified compounds were flavonols, represented by derivatives of kaempferol and quercetin (in glycoside form). Flavone glycosides are known to have strong antineoplastic and antioxidant properties; they are beneficial for patients with cardiovascular disease and transplants [18]. Kaempferol and quercetin glycosides produce a pungent taste in the mouth, and—to a lesser extent—contribute to a bitter taste, which affects the sensory properties of the finished beer product [18]. The flavonol contents in the control wheat beer samples (E0 and L0) were in the range of 2.18–3.15 mg/L, whereas the addition of defatted sea-buckthorn juice led to an increase in the polyphenol concentrations in the finished beer product by an average of 24.18% and by 48.85% in samples with sea-buckthorn juice added at a rate of 5% v/v and 10% v/v, respectively (Table 3). The control wheat beer samples (E0 and L0) were found to contain three compounds, i.e., K-3-O-sophoroside, K-3-O-rutinoside-7-O-glucoside, and K-3-O-glucoside-7-O-glucoside, which possibly originated from the hops added to the wort during the boiling process; their mean contents were 0.79 mg/L, 0.83 mg/L, and 1.06 mg/L, respectively (Table 3; Figures 3–5). The kaempferol-O-glucoside contained in hops is extracted even after 30 min of wort boiling (depending on the dose of the wort [35]). The contents of kaempferol in barley beers were on average in the range of 0.10–1.64 mg/L [36,37]. The wheat beer samples enhanced with defatted sea-buckthorn juice were found to contain the compounds Q-3-O-rutinoside-7-O-glucoside, K-3-O-glucoside, and K-3-O-glucuronide-7-O-glucoside (Figures 3–5), which were extracted from sea-buckthorn juice during fermentation and may have been rearranged into more complex glycoside derivatives; notably, the concentration of the latter compound was on average two times higher in the wheat beer samples enhanced with sea-buckthorn juice at a rate of 10% v/v compared to the samples with a 5% v/v addition of sea-buckthorn juice (Table 3). A study by Guo et al. [38], investigating the fruit of four sea-buckthorn subvarieties, showed the mean contents of Q-3-O-rutinoside and Q-3-O-glucoside of 32.9 mg·100 g^{-1} d.w. and 39.7 mg·100 g^{-1} d.w., respectively. On the other hand, Chen et al. [39] reported the following mean contents of these polyphenols: Q-3-O-rutinoside–52.0 mg·100 g^{-1} d.w. and Q-3-O-glucoside–53.3 mg·100 g^{-1} d.w. (d.w.–dry weight).

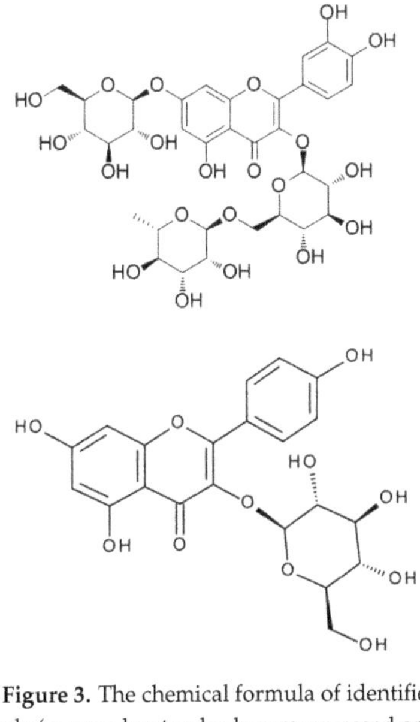

Figure 3. The chemical formula of identified polyphenols—from left—Q-3-O-rut-7-O-glc and K-3-O-glc (source: lgcstandards.com; accessed on 14 June 2022).

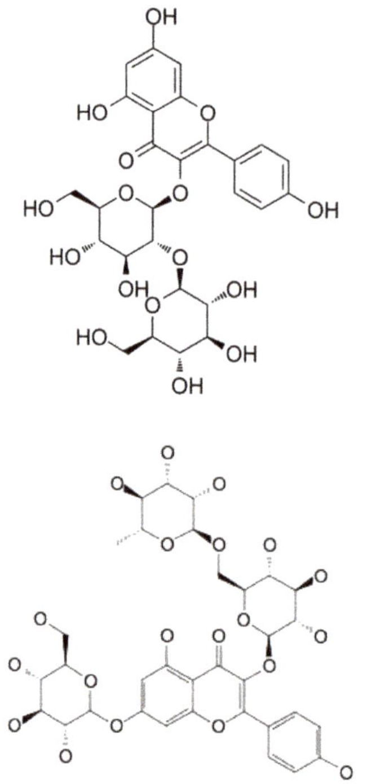

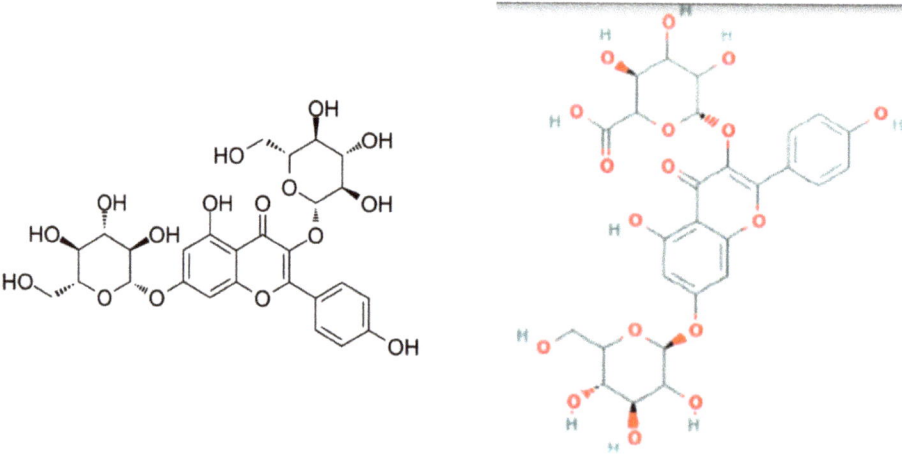

Figure 4. The chemical formula of identified polyphenols—from left—K-3-O-sophoroside and K-3-O-rut-7-O-glc (source: lgcstandards.com; accessed on 14 June 2022).

Figure 5. The chemical formula of identified polyphenols—from left—K-3-O-glc-7-O-glc and K-3-O-gluc-7-O-glc (source: lgcstandards.com; accessed on 14 June 2022).

2.3. Sensory Analysis of Wheat Beers

The sensory qualities of the wheat beer samples enhanced with defatted sea-buckthorn juice determine the specific beer style and contribute to the attractiveness of the beverage for consumers. The results of the sensory assessment of the wheat beers performed by a panel of 11 experts are shown in Table 4 and Figures 6 and 7.

Table 4. Sensory analysis of wheat beer.

	E0	E5	E10	L0	L5	L10
Aroma	4.23 [ab] ± 0.41	4.18 [ab] ± 0.25	3.82 [a] ± 0.60	3.95 [a] ± 0.57	4.54 [b] ± 0.68	4.27 [ab] ± 0.46
Taste	3.73 [a] ± 0.24	4.27 [ab] ± 0.34	3.73 [a] ± 0.28	3.91 [ab] ± 0.44	4.54 [b] ± 0.52	3.77 [a] ± 0.28
Foam stability	3.55 [a] ± 0.13	3.82 [a] ± 0.30	3.41 [a] ± 0.17	3.55 [a] ± 0.33	3.59 [a] ± 0.26	3.50 [a] ± 0.44
Bitterness	4.00 [a] ± 0.17	4.09 [a] ± 0.14	3.82 [a] ± 0.25	4.18 [a] ± 0.37	4.27 [a] ± 0.34	3.64 [a] ± 0.12
Saturation	3.73 [a] ± 0.34	4.00 [a] ± 0.17	3.64 [a] ± 0.27	4.00 [a] ± 0.29	4.00 [a] ± 0.33	3.45 [a] ± 0.38
Overall impression	3.86 [ab] ± 0.47	4.13 [ab] ± 0.37	3.72 [a] ± 0.36	3.91 [ab] ± 0.66	4.28 [b] ± 0.45	3.78 [a] ± 0.44

Data are expressed as mean value (n = 11) ± SD; SD—standard deviation. Mean values within a row with different letters are significantly different ($p < 0.05$). E—'Elixer' variety; L—'Lawina' variety; 0—wheat beer without defatted sea buckthorn juice; 5—wheat beer with 5% v/v defatted sea buckthorn juice; 10—wheat beer with 10% v/v defatted sea buckthorn juice.

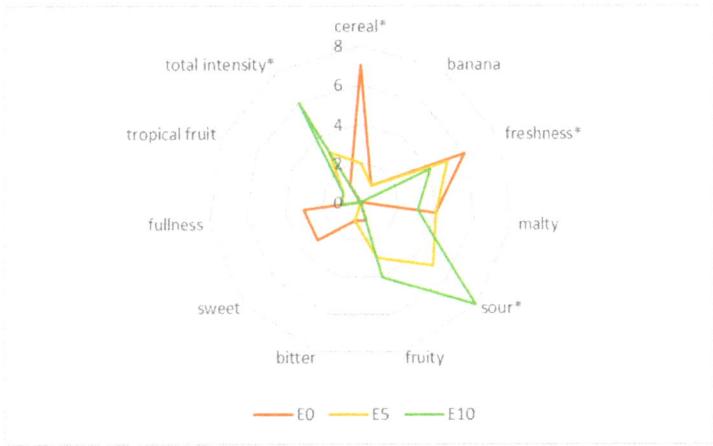

Figure 6. Sensory profile of wheat beer—control sample (E0) and sample with defatted sea-buckthorn juice added at a rate of 5% v/v (E5) as well as sample with defatted sea-buckthorn juice added at a rate of 10% v/v (E10). (* marks the attributes that were statistically different at $p \leq 0.05$).

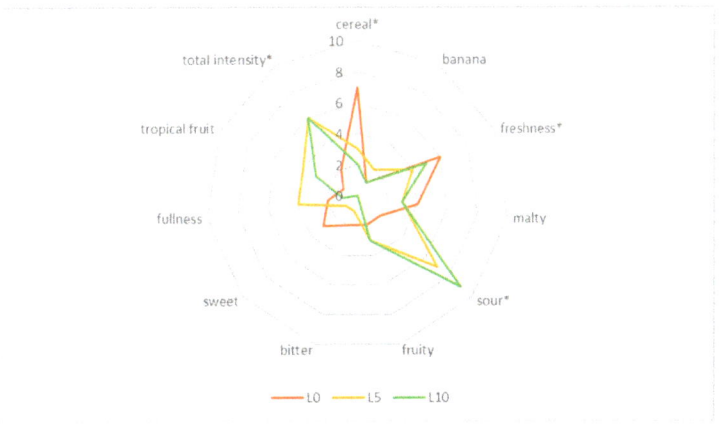

Figure 7. Sensory profile of wheat beer—control sample (L0) and sample with defatted sea-buckthorn juice added at a rate of 5% v/v (L5) as well as sample with defatted sea-buckthorn juice added at a rate of 10% v/v (L10). (* marks the attributes that were statistically different at $p \leq 0.05$).

The wheat beer samples enhanced with defatted sea-buckthorn juice at a rate of 5% v/v were found to have the highest sense of flavour (combination of taste and aroma), bitter taste, and saturation, compared to the other beer samples evaluated. With regard to the control beer samples (E0 and L0), the overall impression evaluated by the expert panel was reflected by a score in the range of 3.86–3.91 on a 5-point scale. The lowest rating was identified in the case of beer samples with a 10% v/v addition of defatted sea-buckthorn juice (Table 4). Out of all the quality properties assessed, the stability of the beer head in the wheat beer samples received the lowest rating, whether or not sea-buckthorn juice was added in the production process. The taste and aroma of the beer are not only affected by the raw materials used but also by the products of the fermentation process (e.g., aldehydes, phenols, and esters), which impact the taste profile of the beer.

The sensory profile of the investigated wheat beer varied; the control samples (with no addition of sea-buckthorn juice) had a grainy and malty flavour produced by such compounds as maltol and furaneol [40]; they also had a refreshing and sweet taste, characteristic for wheat beers (Figures 6 and 7). The sensory assessment showed that the beer samples enhanced with sea-buckthorn juice had a stronger flavour and more refreshing quality as well as an acidic taste, which was more pronounced in the wheat beer samples with a higher proportional share of the juice added. According to the assessing experts, the control samples E10 and L10 would not be accepted by consumers because of their highly acidic taste. The assessing panel expressed an opinion that, irrespective of the wheat variety used to produce the malt, the beer samples had a balanced taste and aroma profile, whether or not sea-buckthorn juice was added during the production process. The interactions taking place during the fermentation and maturation of beer between esters, sulphur compounds, carbonyls, phenolic compounds, alcohols, and organic acids significantly affect the taste of the produced beer [40]. Beer with a distinct fruity flavour, sweet aftertaste, and pleasant aroma is more favoured and desirable for consumers compared to traditional types of beer [24,41].

3. Materials and Methods

3.1. Material

Grains of two varieties of winter wheat, i.e., 'Lawina' and 'Elixer', that were used in the production of wheat beer were obtained from a field experiment conducted in 2021 in the village of Kosina (50°04′17″ N 22°19′46″ E), Podkarpackie Region, Poland. Grain of the winter wheat varieties was harvested after achieving full maturity, and following a resting period, it was used for preparing five-day wheat malts (the methodology of the malting process was described in Belcar et al. [42]). The wheat malt of the 'Elixer' variety had the following characteristics: extract potential—86.0% d.m., total protein content—11.6% d.m., content of soluble protein—4.71% d.m., diastatic power—331 WK, and degree of final attenuation—82.4%, whereas the wheat malt of the 'Lawina' variety had the following characteristics: extract potential—85.1% d.m., total protein content—11.0% d.m., content of soluble protein—4.42% d.m. (d.m.—dry matter), diastatic power—336 WK, and degree of final attenuation—81.7%.

Materials used in the production of beer samples included commercially available barley malt acquired from Viking Malt company (Strzegom, Poland). The barley malt had the following characteristics: extract potential—80.0% d.m., total protein content—11.4% d.m., content of soluble protein—3.75% d.m. (d.m.—dry matter), diastatic power—324 WK, and degree of final attenuation—82.1%. The wheat and barley malts were refined to the required particle size using a Cemotec disc mill manufactured by FOSS. The input material used in the brewing process comprised commercial barley malt at a rate of 60% and wheat malt at a rate of 40%.

Wheat beer samples were enhanced using defatted juice made from sea-buckthorn berries (after sedimentation) that was produced in 2021 by the Szarłat Company (Cibory Gałeckie, Podlaskie Region, Poland). The defatted juice had the following chemical parameters: fat content—0.04 g/100 g, L-ascorbic acid content—44.45 mg/100 mL, extract content—8.44%, and total acidity—3.36 g/100 mL.

3.2. Production of Beer

The production process, based on the infusion method, was carried out in the laboratory of the Department of Agricultural and Food Engineering at the University of Rzeszów. Barley malt with a weight of 3.0 kg and wheat malt with a weight of 2.0 kg were refined and placed in a brew kettle ROYAL RCBM-40N (Expondo; Poland; applied at 80% process efficiency) with 15.0 L of water (3 L of water per each kg of malt). The processes of mashing, boiling with hops, and chilling of beer wort were conducted in line with the methodology described by Gorzelany et al. [8].

All six beer wort samples were found with an extract content of 12.0 °P. The chilled wort samples were poured into 30 L fermentation vessels along with the yeast *Saccharomyces cerevisae* Fermentis Safale US-05 (6×10^9/g), earlier subjected to a dehydration process in line with the manufacturer's instructions (0.58 g d.m./L of wort). The fermentation process was carried out at 21 °C. After the fermentation process had continued for 7 days, defatted sea-buckthorn juice was added to the beer in specified quantities (0, 5, or 10% relative to wort volume) and then the fermentation process continued for the next 14 days. After 21 days, an aqueous solution of sucrose (0.3%) was added and the beer was poured into bottles for refermentation to achieve an adequate level of carbonation. The beer was then kept at 20 °C. Sensory assessment and physicochemical tests were performed one month after the bottling.

Wheat beer produced using malt obtained from the winter wheat variety 'Elixer' and with no addition of defatted sea-buckthorn juice is marked E0, whereas the sample with the 5% v/v addition of defatted sea-buckthorn juice is marked E5 and the sample with the 10% v/v addition of sea-buckthorn juice is marked E10. Wheat beer produced using malt obtained from the winter wheat variety 'Lawina' and with no addition of defatted sea-buckthorn juice is marked L0, whereas the sample with the 5% v/v addition of defatted sea-buckthorn juice is marked L5 and the sample with the 10% v/v addition of sea-buckthorn juice is marked L10. A total of six variants of wheat beer were produced.

3.3. Analysis of Beer Quality Indicators

Alcohol contents [% m/m and % v/v], apparent extract [% m/m], real extract [% m/m], original extract in beer [% m/m], degree of final apparent and real fermentation [%], total acidity [0.1 M NaOH/100 mL], pH, colour [EBC units], carbon dioxide contents [%], contents of bitter substances [IBU units], as well as energy value of beer [kcal/100 mL] were determined following the methodology described by Belcar et al. [34]. The analyses were performed in three replications.

3.4. Contents of Bioactive Compounds in Wheat Beers

The total contents of polyphenols [mg GAE/L] determined using the Folin–Ciocalteu method, as well as the polyphenol profile of the beer samples, were measured in compliance with the methodology described by Gorzelany et al. [8].

Determination of polyphenolic compounds [mg/L] was carried out using the UPLC equipped with a binary pump, column and sample manager, photodiode array detector (PDA), tandem quadrupole mass spectrometer (TQD) with an electrospray ionization (ESI) source working in negative mode (Waters, Milford, MA, USA) according to the method of Żurek et al. [43]. Separation was performed using the UPLC BEH C18 column (1.7 μm, 100 mm × 2.1 mm, Waters) at 50 °C at a flow rate of 0.35 mL/min. The injection volume of the samples was 5 μL. The mobile phase consisted of water (solvent A) and 40% acetonitrile in water v/v (solvent B). The following TQD parameters were used: capillary voltage of 3500 V; con voltage of 30 V; con gas flow 100 L/h; source temperature 120 °C; desolvation temperature 350 °C; and desolvation gas flow rate of 800 L/h. Polyphenolic identification and quantitative analyses were performed based on the mass-to-charge ratio, retention time, specific PDA spectra, fragment ions, and comparison of data obtained with commercial standards and literature findings. The analyses were performed in three replications.

Contents of ascorbic acid [mg/100 mL] in sea-buckthorn berries were assessed in conformity with PN-A-04019:1998 [44]. The analyses were performed in three replications.

3.5. Antioxidant Activity in Wheat Beers

Antioxidant capacity of fruit beers (assessed using DPPH [mM TE/L], FRAP [mM Fe^{2+}/L, and ABTS [mM TE/L] assays) was measured following the methodology described by Gorzelany et al. [8]. The analyses were performed in three replications.

3.6. Sensory Assessment in Beers

Sensory assessment was performed by a panel of 11 experts (4 women and 7 men, aged 30–40 years), in a sensory analysis laboratory in line with the EBC 13.13 method [45]. Beer samples, chilled to a temperature of 10 °C and coded, were served in a random order in transparent plastic cups with a capacity of 250 mL. After each test, the experts were given water to rinse their mouths. Sensory analysis of the beer samples was performed using a 5-point scale, assessing the specific quality characteristics, i.e., aroma (5—very strong, distinctive, and pleasant; 1—imperceptible aroma/ unpleasant smell), taste (5—very good; 1—bad); beer head stability (5—highly stable; 1—unstable), bitter taste (5—weak; 1—very strong), and carbonation (5—high; 1—poor or none). The average score described the general impression (5—excellent; 1—poor) related to the investigated wheat beers. Additionally, evaluation of the beer samples in terms of their taste and aroma applied the sensory profile describing the quality characteristics (malty, fruity, sweet, grainy, strong, full, fresh, phenolic, bitter, and sour) in line with EBC 13.12 [46]. The sensory profile of the fruit beer produced with the addition of defatted sea-buckthorn juice was compared to the control beer (no addition of sea-buckthorn juice).

3.7. Statistical Analysis

The results of the fruit beer evaluation are shown as mean values and standard deviations. The statistical analyses of the results were computed using Statistica 13.3 (TIBCO Software Inc., Tulsa, OK, USA). The results related to physicochemical characteristics, polyphenol contents, and antioxidant activity of fruit beer samples were examined using the two-factor completely randomized ANOVA with a significance level of $\alpha = 0.05$. The mean values were compared using the Tukey HSD test.

4. Conclusions

The study, designed to assess the feasibility of defatted sea-buckthorn juice as an enhancer to be used in the production of fruit wheat beers, showed that the most balanced sensory profile (intensity, perceived bitter flavour, as well as fruity taste and aroma) is found in beer samples enhanced with juice at a rate of 5% v/v. Additionally, these beer samples were shown to have better colour, as well as higher polyphenol contents and antioxidant capacity. Although the addition of defatted sea-buckthorn juice at a rate of 10% v/v positively affected the health-promoting properties of wheat beer (with a mean content of ascorbic acid of 4.5 mg/100 mL), the sensory properties of this type of beer were not acceptable for the assessing panel, mainly due to its highly acidic taste. The enhancement of wheat beer with defatted sea-buckthorn juice at a rate of 5% v/v could effectively be applied to expand the assortment of fruit wheat beers on offer.

Author Contributions: Conceptualization, J.B. and J.G.; methodology, J.B.; validation, J.B. and J.G.; formal analysis, J.B.; data curation, J.B.; writing—original draft preparation, J.B.; writing—review and editing, J.B. and J.G.; visualization, J.B.; supervision, J.G.; project administration, J.B.; funding acquisition, J.B. All authors have read and agreed to the published version of the manuscript.

Funding: This research received no external funding.

Institutional Review Board Statement: Not applicable.

Informed Consent Statement: Not applicable.

Data Availability Statement: Not applicable.

Acknowledgments: The authors wish to express their gratitude to the M. and W. Lemkiewicz 'Szarłat' company from Cibory Gałeckie (Podlaskie Region, Poland) for the material provided for the present study (defatted sea buckthorn juice).

Conflicts of Interest: The authors declare no conflict of interest.

Sample Availability: Samples of the compounds used in the research are available from the authors.

References

1. Kunze, W. *Technology Brewing and Malting*, 4th ed.; VLB Berlin: Berlin, Germany, 2010; pp. 108, 843.
2. Byeon, Y.S.; Lim, S.-T.; Kim, H.-J.; Kwak, H.S.; Kim, S.S. Quality Characteristics of Wheat Malts with Different Country of Origin and Their Effect on Beer Brewing. *J. Food Qual.* **2021**, *2021*, 2146620. [CrossRef]
3. Hu, X.; Jin, Y.; Du, J. Differences in protein content and foaming properties of cloudy beers based on wheat malt content. *J. Inst. Brew.* **2019**, *125*, 235–241. [CrossRef]
4. Wu, X.; Du, J.; Zhang, K.; Ju, Y.; Jin, Y. Changes in protein molecular weight during cloudy wheat beer brewing. *J. Inst. Brew.* **2015**, *121*, 137–144. [CrossRef]
5. He, G.; Du, J.; Zhang, K.; Wei, G.; Wang, W. Antioxidant capability and potableness of fresh cloudy wheat beer stored at different temperatures. *J. Inst. Brew.* **2012**, *118*, 386–392. [CrossRef]
6. Baigts-Allende, D.K.; Pérez-Alva, A.; Ramírez-Rodrigues, M.A.; Palacios, A.; Ramírez-Rodrigues, M.M. A comparative study of polyphenolic and amino acids profiles of commercial fruit beers. *J. Food Compos. Anal.* **2021**, *100*, 103921. [CrossRef]
7. Yang, Q.; Tu, J.; Chen, M.; Gong, X. Discrimination of Fruit Beer Based on Fingerprints by Static Headspace-Gas Chromatography-Ion Mobility Spectrometry. *J. Am. Soc. Brew. Chem.* **2021**, 1946654. [CrossRef]
8. Gorzelany, J.; Michałowska, D.; Pluta, S.; Kapusta, I.; Belcar, J. Effect of Ozone-Treated or Untreated Saskatoon Fruits (*Amelanchier alnifolia* Nutt.) Applied as an Additive on the Quality and Antioxidant Activity of Fruit Beers. *Molecules* **2022**, *27*, 1976. [CrossRef]
9. Patrașcu, L.; Banu, I.; Bejan, M.; Aprodu, I. Quality parameters of fruit beers available on Romanian market. *St. Cerc. St. CICBIA.* **2018**, *19*, 323–335.
10. Nardini, M.; Garaguso, I. Characterization of bioactive compounds and antioxidant activity of fruit beers. *Food Chem.* **2020**, *305*, 125437. [CrossRef]
11. Martínez, A.; Vegara, S.; Martí, N.; Valero, M.; Saura, D. Physicochemical characterization of special persimmon fruit beers using bohemian pilsner malt as a base. *J. Inst. Brew.* **2017**, *123*, 319–327. [CrossRef]
12. Zapałowska, A.; Matłok, N.; Zardzewiały, M.; Piechowiak, T.; Balawejder, M. Effect of Ozone Treatment on the Quality of Sea Buckhorn (*Hippophae rhamnoides* L.). *Plants* **2021**, *10*, 847. [CrossRef]
13. Piłat, B.; Zadernowski, R. Sea buckthorn (*Hippophae rhamnoides* L.) in cancer prevention. *Post. Fitoter.* **2019**, *20*, 111–117. (In Polish)
14. Teleszko, M.; Wojdyło, A.; Rudzińska, M.; Oszmiański, J.; Golis, T. Analysis of lipophilic and hydrophilic bioactive compounds content in Sea buckthorn (*Hippophaë rhamnoides* L.) berries. *J. Argic. Food Chem.* **2015**, *63*, 4120–4129. [CrossRef]
15. Pop, R.M.; Weesepoel, Y.; Socaciu, C.; Pintea, A.; Vincken, J.-P.; Gruppen, H. Carotenoid composition of berries and leaves from six Romanian sea buckthorn (*Hippophae rhamnoides* L.) varieties. *Food Chem.* **2014**, *147*, 1–9. [CrossRef]
16. Bal, L.M.; Meda, V.; Naik, S.N.; Satya, S. Sea buckthorn berries: A potential source of valuable nutrients for nutraceuticals and cosmeceuticals. *Food Res. Int.* **2011**, *44*, 1718–1727. [CrossRef]
17. Piłat, B.; Zadernowski, R. Sea buckthorn berries (*Hippophae rhamnoides* L.)—A rich source of biologically active compounds. *Post. Fitoter.* **2016**, *4*, 298–306. (In Polish)
18. Ma, X.; Yang, W.; Kallio, H.; Yang, B. Health promoting properties and sensory characteristics of phytochemicals in berries and leaves of sea buckthorn (*Hippophaë rhamnoides*). *Crit. Rev. Food Sci. Nutr.* **2022**, *62*, 3798–3816. [CrossRef]
19. Zeb, A. Chemical and Nutritional Constituents of Sea Buckthorn Juice. *Pak. J. Nutr.* **2004**, *3*, 99–106.
20. Depraetere, S.; Delvaux, F.; Coghe, S.; Delvaux, F.R. Wheat Variety and Barley Malt Properties: Influence on Haze Intensity and Foam Stability of Wheat Beer. *J. Inst. Brew.* **2004**, *110*, 200–206. [CrossRef]
21. Geertsen, J.B.H.; Allesen-Holm, D.V.; Byrne, D.; Giacalone, D. Consumer-led development of novel sea buckthorn-based beverages. *J. Sens. Stud.* **2016**, *31*, 245–255. [CrossRef]
22. Mascia, I.; Fadda, C.; Dostálek, P.; Olšovská, J.; Del Caro, A. Preliminary characterization of an Italian craft durum wheat beer. *J. Inst. Brew.* **2014**, *120*, 495–499. [CrossRef]
23. Gasiński, A.; Kawa-Rygielska, J.; Szumny, A.; Czubaszek, A.; Gąsior, J.; Pietrzak, W. Volatile Compounds Content Physicochemical Parameters and Antioxidant Activity of Beers with Addition of Mango Fruit (*Mangifera Indica*). *Molecules* **2020**, *25*, 3033. [CrossRef]
24. Adadi, P.; Kovaleva, E.G.; Glukhareva, T.V.; Shatunova, S.A.; Petrov, A.S. Production and analysis of non-traditional beer supplemented with sea buckthorn. *Agron. Res.* **2017**, *15*, 1831–1845. [CrossRef]
25. Habschied, K.; Košir, I.J.; Krstanović, V.; Kumrić, G.; Mastanjević, K. Beer Polyphenols—Bitterness, Astringency, and Off-Flavors. *Beverages* **2021**, *7*, 38. [CrossRef]

26. Michalak, M.; Podsędek, A.; Glinka, R. Antioxidant potential and polyphenolic compounds of glycolic extracts from *Hippophae rhamnoides* L. *Post. Fitoter.* **2016**, *17*, 33–38. (In Polish)
27. Kapur, A.; Hasković, A.; Čopra-Janičijecić, A.; Klepo, L.; Topčagić, A.; Tahirović, I.; Sofič, E. Spectrophotometric analysis of total ascorbic acid content in various fruits and vegetables. *Bull. Chem. Technol. Bosnia Herzeg.* **2012**, *38*, 39–42.
28. Pimentel, C.É.M.; Santiago, I.L.; Oliviera, S.K.M.; Serudo, R.L. Production of artisan beer with added ascorbic acid from amazonic fruit. *Braz. J. Dev.* **2019**, *5*, 18553–18560. [CrossRef]
29. Bogdan, P.; Kordialik-Bogacka, E. Antioxidant activity of beers produced with the addition of unmalted quinoa and amaranth. *Food Sci. Technol. Qual.* **2016**, *3*, 118–126. (In Polish) [CrossRef]
30. Ditrych, M.; Kordialik-Bogacka, E.; Czyżowska, A. Antiradical and Reducing Potential of Commercial Beer. *Czech J. Food Sci.* **2015**, *33*, 261–266. [CrossRef]
31. Deng, Y.; Lim, J.; Nguyen, T.T.H.; Mok, I.-K.; Piao, M.; Kim, D. Composition and biochemical properties of ale beer enriched with lignans from Schisandra chinensis Baillon (omija) fruits. *Food Sci. Biotechnol.* **2020**, *29*, 609–617. [CrossRef]
32. Gouvintas, I.; Breda, C.; Barros, A.I. Characterization and Discrimination of Commercial Portuguese Beers Based on Phenolic Composition and Antioxidant Capacity. *Foods* **2021**, *10*, 1144. [CrossRef]
33. Mikyška, A.; Dušek, M.; Slabý, M. How does fermentation, filtration and stabilization of beer affect polyphenols with health benefits. *Kvas. Prum.* **2019**, *65*, 120–126. [CrossRef]
34. Belcar, J.; Buczek, J.; Kapusta, I.; Gorzelany, J. Quality and Pro-Healthy Properties of Belgian Witbier-Style Beers Relative to the Cultivar of Winter Wheat and Raw Materials Used. *Foods* **2022**, *11*, 1150. [CrossRef]
35. Mikyška, A.; Dušek, M. How wort boiling process affect flawonoid polyphenols in beer. *Kvas. Prum.* **2019**, *65*, 192–200. [CrossRef]
36. Radonjič, S.; Maraš, V.; Raičevič, J.; Košmerl, T. Wine or Beer? Comparison, Changes and Improvement of Polyphenolic Compounds during Technological Phases. *Molecules* **2020**, *25*, 4960. [CrossRef]
37. Almaguer, C.; Schonberger, C.; Gastl, M.; Arendt, E.K.; Becker, T. *Humulus lupulus*—A story that begs to be told. A review. *J. Inst. Brew.* **2014**, *120*, 289–314. [CrossRef]
38. Guo, R.; Guo, X.; Li, T.; Fu, X.; Liu, R.H. Comparative assessment of phytochemical profiles, antioxidant and antiproliferative activities of sea buckthorn (*Hippophae rhamnoides* L.) berries. *Food Chem.* **2017**, *221*, 997–1003. [CrossRef]
39. Chen, C.; Zhang, H.; Xiao, W.; Yong, Z.-P.; Bai, N. High-performance liquid chromatographic fingerprint analysis for different origins of sea buckthorn berries. *J. Chromatogr. A* **2007**, *1154*, 250–259. [CrossRef]
40. Faltermaier, A.; Waters, D.; Becker, T.; Arendt EGastl, M. Common wheat (*Triticum aestivum* L.) and its use as a brewing cereal—A review. *J. Inst. Brew.* **2014**, *120*, 1–15. [CrossRef]
41. Viejo, G.; Sigfredo, C.; Sigfredo, F.; Damir, T.; Amruta, G.; Frank, D. Chemical characterization of aromas in beer and their effect on consumers liking. *Food Chem.* **2019**, *293*, 479–485. [CrossRef]
42. Belcar, J.; Sekutowski, T.R.; Zardzewiały, M.; Gorzelany, J. Effect of malting process duration on malting losses and quality of wheat malts. *Acta Univ. Cibin. Ser. E Food Technol.* **2021**, *25*, 221–232. [CrossRef]
43. Żurek, N.; Karatsai, O.; Rędowicz, M.J.; Kapusta, I. Polyphenolic Compounds of Crataegus Berry, Leaf, and Flower Extracts Affect Viability and Invasive Potential of Human Glioblastoma Cells. *Molecules* **2021**, *26*, 2656. [CrossRef]
44. *PN-A-04019: 1998*; Food products—Determination of vitamin C content. Polish Committee for Standardization: Warsaw, Poland, 1998.
45. Analytica EBC. 13.13—Sensory Analysis: Routine Descriptive Test Guideline. In *European Brewery Convention*; Hans Carl Getränke-Fachverlag: Nürnberg, Germany, 2004.
46. Analytica EBC. 13.12—Sensory Analysis: Flavour Terminology and Reference Standards. In *European Brewery Convention*; Hans Carl Getränke-Fachverlag: Nürnberg, Germany, 2004.

MDPI
St. Alban-Anlage 66
4052 Basel
Switzerland
www.mdpi.com

Molecules Editorial Office
E-mail: molecules@mdpi.com
www.mdpi.com/journal/molecules

Disclaimer/Publisher's Note: The statements, opinions and data contained in all publications are solely those of the individual author(s) and contributor(s) and not of MDPI and/or the editor(s). MDPI and/or the editor(s) disclaim responsibility for any injury to people or property resulting from any ideas, methods, instructions or products referred to in the content.

www.ingramcontent.com/pod-product-compliance
Lightning Source LLC
LaVergne TN
LVHW070631100526
838202LV00012B/781